METHODS IN MOLECULAR BIOLOGY

Series Editor
John M. Walker
School of Life and Medical Sciences
University of Hertfordshire
Hatfield, Hertfordshire, UK

For further volumes:
http://www.springer.com/series/7651

For over 35 years, biological scientists have come to rely on the research protocols and methodologies in the critically acclaimed *Methods in Molecular Biology series.* The series was the first to introduce the step-by-step protocols approach that has become the standard in all biomedical protocol publishing. Each protocol is provided in readily-reproducible step-by step fashion, opening with an introductory overview, a list of the materials and reagents needed to complete the experiment, and followed by a detailed procedure that is supported with a helpful notes section offering tips and tricks of the trade as well as troubleshooting advice. These hallmark features were introduced by series editor Dr. John Walker and constitute the key ingredient in each and every volume of the *Methods in Molecular Biology series.* Tested and trusted, comprehensive and reliable, all protocols from the series are indexed in PubMed.

RNA Folding

Methods and Protocols

Edited by

Ronny Lorenz

Institute for Theoretical Chemistry, University of Vienna, Vienna, Austria

Humana Press

Editor
Ronny Lorenz
Institute for Theoretical Chemistry
University of Vienna
Vienna, Austria

ISSN 1064-3745 ISSN 1940-6029 (electronic)
Methods in Molecular Biology
ISBN 978-1-0716-3518-6 ISBN 978-1-0716-3519-3 (eBook)
https://doi.org/10.1007/978-1-0716-3519-3

This Humana imprint is published by the registered company Springer Science+Business Media, LLC, part of Springer Nature.
The registered company address is: 1 New York Plaza, New York, NY 10004, U.S.A.

Paper in this product is recyclable.

Preface

Proteins and DNA, along with RNA molecules, are undoubtedly the most essential biopolymers in nature. Together, they serve as the building blocks for all living organisms on earth. DNA primarily functions as a storage medium for genetic code, while proteins play a vital role in maintaining cellular function and carrying out various metabolic processes. RNA, on the other hand, has long been considered a mere messenger that transcribes and passes information from DNA to the protein translation machinery. Although tRNAs and rRNAs were recognized early on as crucial components of the translation process, it was only recently discovered that RNA's functions extend beyond its nucleotide sequence. This fact had been ignored for many years.

About two decades ago, with the emergence of large-scale genome sequencing projects, numerous non-coding RNAs (ncRNAs) were discovered. Interestingly, most RNA transcripts in the human genome are non-coding, which raises questions about their functionality and how they operate.

It's now known that most ncRNAs serve their cellular purpose through intricate structures, created by partially self-complementary nucleotide sequences that enable the RNA to fold back on itself, much like double-stranded DNA helices. These structured and unstructured regions then potentially act as scaffolds or binding sites for other molecules, such as proteins or additional RNAs.

The interaction between RNA-protein or RNA-RNA complexes gives rise to a wide array of functional entities that regulate nearly all cellular processes. Moreover, RNA has an extra hydroxyl group on its ribose, enabling it to catalyze various chemical reactions. It's no wonder that some RNAs fold into specific conformations to function as enzymes, known as ribozymes.

Certain ribozymes are responsible for reactions like self-cleavage or ligation of the RNA strand, while others introduce post-transcriptional modifications, like adding or removing chemical groups to RNA nucleotides. In instances like tRNAs, these modifications are crucial for helping the RNAs fold into their proper shape or for revealing the correct binding sites for other ligands or RNAs.

Protein-coding mRNA transcripts also contain essential RNA structures. In eukaryotes, for instance, the folding of 5' and 3' untranslated regions (UTRs) in processed mRNA transcripts can influence their lifespan, i.e., the time they are available for translation. Structured elements like tightly bound helices or G-quadruplexes at strand ends can prevent exonucleases from degrading the mRNA.

The accessibility of binding sites for microRNAs (miRNAs) in an mRNA's 3' UTR determines how effectively the transcripts can be regulated through miRNA degradation. RNA structures within the coding part of mRNA sequences serve as recognition and binding sites for transport proteins, which direct these transcripts to specific cell regions or even distant organelles in plant phloem.

In bacteria and archaea, numerous riboswitches have been identified. These upstream elements of protein-coding genes regulate gene expression at transcription or translation levels by conformational changes. Riboswitches adopt different RNA structures based on transcription speed, translation speed, or the presence or absence of ligands, switching genetic programs on or off accordingly.

For single-stranded RNA (ssRNA) viruses, structural conformations play a significant role in their lifecycle. Well-known examples of ssRNA viruses with critical structural elements include flaviviruses, the influenza virus, and coronaviruses, such as the one responsible for the SARS-CoV-2 pandemic. RNA folding gained more attention in RNA design following the COVID-19 outbreak, as large-scale treatment with mRNA vaccines required a deep understanding of potential mRNA structures.

Insights into favorable or unfavorable structures can be obtained through a combination of experiments, prediction tools, and algorithms. Like many post-transcriptionally modified natural RNAs, artificial mRNA vaccines use non-canonical bases to avoid early detection by cellular RNA degradation mechanisms. However, more focus is needed on prediction tool development and parameterization, as modified bases are currently unsupported by most prediction tools. Many tools lack the flexibility to include them in sequence input, and only a few energy parameters for a limited number of RNA modifications are available to support RNA structure prediction with modified bases.

In summary, RNA folding plays a crucial role across numerous aspects of biology. To understand and predict the behavior and function of RNAs, sophisticated tools are needed to simulate and analyze their potential for forming structures. Most chapters in this book discuss RNA structures at the secondary structure level. This should not be seen as a limitation, but rather a recognition that efficient algorithms for RNA tertiary structure prediction are not yet available. Furthermore, predicting the spatial arrangement of an RNA molecule typically necessitates a solid understanding of its secondary structure. This book aims to illuminate the various levels of prediction and the most significant algorithmic approaches.

The book is organized as follows. Chapters 1 and 2 cover the energy parameters of the nearest-neighbor (NN) energy model, which underpins numerous RNA structure prediction tools. Chapter 1 discusses the analysis of folding free energy changes, while Chap. 2 delves into the optimization and salt correction of NN parameters for predicting RNA duplex melting temperatures.

Chapter 3 provides an overview of key thermodynamic properties that can be predicted for RNA molecules, particularly when interacting with small ligands or proteins.

Chapters 4 and 5 support RNA secondary structure prediction using experimental RNA structure probing data. Chapter 4 presents experimental setup and in silico analysis for integrating data from multiple structure probing agents, while Chap. 5 thoroughly investigates how parameters associated with the integration of SHAPE-directed probing data impact prediction outcomes.

Chapter 6 explores classified dynamic programming, which helps address the exponential growth of candidate structures an RNA molecule may fold into. Chapter 7 presents an approach for identifying and relating common RNA structure motifs with potential non-canonical base pairs based on available RNA 3D structures and multiple sequence alignments.

Chapter 8 demonstrates an approach incorporating as much data on base modifications' effects on secondary structure prediction as possible, using a heavily modified tRNA as an example. Chapter 9 illustrates methods and application scenarios for predicting RNA-RNA interactions.

Chapters 10 and 11 focus on sequence evolution and multiple RNA sequences with potentially conserved structures. Chapter 10 presents an efficient tool for simultaneous RNA folding and alignment, which allows for better detection of conserved RNA structures compared to ab initio alignment-based methods. Chapter 11 provides an overview of covariance models that help construct RNA families, identify members of a common RNA family, and compare RNA families.

Chapter 12 takes RNA sequence design to the next level with a novel, versatile framework capable of handling both positive and negative design objectives.

Chapters 13 and 14 discuss kinetic folding approaches, addressing the dynamic nature of RNA folding. Chapter 13 introduces a fast heuristic approach for co-transcriptional RNA secondary structure prediction, highlighting differences and drawbacks of "classical" thermodynamic methods. Chapter 14 focuses on quality control for co-transcriptional structure prediction, presenting a user-friendly pipeline that simulates RNA folding as a Markov process and helps minimize issues related to time- and memory-consuming tasks.

Finally, Chap. 15 goes beyond secondary structures, using secondary structure models and small-angle X-ray scattering (SAXS) data to demonstrate how a predicted preliminary coarse-grained RNA 3D structure can be refined and converted back to all-atom coordinates.

I would like to express my deepest gratitude to everyone involved in the creation of this book. I am truly thankful for the authors' thorough and up-to-date contributions, as well as the hard work they put into writing their chapters. I also want to extend my appreciation to Series Editor John Walker and Associate Editor Anna Rakovsky for their patience and unwavering encouragement. Last but not least, I am indebted to my family for their support and the time they allowed me to dedicate to the completion of this project.

Vienna, Austria *Ronny Lorenz*

Contents

Contributors

DELPHINE ALLOUCHE • *CiTCOM, Cibles Thérapeutiques et conception de médicaments, UMR8038 CNRS, Université de PARIS, Paris, France; Sanofi mRNA center of excellence 1541, Marcy-l'Etoile, France*
STEFAN BADELT • *Department of Theoretical Chemistry, University of Vienna, Vienna, Austria*
SARGUEIL BRUNO • *CiTCOM, Cibles Thérapeutiques et conception de médicaments, UMR8038 CNRS, Université de PARIS, Paris, France*
GRÉGOIRE DE BISSCHOP • *CiTCOM, Cibles Thérapeutiques et conception de médicaments, UMR8038 CNRS, Université de PARIS, Paris, France; Institut de Recherches Cliniques de Montréal (IRCM), Montréal, QC, Canada*
FRANCOIS-XAVIER LYONNET DU MOUTIER • *CiTCOM, Cibles Thérapeutiques et conception de médicaments, UMR8038 CNRS, Université de PARIS, Paris, France*
FLORIAN EGGENHOFER • *Bioinformatics Group, Department of Computer Science University of Freiburg, Freiburg, Germany*
IZABELA FERREIRA • *Departamento de Física, Universidade Federal de Minas Gerais, Belo Horizonte-MG, Brazil*
SVEN FINDEIß • *Bioinformatics Group, Department of Computer Science, and Interdisciplinary Center for Bioinformatics, Leipzig University, Leipzig, Germany*
TORIN GREENWOOD • *North Dakota State University, Fargo, ND, USA*
PIERRE HARDOUIN • *CiTCOM, Cibles Thérapeutiques et conception de médicaments, UMR8038 CNRS, Université de PARIS, Paris, France*
CHRISTINE E. HEITSCH • *Georgia Institute of Technology, Atlanta, GA, USA*
IVO L. HOFACKER • *Department of Theoretical Chemistry, Faculty of Chemistry, University of Vienna, Vienna, Austria; Research Group Bioinformatics and Computational Biology, Faculty of Computer Science, University of Vienna, Vienna, Austria*
CHRISTIAN HÖNER ZU SIEDERDISSEN • *Bioinformatics Group, Department of Computer Science, University of Leipzig, Leipzig, Germany; Interdisciplinary Center for Bioinformatics, University of Leipzig, Leipzig, Germany; Bioinformatics/High-Throughput Analysis, Faculty of Mathematics and Computer Science, Friedrich Schiller University Jena, Jena, Germany*
FELIX KÜHNL • *Bioinformatics Group, Department of Computer Science, and Interdisciplinary Center for Bioinformatics, Leipzig University, Leipzig, Germany*
RONNY LORENZ • *Department of Theoretical Chemistry, University of Vienna, Vienna, Austria*
DAVID H. MATHEWS • *Department of Biochemistry & Biophysics and Center for RNA Biology, University of Rochester Medical Center, Rochester, NY, USA; Department of Biostatistics & Computational Biology, University of Rochester Medical Center, Rochester, NY, USA*
MILAD MILADI • *Bioinformatics Group, Department of Computer Science, University of Freiburg, Freiburg, Germany*
SIMÓN POBLETE • *Instituto de Ciencias Físicas y Matemáticas, Universidad Austral de Chile, Valdivia, Chile; Computational Biology Lab, Fundación Ciencia & Vida, Santiago, Chile;*

Facultad de Ingeniería, Arquitectura y Diseño, Universidad SanSebastián, Santiago, Chile

YANN PONTY • *CNRS UMR 7161, LIX, Ecole Polytechnique, Palaiseau, France; LIX, CNRS UMR 7161, Ecole Polytechnique, Institut Polytechnique de Paris, Palaiseau, France*

MARTIN RADEN • *Bioinformatics Group, Department of Computer Science, University of Freiburg, Freiburg, Germany*

VLADIMIR REINHARZ • *Department of Computer Science, Université du Québec à Montréal, Montreal, QC, Canada*

AFAF SAAIDI • *Georgia Institute of Technology, School of Mathematics, Atlanta, GA, USA*

ROMAN SARRAZIN-GENDRON • *School of Computer Science, McGill University, Montreal, QC, Canada*

PETER F. STADLER • *Bioinformatics Group, Department of Computer Science, Interdisciplinary Center of Bioinformatics, German Centre for Integrative Biodiversity Research (iDiv) Halle-Jena-Leipzig, Competence Center for Scalable Data Services and Solutions, and Leipzig Research Center for Civilization Diseases, Leipzig University, Leipzig, Germany; Max Planck Institute for Mathematics in the Sciences, Leipzig, Germany; Institute for Theoretical Chemistry, University of Vienna, Vienna, Austria; Facultad de Ciencias, Universidad Nacional de Colombia, Bogotá, D.C., Colombia; Santa Fe Institute, Santa Fe, NM, USA*

BERNHARD C. THIEL • *Department of Theoretical Chemistry, Faculty of Chemistry, University of Vienna, Vienna, Austria*

YULIIA VARENYK • *Department of Theoretical Chemistry, University of Vienna, Vienna, Austria*

BJÖRN VOß • *RNA Biology and Bioinformatics, Institute of Biomedical Genetics, University of Stuttgart, Stuttgart, Germany*

JÉRÔME WALDISPÜHL • *School of Computer Science, McGill University, Montreal, QC, Canada*

GERALD WEBER • *Departamento de Física, Universidade Federal de Minas Gerais, Belo Horizonte-MG, Brazil*

SEBASTIAN WILL • *LIX, CNRS UMR 7161, Ecole Polytechnique, Institut Polytechnique de Paris, Palaiseau, France*

HUA-TING YAO • *LIX, CNRS UMR 7161, Ecole Polytechnique, Institut Polytechnique de Paris, Palaiseau, France; School of Computer Science, McGill University, Montreal, Canada; Department of Theoretical Chemistry, University of Vienna, Vienna, Austria*

JEFFREY ZUBER • *Department of Biochemistry & Biophysics and Center for RNA Biology, University of Rochester Medical Center, Rochester, NY, USA*

Chapter 1

Estimating RNA Secondary Structure Folding Free Energy Changes with *efn2*

Jeffrey Zuber and David H. Mathews

Abstract

A number of analyses require estimates of the folding free energy changes of specific RNA secondary structures. These predictions are often based on a set of nearest neighbor parameters that models the folding stability of a RNA secondary structure as the sum of folding stabilities of the structural elements that comprise the secondary structure. In the software suite RNAstructure, the free energy change calculation is implemented in the program *efn2*. The *efn2* program estimates the folding free energy change and the experimental uncertainty in the folding free energy change. It can be run through the graphical user interface for RNAstructure, from the command line, or a web server. This chapter provides detailed protocols for using efn2.

Key words RNAstructure, efn2, Folding free energy, Equilibrium constant, Pseudoknot, RNA secondary structure

1 Introduction

Noncoding RNAs (ncRNA), functional RNAs that are not transcribed into protein, have functions that range from enzymatic catalysis (ribozymes) [1] to regulation of gene expression (siRNA, miRNA, and riboswitches) [2, 3] and to target identification (guide RNAs) [4]. Many of the functions of these ncRNA sequences are mediated by the structures that they form, either within the sequence or with a second nucleic acid sequence.

RNA structure is hierarchical [5]. The primary structure of RNA is the sequence of covalently linked nucleotides. The secondary structure of RNA is the set of canonical base pairs in the RNA, stacks of which form helices (Fig. 1). These helices are flanked by loops, regions of nucleotides that are not canonically base paired, although often the nucleotides are noncanonically paired [6]. Loops include hairpin loops (with one exiting helix), bulge loops (with two exiting helices and all unpaired nucleotides on one

Ronny Lorenz (ed.), *RNA Folding: Methods and Protocols*, Methods in Molecular Biology, vol. 2726,
https://doi.org/10.1007/978-1-0716-3519-3_1,

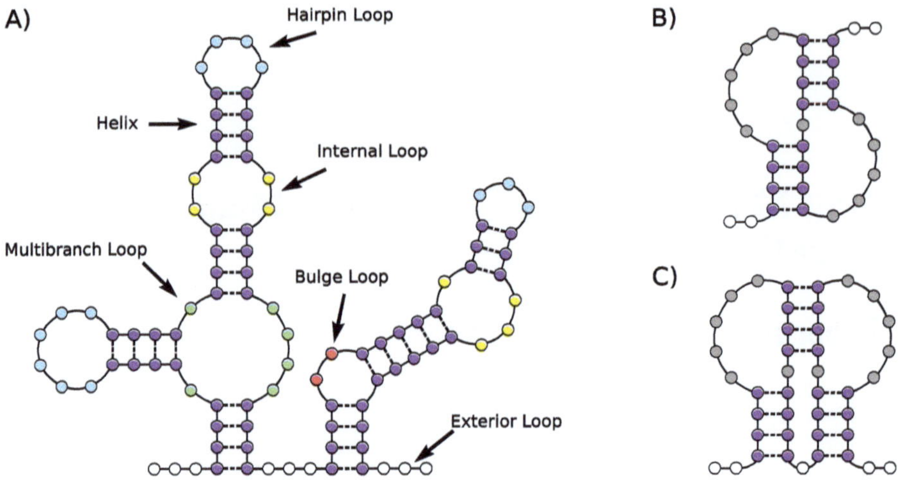

Fig. 1 RNA secondary structure motifs. The circles represent nucleotides, solid lines represent covalent bonds between nucleotides, and dashed lines represent base pairing interactions. (**a**) RNA secondary structure is defined as a set of base-paired nucleotides, which form helices (purple circles). These helices and unpaired nucleotides combine to form loops, such as hairpins (blue circles), internal loops (yellow circles), bulge loops (red circles), multibranch loops (green circles), and exterior loops (white circles). (**b** and **c**) Examples of pseudoknotted RNA structures (grey circles). (**b**) This is an H-type pseudoknot where nucleotides in a hairpin loop base pair with nucleotides outside the closing stem. (**c**) This is the kissing loop motif

strand of the loop), internal loops (with two exiting helices and unpaired nucleotides on both strands of the loop), multibranch loops (with 3 or more exiting helices), and exterior loops (which contain the 5′ and 3′ ends of the RNA molecule). The tertiary structure of RNA includes additional contacts and is defined by the coordinates of all atoms in the RNA. The secondary structure generally forms faster [7] and is more thermostable [8, 9] than tertiary structure, allowing RNA secondary structure to be predicted independently of tertiary structure [5].

To estimate the free energy change of folding (*See* **Note 1**) to a secondary structure from random coil, a set of nearest-neighbor thermodynamic parameters can be used [10, 11]. To use the nearest neighbor parameters, an RNA secondary structure is decomposed into the individual motifs that make up the structure, including stacks of two sequential base pairs and the loops. The predicted folding free energy change for the secondary structure is then calculated by adding the contributions of each motif to the folding free energy change, where the contributions are parameterized in lookup tables.

The nearest neighbor parameters were derived using linear regression on a database of folding stabilities determined by optical melting data of small model RNA structures [12]. Parameters for stacks of Watson-Crick-Franklin base pairs were determined from a set of folding stabilities for helices composed only of Watson-Crick-Franklin base pairs [11]. These parameter values were then fixed in value for subsequent estimates of the remaining parameters. The free energy change contributions of GU-containing stacks were then determined in an additional linear regression. The folding free energy contributions of terminal mismatches and dangling end nucleotides were determined from the difference in folding free energy change for duplexes with and without the mismatch or dangling end. Next, the loop stabilities were determined from optical melting experiments on small model systems by subtracting the stability of the helical components from the stabilities of systems with a single loop. Parameters for hairpin loops, internal loops, bulge loops, and multibranch loops parameters are fit in six separate linear regressions.

The nearest neighbor parameters are termed the "Turner Rules" because Doug Turner and coworkers systematically studied helix and loop stabilities and also determined the set of parameters. The most recent complete set of Turner rules, assembled in 2004 [10, 13], includes estimated stabilities for loops and GU base pairs and the set of Watson-Crick-Franklin base pair terms determined in 1998 [11]. Enthalpy parameters were derived to extrapolate folding stabilities to temperatures other than the default of 37 °C [14]. The 2004 Turner rules are enumerated in the nearest neighbor database (NNDB), along with examples of their use [13]. Additionally, a method was recently implemented to estimate the experimental uncertainty in a folding free energy change prediction [15]. To estimate the uncertainty, an accounting is required of the substantial correlation between parameters [16].

A number of software packages use the nearest neighbor rules to predict RNA secondary structures [17–20], design sequences that will fold to a structure [21–26], or estimate accessibility to oligonucleotide [27–30] or small molecule binding [31]. For many of these applications, the parameters are used with dynamic programming algorithms to predict lowest free energy structures [32] or partition functions [33] for structures. Additionally, a number of methods that infer folding parameters from known secondary structures use the terms to parameterize the favorability of the structure motifs [34–36].

RNAstructure is a software suite developed by our group that uses the nearest neighbor thermodynamic parameters to perform analyses on RNA sequences [18]. RNAstructure provides tools to predict RNA secondary structures, base pair probabilities, conserved secondary structures for a set of homologs, interactions between RNA strands, and to design RNA sequences that fold into a desired structure. One of the tools provided in RNAstructure is *efn2*, a utility to predict the folding free energy change for a specific RNA secondary structure. Recently, we extended efn2 to also estimate the uncertainty in the free energy calculation (*See* **Note 2**) [15, 16].

Folding free energy changes, as calculated by *efn2*, are useful for many reasons. Comparing folding free energy change values can be used to calculate the relative populations of specific structures or to compare two proposed structures [37]. The dynamic programming algorithms [17, 18, 38] used to predict RNA secondary structures do not use the full energy model for multibranch loops described in the Turner 2004 rules [39]; instead they use simplified energy models [40]. Using *efn2* on the full set of suboptimal secondary structures for a given sequence generated by the secondary structure prediction program *Fold* will validate that the full energy model still ranks the structures like the simple energy model. Additionally, estimated folding free energies can be used to generate hypotheses for mechanisms of action [41], predicting thermodynamics of specific unfolding events [42], and to predict the thermodynamic impact of incorporating base substitutions [43].

This chapter provides an overview of nearest neighbor analysis, a protocol for using the command-line software *efn2* for estimating folding free energy changes and errors, a protocol for using the RNAstructure webserver for estimating folding free energy change, a protocol for using the RNAstructure graphical interface for estimating folding free energy change, and a protocol for using the RNAstructure structure editor for estimating folding stability.

2 Materials

The protocols use the RNAstructure software package, which is freely available at: https://rna.urmc.rochester.edu/RNAstructure.html. It is available as either source code or as precompiled programs. Instructions for installing the software are available in the online help manual at: https://rna.urmc.rochester.edu/RNAstructureHelp.html. The protocol for using the command line program *efn2* assumes familiarity with the command line (either the Unix command line for Linux, Mac OS X, or Windows 10 Subsystem for Linux or the Microsoft DOS command line for Windows). The graphical interface (GUI) installs and runs on Microsoft Windows, Mac OS, or Unix /Linux platforms. The webserver uses a web browser and is known to work with the common web browsers (Chrome, Safari, Firefox, etc.) we have tested.

3 Methods

3.1 Nearest Neighbor Thermodynamic Parameters

The nearest neighbor thermodynamic rules approximate the folding free energy change for a specific secondary structure as the linear combination of the folding free energy changes for each individual secondary structure motif, expressed in the following equation:

$$\Delta G = \sum_{i}^{\text{NNs}} P_i \times \text{NN}_i \tag{1}$$

where P_i is the parameter coefficient (the number of times a specific secondary structure element appears in a secondary structure) and NN_i is the value of the *i*th nearest neighbor parameter (the folding free energy change value for a secondary structure element).

The parameter values for the 2004 Turner rules are derived from a set 802 optical melting experiments [10, 44]. These optical melting experiments were performed on small oligonucleotides that were designed to test the stability of specific RNA secondary structures and sequences. These parameters include energy models for helices, terminal mismatches (with unpaired nucleotides stacking on both the 5′ and 3′ sides of a helix end), dangling ends (with an unpaired nucleotide staking on either the 5′ or 3′ side of a helix end), hairpin loops, internal loops, bulge loops, mismatch-mediated coaxial stacking, and multibranch loop initiation. For multibranch and exterior loops, the energy of the optimal configuration of terminal mismatches, dangling ends, and coaxial stacks is also added. An example calculation for a simple hairpin loop structure is shown in Fig. 2. Refer to the nearest neighbor database (NNDB) for a detailed explanation of each of these energy models and their parameters [13]. The NNDB is available online at: https://rna.urmc.rochester.edu/NNDB/.

3.2 Pseudoknot Free Energy Changes

An RNA pseudoknot is present in a structure when the base pairs form nonnested structures. This occurs when there exists base pairs between nucleotides i-j and k-l that have the relation $i < k < j < l$. This is in contrast to nonpseudoknotted structures, where all base pairs are nested, that is, all base pairs are entirely between or entirely outside of every other base pair ($i < j < k < l$ or $i < k < l < j$). Figure 1b, c shows two examples of pseudoknots. The 2004 Turner rules did not originally provide parameters for estimating pseudoknot stability. There is less experimental guidance for the stability of pseudoknots than for other motifs [45–47]. We extended the 2004 Turner model to estimate pseudoknot stabilities [48] using a functional form inspired by polymer physics [49]. This functional form is incorporated into *efn2*. Aside from our pseudoknot parameters, other groups have also worked on this problem in support of RNA secondary structure prediction [50–52]. A more definitive set of parameters will likely require additional experimental measurements.

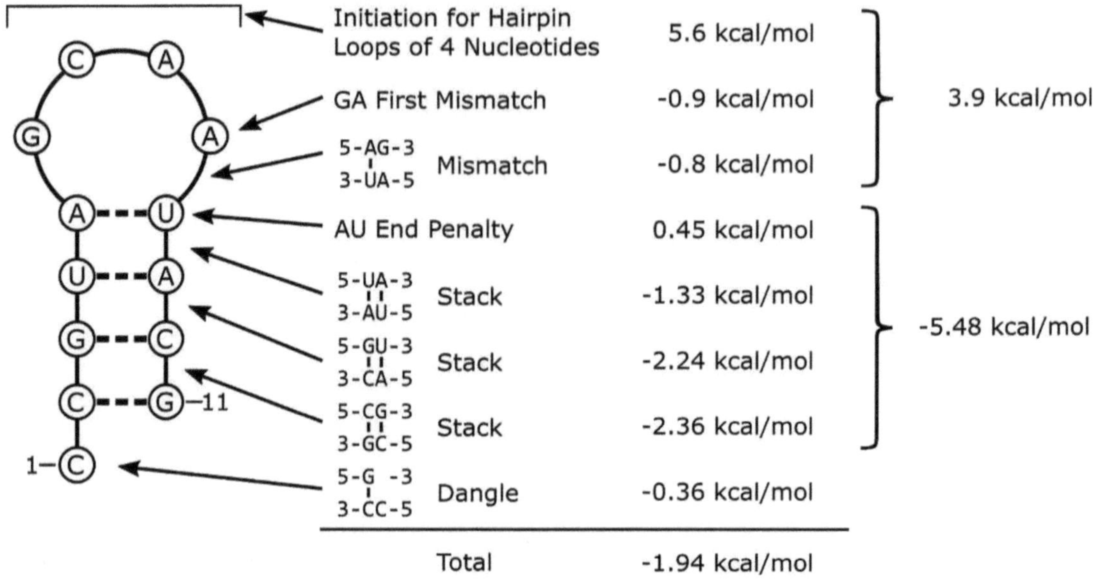

Fig. 2 Example calculation of RNA folding stability using the nearest neighbor parameters. The dangling end stability is sequence specific. Note that the stability of the helix is estimated using stacks of adjacent base pairs. Each end of a helix that terminates in an AU base pair is penalized by 0.45 kcal/mol. The hairpin loop has sequence-dependent terms, which are a stability for a mismatch adjacent to the helix end and a bonus for having a GA mismatch in the first position. The hairpin loop also has a sequence-independent term for closing a loop of 4 unpaired nucleotides. The total stability is the sum of the terms

3.3 Using the Command Line Program efn2

Folding free energy changes can be predicted using the command line program *efn2*. Using the command line program provides the most options as compared to the GUI tools. Before using the command line utility, an environment variable DATAPATH should be defined to specify the location of the thermodynamic parameter data tables. This can be done (in the BASH shell) by running the command:

```
$ export DATAPATH=<PATH-TO-RNAstructure>/data_tables/
```

where <PATH-TO-RNAstructure> needs to be replaced with the location of the RNAstructure directory. In its simplest usage, *efn2* requires two arguments:

```
$ efn2 <structure file> <output file>
```

The first argument is the file containing the RNA secondary structure in a CT or dot-bracket format, which can contain multiple secondary structures for the same sequence. The CT and dot-bracket formats are defined in the online help at https://rna.urmc.rochester.edu/Text/File_Formats.html. The second argument is the output file path where the energies will be written as a plain text file. Example usage and output are shown in Fig. 3. When calculating the folding free energy changes for RNA sequences, the

```
$ efn2 ~/Examples/RA7680.ct RA7680.out
Initializing nucleic acids...done.
Calculating free energies...done.
efn2 complete.

$ cat RA7680.out
Structure: 1   Energy = -31.8 ± 0.7
Structure: 2   Energy = -32.5 ± 0.9
Structure: 3   Energy = -31.4 ± 1.0
Structure: 4   Energy = -29.2 ± 0.8
```

Fig. 3 Sample output of the *efn2* command line utility

experimental uncertainties in the predictions are calculated [15]. These uncertainties are a reflection of the experimental uncertainties in the optical melting experiments from which the nearest neighbor parameter values are derived.

In order to suppress the output of uncertainty values (which might be necessary to ensure compatibility with user scripts written for previous versions of RNAstructure), a flag (−ne) is available for the command line utility. To export the parameter usage count vector, another flag (−c) is available. The resulting text file will contain the parameter usage counts, ordered as in the supplementary information provided in [16]. The parameter usage counts are the cumulative parameter usage counts for all structures in the structure file.

Another important option, flag -S, -s, or −simple, is to use the simplified free energy change model for multibranch loops. When this flag is used, the energy model is identical to that used in RNAstructure's dynamic programming algorithms for structure prediction (*AllSub*, *Fold*, *partition*, and *stochastic*). By default, a logarithmic dependence on the number of unpaired nucleotides (based on Jacobson-Stockmayer theory [53]) and a penalty for asymmetry in the distribution of unpaired nucleotides is used [39]. These aspects of the model might provide more accurate estimates for folding free energy changes, but are difficult to include in dynamic programming algorithms [40, 54]. The flag -t, -T, or --temperature can be used to estimate folding free energy change at a temperature other than 37 °C. This flag should be followed by the absolute temperature in Kelvin. For a full explanation of all of the options available with the command line utility, visit: http://rna.urmc.rochester.edu/Text/efn2.html.

As an alternative to the standard *efn2* output, a thermodynamic details file can also be generated. The thermodynamic details file provides the free energy contribution of each stack and each loop in the secondary structures. This output is enabled using the -w, -W, or --writedetails flags in the command.

3.4 Using efn2 as Part of the Graphical User Interface

As an alternative to the command line utility, folding free energy changes can be predicted with the RNAstructure graphical user interface (GUI). After installation, open the RNAstructure program. The *efn2* utility can be accessed from the RNA or DNA menu (where DNA folding free energy parameters would be used for DNA efn2). In the *efn2* sub-window, there are data fields for the structure file and output file arguments (Fig. 4). After choosing the input structure file, the output file name is automatically inserted, but can be changed by clicking the adjacent "..." button. After clicking START, the *efn2* utility will run and create the output file. The output file is plain text and the contents can be viewed with any text editor. On Windows, Notepad or WordPad can be used. On the Mac, TextEdit can be used. At this time, the RNAstructure GUI does not calculate the experimental uncertainty in the folding free energy predictions. Additionally, the only available option is the flag for writing the thermodynamic details file.

Another GUI program that can be used to calculate the folding free energy change of a secondary structure is the Structure Editor program distributed with RNAstructure. This is primarily a software program for editing structures and creating publication-

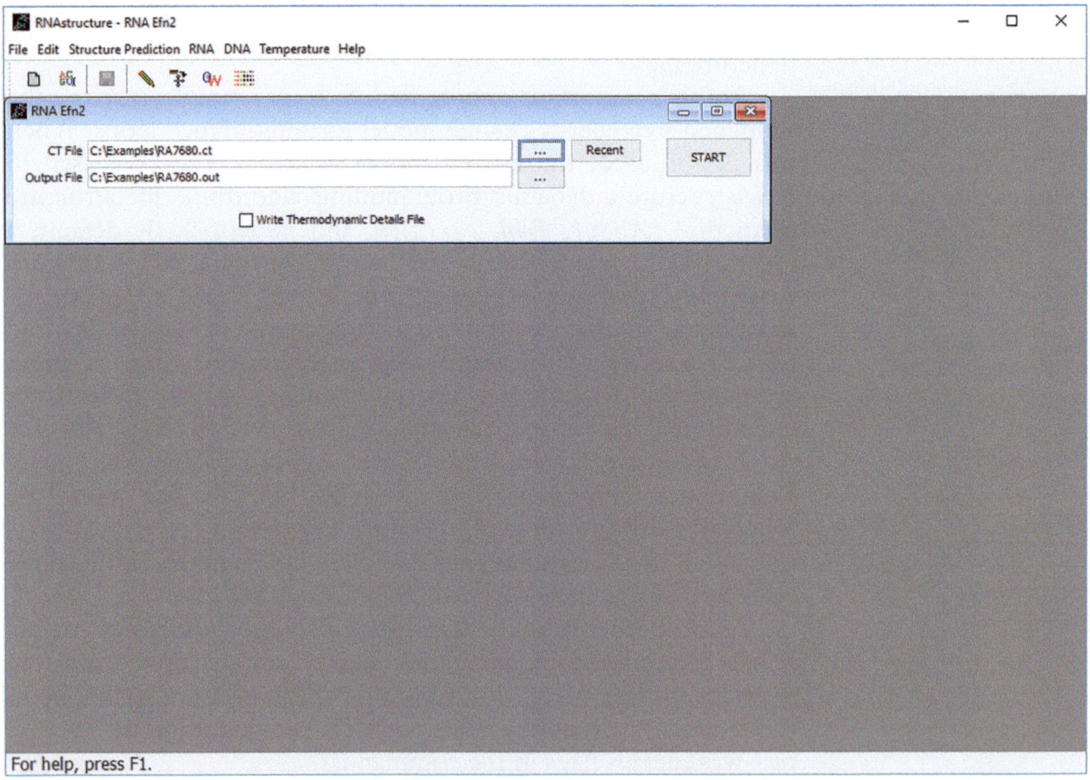

Fig. 4 Running *efn2* through the RNAstructure GUI. This shows the input form as it appears on Windows. The user has already selected the input CT file

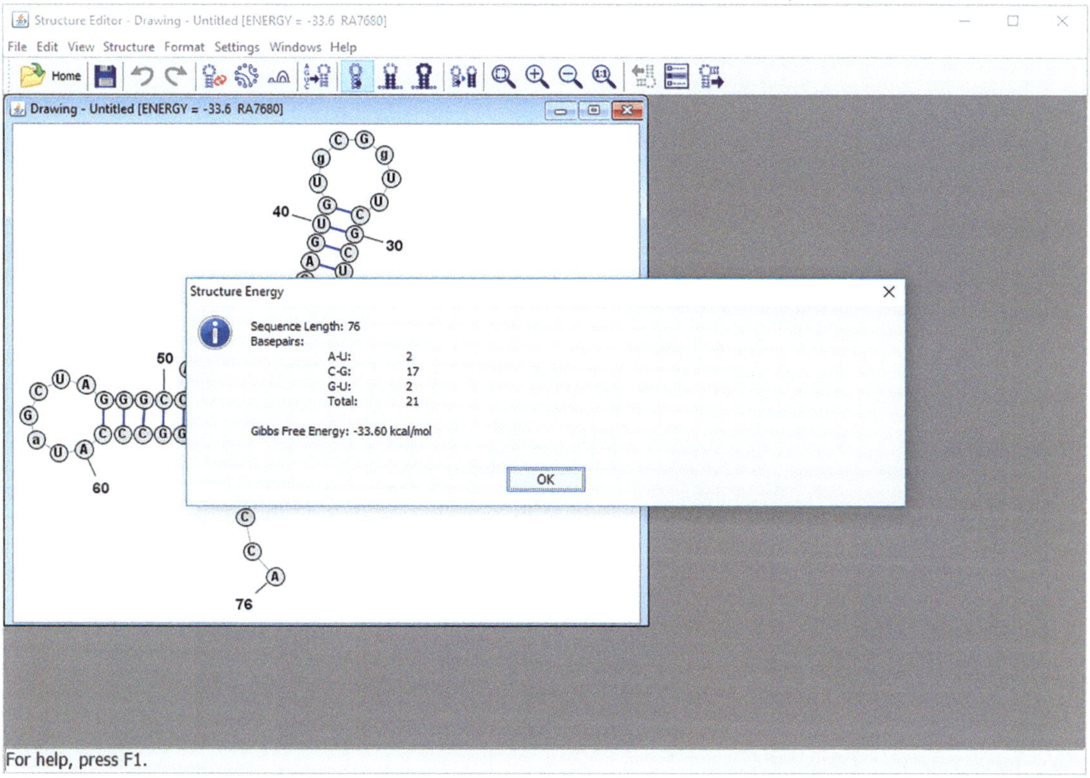

Fig. 5 Evaluating fold free energy change using Structure Editor. In this view, the structure energy window is superimposed above the structure

quality structure figures, but it is also capable of predicting structures and estimating folding free energy changes. To access *efn2*, start the Structure Editor program. Then open the file containing the secondary structure(s), which can be in CT or dot-bracket format. Select the secondary structure using commands in the View menu. Then select "Calculate Energy" from the View menu, and the estimated folding free energies are shown on the screen. An example output is shown in Fig. 5. Note that the energies calculated with this program use the simple multibranch loop energy function (equivalent to using the '-S' flag on the command line utility).

3.5 Using the Web Server

As an alternative to installing the program on a local computer, *efn2* is also available on the Mathews Lab web servers. To access the web server, navigate to the web page https://rna.urmc.rochester.edu/RNAstructureWeb/index.html and click the link to the *efn2* server, which will take the user to the job submission page (Fig. 6).

From that page, the user can upload a structure file and select nucleic acid type (RNA or DNA). Optionally, the temperature can be adjusted from 310.15 K (37 °C), a thermodynamic details file can be chosen as the output, and an e-mail address can be provided. Clicking the 'Submit Query' button will add the calculation to the

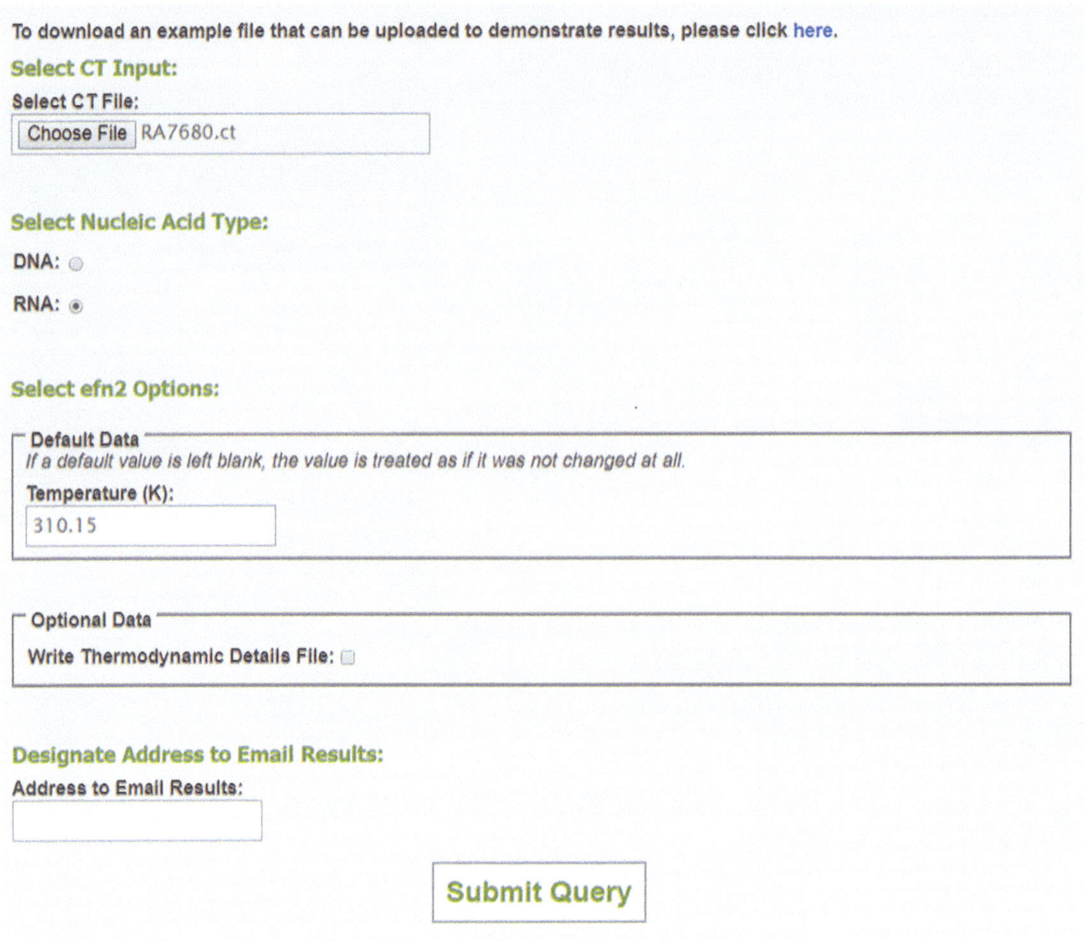

Fig. 6 Running *efn2* through the RNAstructure web server. This view shows the input form as seen in a web browser

web server's job queue. After the calculation is complete, the results will appear on the following page. If an email address was provided, an email containing a link to the results page will be sent after the job is completed.

4 Notes

4.1 Meaning of Free Energy Change

The free energy change quantifies the stability of a structure relative to the random coil, the unstructured state. For unimolecular secondary structure, a structure involving a single strand, the ratio of concentration of strands in a given structure to the concentration of strands in the random coil is the equilibrium constant for that structure:

$$K = \frac{[S]}{[\mathrm{RC}]}$$

where K is the equilibrium constant, $[S]$ is the concentration of the structure with equilibrium constant K, and [RC] is the concentration of the random coil. The free energy change determines the equilibrium constant:

$$K = e^{-\Delta G^\circ/RT}$$

where ΔG° is the folding free energy change, R is the gas constant, and T is the absolute temperature. Therefore, the lower the folding free energy change for a structure (the more negative the value), the larger the equilibrium constant, and the more favorable the structure compared to alternative structures.

4.2 Uncertainty in Folding Free Energy Prediction

For an equation of the form of Eq. 1, the variance in the output can be determined with the following matrix operations:

$$\sigma^2_{\Delta G} = \boldsymbol{P} \times \boldsymbol{\Sigma}_{\boldsymbol{P}} \times \boldsymbol{P}^\top \qquad (4)$$

where $\boldsymbol{P}$ is the parameter coefficient vector (the number of uses of each parameter), $\boldsymbol{\Sigma}_{\boldsymbol{P}}$ is the variance-covariance matrix (not to be confused with the summation operator), and $\boldsymbol{P}^\top$ is the transpose of the parameter coefficient vector. We estimated the variance-covariance matrix by observing the parameter variance and covariance for the free energy parameters in 100,000 sets of nearest neighbor parameter values derived from perturbed optical melting data, where the free energy change values for each experiment was varied within experimental uncertainty [16]. This accounts for all the explicit and implicit relationships between all the parameters in the nearest neighbor thermodynamic model.

Acknowledgment

This protocol was written with the support of the National Institutes of Health grant R35GM145283 to D.H.M.

References

1. Doudna JA, Cech TR (2002) The chemical repertoire of natural ribozymes. Nature 418(6894):222–228
2. Wu L, Belasco JG (2008) Let me count the ways: mechanisms of gene regulation by miRNAs and siRNAs. Mol Cell 29(1):1–7
3. Serganov A, Nudler E (2013) A decade of riboswitches. Cell 152(1–2):17–24
4. Yu YT, Meier UT (2014) RNA-guided isomerization of uridine to pseudouridine--pseudouridylation. RNA Biol 11(12):1483–1494
5. Tinoco I Jr, Bustamante C (1999) How RNA folds. J Mol Biol 293(2):271–281
6. Leontis NB, Stombaugh J, Westhof E (2002) The non-Watson-Crick base pairs and their associated isostericity matrices. Nucleic Acids Res 30(16):3497–3531

7. Woodson SA (2000) Recent insights on RNA folding mechanisms from catalytic RNA. Cell Mol Life Sci 57(5):796–808
8. Crothers DM, Cole PE, Hilbers CW, Shulman RG (1974) The molecular mechanism of thermal unfolding of Escherichia coli formylmethionine transfer RNA. J Mol Biol 87(1):63–88
9. Onoa B, Tinoco I Jr (2004) RNA folding and unfolding. Curr Opin Struct Biol 14(3):374–379
10. Mathews DH, Disney MD, Childs JL, Schroeder SJ, Zuker M, Turner DH (2004) Incorporating chemical modification constraints into a dynamic programming algorithm for prediction of RNA secondary structure. Proc Natl Acad Sci U S A 101(19):7287–7292
11. Xia T, SantaLucia J Jr, Burkard ME, Kierzek R, Schroeder SJ, Jiao X, Cox C, Turner DH (1998) Thermodynamic parameters for an expanded nearest-neighbor model for formation of RNA duplexes with Watson-crick base pairs. Biochemistry 37(42):14719–14735
12. Andronescu M, Condon A, Turner DH, Mathews DH (2014) The determination of RNA folding nearest neighbor parameters. Methods Mol Biol 1097:45–70
13. Turner DH, Mathews DH (2010) NNDB: the nearest neighbor parameter database for predicting stability of nucleic acid secondary structure. Nucleic Acids Res 38(Database issue):D280–D282
14. Lu ZJ, Turner DH, Mathews DH (2006) A set of nearest neighbor parameters for predicting the enthalpy change of RNA secondary structure formation. Nucleic Acids Res 34(17):4912–4924
15. Zuber J, Mathews DH (2019) Estimating uncertainty in predicted folding free energy changes of RNA secondary structures. RNA 25(6):747–754
16. Zuber J, Cabral B, McFadyen I, Mauger DM, Mathews DH (2018) Analysis of RNA nearest neighbor parameters reveals interdependencies and quantifies the uncertainty in RNA secondary structure prediction. RNA 24(11):1568–1582
17. Lorenz R, Bernhart SH, Honer Zu Siederdissen C, Tafer H, Flamm C, Stadler PF, Hofacker IL (2011) ViennaRNA Package 2.0. Algorithms Mol Biol 6:26
18. Reuter JS, Mathews DH (2010) RNAstructure: software for RNA secondary structure prediction and analysis. BMC Bioinformatics 11:129
19. Markham NR, Zuker M (2008) UNAFold: software for nucleic acid folding and hybridization. Methods Mol Biol 453:3–31
20. Andronescu M, Aguirre-Hernandez R, Condon A, Hoos HH (2003) RNAsoft: a suite of RNA secondary structure prediction and design software tools. Nucleic Acids Res 31(13):3416–3422
21. Zadeh JN, Wolfe BR, Pierce NA (2011) Nucleic acid sequence design via efficient ensemble defect optimization. J Comput Chem 32(3):439–452
22. Andronescu M, Fejes AP, Hutter F, Hoos HH, Condon A (2004) A new algorithm for RNA secondary structure design. J Mol Biol 336(3):607–624
23. Wolfe BR, Pierce NA (2015) Sequence design for a test tube of interacting nucleic acid strands. ACS Synth Biol 4(10):1086–1100
24. Bellaousov S, Kayedkhordeh M, Peterson RJ, Mathews DH (2018) Accelerated RNA secondary structure design using preselected sequences for helices and loops. RNA 24(11):1555–1567
25. Lee J, Kladwang W, Lee M, Cantu D, Azizyan M, Kim H, Limpaecher A, Yoon S, Treuille A, Das R, Ete RNAP (2014) RNA design rules from a massive open laboratory. Proc Natl Acad Sci U S A 111(6):2122–2127
26. Garcia-Martin JA, Clote P, Dotu I (2013) RNAiFOLD: a constraint programming algorithm for RNA inverse folding and molecular design. J Bioinforma Comput Biol 11(2):1350001
27. Lu ZJ, Mathews DH (2008) Efficient siRNA selection using hybridization thermodynamics. Nucleic Acids Res 36(2):640–647
28. Mathews DH, Burkard ME, Freier SM, Wyatt JR, Turner DH (1999) Predicting oligonucleotide affinity to nucleic acid targets. RNA 5(11):1458–1469
29. Bernhart SH, Muckstein U, Hofacker IL (2011) RNA accessibility in cubic time. Algorithms Mol Biol 6(1):3
30. Long D, Lee R, Williams P, Chan CY, Ambros V, Ding Y (2007) Potent effect of target structure on microRNA function. Nat Struct Mol Biol 14(4):287–294
31. Disney MD, Winkelsas AM, Velagapudi SP, Southern M, Fallahi M, Childs-Disney JL (2016) Inforna 2.0: a platform for the sequence-based design of small molecules targeting structured RNAs. ACS Chem Biol 11(6):1720–1728
32. Zuker M, Stiegler P (1981) Optimal computer folding of large RNA sequences using

thermodynamics and auxiliary information. Nucleic Acids Res 9(1):133–148

33. McCaskill JS (1990) The equilibrium partition function and base pair binding probabilities for RNA secondary structure. Biopolymers 29(6–7):1105–1119
34. Andronescu M, Condon A, Hoos HH, Mathews DH, Murphy KP (2010) Computational approaches for RNA energy parameter estimation. RNA 16(12):2304–2318
35. Do CB, Woods DA, Batzoglou S (2006) CONTRAfold: RNA secondary structure prediction without physics-based models. Bioinformatics 22(14):e90–e98
36. Rivas E, Lang R, Eddy SR (2012) A range of complex probabilistic models for RNA secondary structure prediction that includes the nearest-neighbor model and more. RNA 18(2):193–212
37. Fang R, Moss WN, Rutenberg-Schoenberg M, Simon MD (2015) Probing Xist RNA structure in cells using targeted structure-seq. PLoS Genet 11(12):e1005668
38. Zuker M (2003) Mfold web server for nucleic acid folding and hybridization prediction. Nucleic Acids Res 31(13):3406–3415
39. Mathews DH, Turner DH (2002) Experimentally derived nearest-neighbor parameters for the stability of RNA three- and four-way multibranch loops. Biochemistry 41(3):869–880
40. Ward M, Datta A, Wise M, Mathews DH (2017) Advanced multi-loop algorithms for RNA secondary structure prediction reveal that the simplest model is best. Nucleic Acids Res 45(14):8541–8550
41. Whipple JM, Lane EA, Chernyakov I, D'Silva S, Phizicky EM (2011) The yeast rapid tRNA decay pathway primarily monitors the structural integrity of the acceptor and T-stems of mature tRNA. Genes Dev 25(11): 1173–1184
42. Mustoe AM, Busan S, Rice GM, Hajdin CE, Peterson BK, Ruda VM, Kubica N, Nutiu R, Baryza JL, Weeks KM (2018) Pervasive regulatory functions of mRNA structure revealed by high-resolution SHAPE probing. Cell 173(1):181–195 e118
43. Payea MJ, Sloma MF, Kon Y, Young DL, Guy MP, Zhang X, De Zoysa T, Fields S, Mathews DH, Phizicky EM (2018) Widespread temperature sensitivity and tRNA decay due to mutations in a yeast tRNA. RNA 24(3):410–422
44. Zuber J, Sun H, Zhang X, McFadyen I, Mathews DH (2017) A sensitivity analysis of RNA folding nearest neighbor parameters identifies a subset of free energy parameters with the greatest impact on RNA secondary structure prediction. Nucleic Acids Res 45(10):6168–6176
45. Liu B, Mathews DH, Turner DH (2010) RNA pseudoknots: folding and finding. F1000 Biol Rep 2:8
46. Liu B, Shankar N, Turner DH (2010) Fluorescence competition assay measurements of free energy changes for RNA pseudoknots. Biochemistry 49(3):623–634
47. Wyatt JR, Puglisi JD, Tinoco I Jr (1990) RNA pseudoknots. Stability and loop size requirements. J Mol Biol 214(2):455–470
48. Hajdin CE, Bellaousov S, Huggins W, Leonard CW, Mathews DH, Weeks KM (2013) Accurate SHAPE-directed RNA secondary structure modeling, including pseudoknots. Proc Natl Acad Sci U S A 110(14):5498–5503
49. Aalberts DP, Nandagopal N (2010) A two-length-scale polymer theory for RNA loop free energies and helix stacking. RNA 16(7): 1350–1355
50. Rivas E, Eddy SR (1999) A dynamic programming algorithm for RNA structure prediction including pseudoknots. J Mol Biol 285(5): 2053–2068
51. Andronescu MS, Pop C, Condon AE (2010) Improved free energy parameters for RNA pseudoknotted secondary structure prediction. RNA 16(1):26–42
52. Dirks RM, Pierce NA (2003) A partition function algorithm for nucleic acid secondary structure including pseudoknots. J Comput Chem 24(13):1664–1677
53. Jacobson HS, Stockmayer WH (1950) Intramolecular reaction in polycondensations. I. The theory of linear systems. J Chem Phys 18:1600–1606
54. Ward M, Sun H, Datta A, Wise M, Mathews DH (2019) Determining parameters for non-linear models of multi-loop free energy change. Bioinformatics 35(21):4298–4306

Chapter 2

VarGibbs Usage in the Optimization of Nearest-Neighbor Parameters and Prediction of Melting Temperature of RNA Duplexes

Izabela Ferreira and Gerald Weber

Abstract

The nearest-neighbor (NN) model is a general tool for the evaluation for oligonucleotide thermodynamic stability. It is primarily used for the prediction of melting temperatures but has also found use in RNA secondary structure prediction and theoretical models of hybridization kinetics. One of the key problems is to obtain the NN parameters from melting temperatures, and VarGibbs was designed to obtain those parameters directly from melting temperatures. Here we will describe the basic workflow from RNA melting temperatures to NN parameters with the use of VarGibbs. We start by a brief revision of the basic concepts of RNA hybridization and of the NN model and then show how to prepare the data files, run the parameter optimization, and interpret the results.

Key words RNA hybridization, Nearest-neighbor model, RNA melting temperature, Salt correction, Multi-parameter optimization

1 Introduction

The nearest-neighbor model (NN) was developed in the early 1970s [1], based on observations that CG content alone is insufficient for providing a satisfactory prediction of oligonucleotide melting temperatures [2]. It is now widely used to describe and predict the thermodynamic features of RNA through enthalpy and entropy NN parameters [3–6]. Of its numerous theoretical applications in molecular biology, perhaps the best known is secondary structure prediction [7–12]. Moreover, sequence-dependent transcription elongation models also require detailed knowledge for NN parameters [13] as well as target affinity [14]. The derived NN enthalpy parameters are also used in mesoscopic theoretical models for the calculation of thermodynamic properties of oligonucleotides [15, 16]. Note that the NN model differs from the model of independent short sequences (ISS) employed by Gray

Ronny Lorenz (ed.), *RNA Folding: Methods and Protocols*, Methods in Molecular Biology, vol. 2726,
https://doi.org/10.1007/978-1-0716-3519-3_2,

Gray [17, 18] through the use of initiation factors that will be discussed later.

The data on which the NN model is based usually involves the measurements of melting temperatures T_m at a range of total strand concentrations C_t to form a Van't Hoff plot $T_m^{-1} \times \log C_t$, from which the total sequence enthalpy and entropy variations, ΔH^{tot} and ΔS^{tot}, are then interpolated [19]. With these total sequence enthalpy and entropy variations at hand, it is possible to calculate the NN parameters with algebraic methods [20]. However, apart from the additional uncertainty in the final NN parameters, this procedure is not feasible when the total thermodynamic parameters are not available or cannot be calculated using algebraic methods.

The `VarGibbs` software package was developed out of the need to calculate NN parameters from melting temperature data that cannot be handled by ordinary algebraic methods such as singular value decomposition, and is specifically aimed towards optimizing melting temperature predictions [21]. We also showed that we could calculate the NN parameters using a data set for which total sequence enthalpy and entropy parameters are unknown [21].

The `VarGibbs` can be used for three basic applications: as a tool for NN analysis, to predict or to salt adjust melting temperatures, and for NN parameter calculations which is its main purpose. Typically, it is possible to apply `VarGibbs` in the following practical situations:

1. To perform a detailed analysis of NN combinations for a set of RNA sequences
2. To calculate melting temperature predictions for existing or new parameters
3. To recalculate melting temperatures at different salt concentrations
4. To calculate new NN parameters for sets of RNA melting temperatures
5. To evaluate the impact of experimental uncertainties on NN parameters

Note that for simple melting temperature calculations there are other options such as online calculators such as https://dna-utah.org/tm/ and other programs [22–25].

Herein, we will provide a workflow guideline that may be useful to treat RNA sequences in some of the situations described above, such as simple temperature or enthalpies/entropies prediction, optimization of parameters and predictions, as well as their uncertainty estimations. For other types of calculations using `VarGibbs` or other types of applications, we recommend a reading of the user guide included in the software package.

2 Materials

VarGibbs is available for Linux systems; *see* Subheading 3.1 for installation instructions. System requirements for VarGibbs are very modest; for all the applications discussed here, the following system requirements are sufficient:

- Processor speed: Minimum 1 GHz; recommended 2 GHz.
- Memory (RAM): Minimum 100 MiB; recommended 1 GiB for large data sets.
- Disk space: Minimum 100 MiB; recommended 1 GiB for large data sets.

For the applications and scripts discussed here the processing time is of the order of few seconds to a few minutes.

The prediction mode -calc=prediction is also available as an online interface in the address (https://bioinf.fisica.ufmg.br/app/comparetm.pl).

3 Methods

3.1 Installing

Pre-compiled packages for 31 Linux distributions can be found at (https://bioinf.fisica.ufmg.br/software) or in the OpenSuse repository (https://build.opensuse.org/package/show/home:drgweber/VarGibbs). The easiest way to install is to visit one of these sites and search VarGibbs for your Linux distribution and download the appropriate packages. Install them using your systems package manager which should take care of obtaining the correct libraries which are needed for VarGibbs to run.

For instance OpenSuse Leap 15.3, download the package and public key at the command line

```
wget https://bioinf.fisica.ufmg.br/software/var-
gibbs-3.1/\
openSUSE_Leap_15.3/VarGibbs-3.1-lp153.2.1.x86_64.
rpm
wget https://build.opensuse.org/projects/home:
drgweber/public_key
```

as superuser, install first the key then the public key, then the .rpm package

```
rpm --import public_key
zypper install VarGibbs-3.1-lp153.2.1.x86_64.rpm
```

Note that you can skip the public key installation, but then will receive a warning regarding *Signature verification failed* which you can bypass by adding

```
-no-gpg-checks.
```

```
zypper --no-gpg-checks install VarGibbs-3.1-lp153.2.1.x86_64.rpm
```

If you do not have a Linux system, we recommend the installation of a distribution in a virtual machine and then the installation of the appropriate `VarGibbs` packages. After installation, there will typically be a binary executable file located in

```
/usr/bin/vargibbs
```

Model parameter files and existing melting temperature sequences are located in

```
/usr/share/VarGibbs/data
```

and example scripts are in

```
/usr/share/VarGibbs/examples
```

The documentation should be located at

```
/usr/share/doc/packages/VarGibbs
```

Note that the exact location will depend on the Linux distribution that you are using. The source code is also available and should compile and run under any Unix type system, in which case you need to copy the files manually to your system folders.

3.2 Running

`VarGibbs` is run exclusively in a command line, such as `bash` or `csh`. Here, all examples are for the `bash` environment. Running `VarGibbs` without any arguments, that is, just typing

```
vargibbs
```

will show you the information about version, which compiler was used and date of build, as well as all available options and then exit. All messages are printed to `stdout` and error messages to `stderr`. The options should be given in form `-option=value`. `VarGibbs` has an extensive and growing range of options; here we will discuss only those regarding RNA parameter optimization; for other options please see the documentation that is included in the package.

`VarGibbs` requires at least the definition of output file names, `-o=`, and the type of action to be performed through the option `-calc=`

1. `-calc=nncheck` reads the sequence and displays how `VarGibbs` will break up the sequences into base pairs, nearest-neighbors, and trimers. No calculations will be performed at this stage.
2. `-calc=prediction` reads the sequence and a parameter file and calculates melting temperature predictions.
3. `-calc=minlocal` and `-calc=minglobal` perform local or global parameter optimizations.

3.3 Nearest-Neighbor (NN) Model

`VarGibbs` implements the nearest-neighbor (NN) model where the enthalpy and entropy variations of a given ith sequence, ΔH_i^{tot} and ΔS_i^{tot}, are subdivided into contributions of consecutive base pairs [1, 20]. The model commonly uses two types of parameters: context-dependent parameters, $\Delta H_\alpha, \Delta S_\alpha$, and initiation parameters, $\Delta H^c, \Delta S^c$, which are typically correction factors such as symmetry correction (ΔS^{symm}) or terminal initiation (ΔH^{term}). These parameters are then summed

$$\Delta H_i^{\text{tot}} = \sum_\alpha n_{i\alpha} \Delta H_\alpha + \sum_c n_i^c \Delta H^c \tag{1}$$

$$\Delta S_i^{\text{tot}} = \sum_\alpha n_{i\alpha} \Delta S_\alpha + \sum_c n_i^c \Delta S^c \tag{2}$$

where $n_{i\alpha}$ are the occurrences of nearest-neighbor types α in the ith sequence. Note that this representation is very straightforward and that there are similar NN schemes using a reduced representation such as by Licinio and Guerra [26]. The initiation parameters are not standardized and vary considerably among the various authors. For instance, SantaLucia et al. [27] use a specific initiation factor when terminal 5′-TA-3′ are present in the duplex, which are not used by other authors. Here, we will refer collectively to the parameters ΔH_α, ΔS_α, ΔH^c, and ΔS^c as the set $\{P\}$, that is,

$$\{P\} = \{p_1, p_2, p_3, \ldots\}, \quad p = \Delta H_\alpha, \Delta S_\alpha, \Delta H^c, \Delta S^c \tag{3}$$

and the predicted melting temperature T_i' (in kelvin) is a function of these parameters

$$T_i'(\{P\}) = \frac{\Delta H_i^{\text{tot}}}{\Delta S_i^{\text{tot}} + R \ln \dfrac{C_t}{f C_0}} \tag{4}$$

where $f = 1$ for self-complementary sequences and $f = 4$ otherwise. C_t is the total strand concentration and C_0 is a reference concentration taken as 1 μM. Note that `VarGibbs` uses Eq. (4) but outputs the temperatures in °C.

3.3.1 Units and Gibbs Free Energy

Note that most authors omit the reference concentration C_0 in Eq. (4), in which case you should enter C_t always in μM. Salt concentration is always given in mM. R is the molar gas constant taken as 1.987 cal/(K·mol) and we use the thermochemical calorie which corresponds to 4.184 J [28]. Note that older texts sometimes refer to cal/(K· mol) as eu or e.u. (entropy unit) [4]. All temperatures are in ° C; when necessary it is internally converted to (from) kelvin by adding (subtracting) 273.15 [28]. The Gibbs free energy ΔG_i at 37 °C is given by

$$\Delta G_i(37\ ^\circ\mathrm{C}) = \Delta H_i - (273.15 + 37)\Delta S_i \tag{5}$$

3.3.2 Self-Complementarity Versus Non-self-Complementarity

When a single RNA is complementary to itself, that is, when two identical strands X can hybridize to form an anti-parallel duplex X·X

$$2\mathrm{X} \leftrightarrow \mathrm{X} \cdot \mathrm{X} \tag{6}$$

then this sequence is called a *self-complementary* (SC) sequence [29]. In contrast, for an awkwardly named non-self-complementary (NSC) sequence, a duplex is formed by different strands, X and Y,

$$\mathrm{X} + \mathrm{Y} \leftrightarrow \mathrm{X} \cdot \mathrm{Y} \tag{7}$$

It is important to realize that the implications of this difference go far beyond the simple f factor in Eq. (4). Strictly speaking, the enthalpy and entropy variations of a SC duplex are not equivalent to those of an NSC duplex, that is, $\Delta H^{\mathrm{SC}} \neq \Delta H^{\mathrm{NSC}}$ and $\Delta S^{\mathrm{SC}} \neq \Delta S^{\mathrm{NSC}}$. The question is then: how to make SC and NSC enthalpies and entropies comparable? The form of Eq. (4) may mislead the reader to think that the answer is to measure the temperature at a total strand concentration NSC C_t four times larger than for SC sequences and indeed it is not uncommon to see authors showing their results in $C_t^{\mathrm{NSC}} = 4C_t^{\mathrm{SC}}$ [27]. However, if one thinks in terms of Boltzmann statistical physics, it becomes obvious that there is a reduction of 2^N in the available configurations due to the symmetry of the SC duplex X·X in comparison to the NSC duplex X·Y. This implies in a reduction in entropy for SC sequences

$$\Delta S^{\mathrm{symm}} = -Nk_B \ln(2) = -R\ln(2) \approx -1.4\ \mathrm{cal/K.mol} \tag{8}$$

where k_B is the Boltzmann constant. This is the origin of the symmetry correction factor which is ubiquitous in the NN models [4, 5]. For RNA, we allowed this factor to vary with sodium concentration and found that it can change substantially [6] from this value.

Some authors [5] go a step further and add the symmetry correction straight into Eq. (4) by modifying the entropy $\Delta S = \Delta S' - R\ln(2)$

$$T^{\mathrm{SC}} = \frac{\Delta H}{\Delta S - R\ln(2) + R\ln\frac{C_t}{C_0}} = \frac{\Delta H}{\Delta S' + R\ln\frac{C_t}{2C_0}} \tag{9}$$

and as a result display their melting temperatures using $C_t^{\mathrm{NSC}} = 2C_t^{\mathrm{SC}}$ [5]. Note that here, we will use exclusively Eq. (4) and when necessary recourse to the symmetry correction factor ΔS^{symm}.

3.3.3 NN Representation

First, we will explain, by means of practical examples, how VarGibbs takes a duplex sequence and breaks it down into its NN components. This is a crucial aspect of the software as it is used to evaluate Eqs. (1) and (2), and understanding it may avoid some common pitfalls.

The notation for RNA canonical base pairs (BP) in VarGibbs uses double characters like AU and CG. In the same way, nearest neighbors (NN) are interpreted as two sequential base pairs (BP) joined with an underscore character _ as in BP_BP. The convention used by VarGibbs is of writing AU_CG, where AU is the first base pair and CG the second base pair such that

$$\begin{matrix} 5' - \mathrm{AC} - 3' \\ 3' - \mathrm{UG} - 5' \end{matrix}, \quad \rightarrow \quad \mathrm{AU\_CG}$$

NN Symmetry Due to the double-stranded symmetry of anti-parallel duplexes, the notation GC_UA is equivalent to AU_CG, that is, they both represent the same physical molecule. To understand this one should perform two rotations

$$\begin{matrix} 5' - \mathbf{GU} - 3' \\ 3' - \mathrm{CA} - 5' \end{matrix} \longrightarrow \begin{matrix} 3'-\mathrm{CA}-5' \\ 5'-\mathbf{GU}-3' \end{matrix} \longrightarrow \begin{matrix} 5' - \mathrm{AC} - 3' \\ 3' - \mathbf{UG} - 5' \end{matrix} \tag{10}$$

where we flip the duplex first around its horizontal axis and then around its vertical axis such that the upper strand is again in $5' \rightarrow 3'$ direction (the $5'$ G was highlighted to guide the eye). This is also the reason for the adopted representation: to work out the symmetric equivalent of GC_UA all you have to do is to read it backward. In this sense, self-complementary NNs such as AU_UA or GC_CG are always represented by palindromes. Whenever there are two equivalent representations of a NN, VarGibbs keeps the one that precedes lexically, that is, it will use only AU_CG but not GC_UA. If for some reason GC_UA is given, it will silently be converted to AU_CG.

3.3.4 Parameter Format

`VarGibbs` understands two types of parameters, those that are NN-dependent, like ΔH_α, and ΔS_α which are formatted as in the following example

$$\Delta H_{\mathrm{AU\_AU}} \rightarrow \texttt{AU\_AU : enthalpy}$$
$$\Delta S_{\mathrm{AU\_AU}} \rightarrow \texttt{AU\_AU : entropy}$$

that is, first the context NN, followed by a colon and then the parameter name. If the parameter applies to two or more different NN contexts, we list all contexts separated by colons as in

```
AU_CG:AU_GC:enthalpy
```

which means that the same parameter applies to `AU_CG` and `AU_GC`. It is not necessary to indicate the symmetric equivalent of an NN, that is, it is ***redundant*** to write

```
AU_CG:GC_UA:enthalpy
```

However if you do, `VarGibbs` will not complain as internally `AU_CG` and `GC_UA` mean exactly the same. On the other hand, sometimes it may be useful to maintain this redundancy for documentation purposes as people unfamiliar with `VarGibbs` inner workings may be misled into thinking that certain NNs are missing from your parameter files.

If the parameter applies to any NN context, use an asterisk $*$ as in

```
*:enthalpy
```

The initiation parameters have specific names which are shown in Table 1, but all of them end in `enthalpy` or `entropy`. Sodium concentration is formatted by the ion type, a colon, and by the word `concentration`

$$[\mathrm{Na}^+] \rightarrow \texttt{Na+ : concentration}$$

This parameter syntax is used in parameter files as well as in the command line.

3.3.5 Reading Duplexes

For whole duplexes, `VarGibbs` always takes the main strand in $5' \rightarrow 3'$ and the complementary $3' \rightarrow 5'$ direction. For instance

$$5' - \mathrm{GACGUC} - 3'$$
$$3' - \mathrm{CUGCAG} - 5'$$

will be represented by the main strand GACGUC and complementary CUGCAG. The reason for this is that when the main and complementary strands are written one atop the other, as shown above, you will immediately be able to work out the Watson-Crick

Table 1
Summary of initiation factors relevant for RNA. Entropies are given in cal/(K·mol) and enthalpies in kcal/mol. References are those where to our knowledge these types of initiation factors were first introduced. For other non-RNA initiations please see VarGibbs documentation

Parameter	Description	Ref.
initiation_CG_entropy	If the sequence contains a *at least* one CG base pair	[27]
initiation_CG_enthalpy	If the sequence contains a *at least* one CG base pair	
terminal_CG_enthalpy	For each terminal CG base pair	[30]
terminal_CG_entropy	For each terminal CG base pair	[30]
terminal_AU_enthalpy	For each terminal AU base pair	
terminal_AU_entropy	For each terminal AU base pair	
symmetry_correction_entropy	Correction applied to self-complementary sequences; see also discussion in Subheading 3.3.2	[27, 30]
symmetry_correction_enthalpy	Correction applied to self-complementary sequences	
initiation_entropy	A fixed correction applied to all sequences	
initiation_enthalpy	A fixed correction applied to all sequences	

base pairs. Single duplexes can be passed on via the command line -seq=GACGUC and -cseq=CUGCAG or read from a file in similar format

```
temperature
GACGUC CUGCAG 0 0 0
```

See Subheading 3.4.1 for further information regarding formatting. Here we simply set the C_t, measured T_m and salt concentrations (last three columns) to zero.

At the command line, if the -seq= is given, the complementary -cseq can in principle be omitted, and then VarGibbs works out the complementarity. In this case please also provide -duplextype=RNA to ensure that VarGibbs uses AU and not AT as complementary pair. Note that the duplex does not have to be formed solely of AU and CG base pairs; any other type of base pair is accepted, for example,

-seq=GACGUC -cseq–CUGUAC

has a GU base pair at is fourth position. However, it is also necessary to provide the corresponding parameters involving GU base pairs. It is also possible to choose completely different alphabetical characters, for example, X or Z, in case you wish to represent different types of nucleotides or even to represent a gap. For instance, we have used X to represent a missing nucleotide for bulges in RNA [31] or to represent a Cy3 fluorophore attached to DNA [32]. In these cases you need to inform `VarGibbs` that you are expanding the set of known nucleotide representations through the `-dict=` option; please see the package documentation for details. In all cases, make sure that you also provide the corresponding parameters consistent with your expanded representation.

3.4 Preparing the Input Files

`VarGibbs` requires two types of input files for any application: a file containing the RNA sequences and another file containing the model parameters. The distribution package comes with a number of sequence files which can be found in the folder `/usr/share/VarGibbs/data`, a summary of existing parameter files is given in Table 2.

3.4.1 Sequence File with Temperatures

For parameter optimization, Subheading 3.7, it is necessary to provide the primary sequences as well as their melting temperatures. Files formatted in this way should start with a single word `temperature` ① in the first line, as in the following example:

Table 2
RNA parameters provided with `VarGibbs`. The filenames in the `VarGibbs` package are given by the parameter code and the `.par` suffix

Parameter code	Description	Year	Reference
`P-FR86`	One of the oldest RNA parameters, mainly of historic interest, for $[Na^+]$ 1000 mM	1986	[4, 8]
`P-XIA98`	Canonical RNA, one of the most frequently used parameters in the literature, for $[Na^+]$ 1000 mM	1998	[5]
`P-CH12`	GU mismatches in RNA, for $[Na^+]$ 1000 mM	2012	[33]
`AOP-FIFRW-[Na+]`	RNA for sodium concentrations `[Na+]`=71, 121, 221, 621, and 1021 mM, with initiation parameters from `P-XIA98`	2019	[6]
`AOP-VIFRW-[Na+]`	RNA for sodium concentrations `[Na+]`=71, 121, 221, 621, and 1021 mM, with optimized initiation parameters	2019	[6]
`P-IP-VIF-Na[Na+]`	Interpolated `AOP-VIFRW` parameters for sodium concentrations between 10 and 1021 mM	2019	[6]

```
temperature① #melting temperatures②
GCGCGC③ CGCGCG④ 55.3⑤ 71⑥ 100⑦
ACCGGU③ UGGCCA④ 43.7⑤ 71⑥ 100⑦
AGCGCU③ UCGCGA④ 41.3⑤ 71⑥ 100⑦ #short sequence②
```

You may add comments in the first line after a space and a # sign ②. The following lines should contain the main strand from 5 ′ to 3 ′③, the complementary strand 3 ′ to 5 ′④, the measured melting temperature ⑤, salt concentration ⑥, and the total strand concentration C_t ⑦. Comments can be added after a # sign. Note that older versions of `VarGibbs` may not work with comments in these types of files.

Sequence files are passed to `VarGibbs` through the `-data=` option, and multiple files can be provided by a comma separated list without spaces as in

```
-data=file1.dat,file2.dat,file3.dat
```

The file extensions are irrelevant and file paths should be given relative to the folder where `VarGibbs` is being run.

3.4.2 Sequence File with Enthalpies and Entropies

Another type of file uses total sequence enthalpies ΔH^{tot} and entropies ΔS^{tot}. The following example shows a file of this type:

```
gibbs①
AGAUAUCU② UCUAUAGA③ -64.5④ -186.8④ 1000⑥
GUCGAC② CAGCUG③ -53.6④ -150.1④ 1000⑥
```

where the file should start with a single line and the word `gibbs` ①, and from the second line on should list the main sequence main strand from 5 ′ to 3 ′②, followed by the complementary strand 3 ′ to 5 ′③. Next should come the total enthalpy variation ④, total entropy variation ⑤, and ending with the salt concentration ⑥. Note that some authors display their tables in $-\Delta H$ and $-\Delta S$ as in Freier et al. [4]; therefore, be sure not to omit the negative sign of the enthalpy and entropy when it is the case. All values should be in units outlined in Subheading 3.3.1. Note that this type of file also requires the option `-ct` to provide the total strand concentration C_t.

3.4.3 Parameter File

A parameter file is required for temperature prediction and parameter optimization. `VarGibbs` has several ready-to-use files which may be used as input parameters in new minimizations or for simple predictions. A summary of parameter files for RNA is given in Table 2. The parameter format follows that described in

Subheading 3.3.4 and numerical values should be given in the units described in Subheading 3.3.1.

The file should start with an arbitrary identification ① in the first line, followed by one type of parameter per line and its numerical value as in

```
xia98-original-table4 ①
#xia98 doi:10.1021/bi9809425 ②
Na+:concentration 1000
AU_AU:enthalpy     -6.82 #5AA3/3UU5 ③
AU_AU:entropy      -19.0 #5AA3/3UU5
terminal_AU_enthalpy 3.72  #per terminal AU pair
terminal_AU_entropy 10.5   #per terminal AU pair
```

lines starting with # ② are ignored as well as any comment after the numerical value ③. The complete file can be found as `P-XIA98.par` in the package `data` folder.

Sometimes one may have only an approximate knowledge of parameter values; this is typically the case for parameter optimization. In this case one can provide a file with generic values ① as in

```
test
*:entropy ①      -23.6 #my best guess
*:enthalpy ①     -8.4  #my best guess
initiation_CG_entropy    -5.9
initiation_AT_entropy    -20
terminal_5TA3_enthalpy    0.4
symmetry_correction_entropy      -20
```

Generic Versus Specific Parameters `VarGibbs` will only use a generic parameter if it cannot find the one for a specific NN context it needs. When this happens a warning like this one is echoed

```
no specific Enthalpy or Entropy values, falling back to generic values
```

Parameter files should be provided through the `-par=` option. Multiple parameter files can be given in a comma separated list with no spaces as in

```
-par=file1.par,file2.par,file3.par
```

The order in which the files are given can make a difference if the files hold differing values for the same parameter. In this case `VarGibbs` uses that **last** that was read and issues a warning like this

```
Warning:replacing AT_AT:enthalpy old value=-9.3522 new value=-8.41043
Warning:replacing AT_AT:entropy old value=-26.487 new value=-18.8472
```

Verifying Parameter Reading If you are providing multiple parameter files, it can become quite complex to understand how exactly VarGibbs interpreted them. You can use the option -printusedpar=1 which will generate a file with extension .usedpar that will show you where VarGibbs has got its parameters from as in this example:

```
parameters
#parameter map has 60 elements
AU_AU:enthalpy -6.82 [-6.820000:P-XIA98.par ①] (1)
AU_AU:entropy -19 [-19.000000:P-XIA98.par ①] (1)
AU_GU:enthalpy -3.96 [-3.960000:P-CH12.par ②] (1)
AU_GU:entropy -11.6 [-11.600000:P-CH12.par ②] (1)
```

where the AU_AU came from P-XIA98.par ① and AU_GU from P-CH12.par ②. Note that the file generated by -printusedpar=1 can be reused in -par= if needed.

3.5 Application: Sequence Analysis

-calc=nncheck

Before any other type of calculation, VarGibbs processes the sequences to work out the nearest-neighbor context following the rules described in Subheading 3.3.3. Whenever you prepare new sequences files, it is strongly advised to verify how they are being read by VarGibbs, for this use the option -calc=nncheck which results in a very detailed report. The report file which ends in .nncheck is machine readable and can be parsed with simple scripts for other uses. This is the only application that does not require a parameter file, only a sequence file or passing the sequence at the command line

```
vargibbs -o=analysis -seq=GACGUC -cseq=CUGCAG
-calc=nncheck
```

For this example the sequence GACGUC/CUGCAG, -calc=nncheck will generate a report analysis.nncheck showing how it is broken down into NNs

```
Exact_neighbours   : GC_AU AU_CG CG_GC GC_UA UA_CG  (5)
```

Then the symmetry operation Eq. (10) is applied

```
Reduced_neighbours : GC_AU AU_CG CG_GC AU_CG GC_AU (5)
```

meaning that actually this sequence has only three different types of NN

```
Reduced_neighbours_set      : (3) AU_CG=2 CG_GC=1 GC_AU=2
```

VarGibbs also analyzes the terminal characteristics

```
Terminal base pairs at 5: GC
Terminal base pairs at 3: CG
Terminal base pairs (BP symmetry reduced) at 5: CG
Terminal base pairs (BP symmetry reduced) at 3: CG
Terminal_5TA3: 0 Terminal_AU: 0 Terminal_AT: 0 Terminal_CG: 2
Terminal_dArU: 0 Terminal_dTrA: 0 Terminal_dCrG: 0 Terminal_dGrC: 0
```

which is used to determine which of the terminal parameters of Table 1 apply to this sequence.

If a sequence file is given ①

```
vargibbs -o=analysis -data=D-XIA98.dat ① -calc=nncheck
```

then in addition to sequence-by-sequence information, a summary of all sequences will be given:

```
10 unique nearest-neighbours in dataset  (NP+P, NN symmetry reduced)
Neighbours set      : AU_AU=55 ① AU_CG=67 AU_GC=83 AU_UA=40 CG_AU=85 CG_CG=51
↪  CG_GC=60 GC_AU=84 GC_CG=74 UA_AU=33
```

where the numbers ① indicate the occurrence of NN combinations in the file.

Please see the package documentation for a full description of all results produced by -calc=nncheck.

3.6 Application: Temperature Prediction

-calc=prediction

Given a duplex, or a set of duplexes, temperature prediction can be performed by selecting existing parameter sets. Prepare the sequence files as described in Subheading 3.4.1, or chose existing files, and select the appropriate parameter. Parameters available for RNA are shown in Table 2 and have the form described in Subheading 3.4.3. Consider the following example:

```
vargibbs -data=D-XIA98.dat ① -o=simple-prediction ② \
-par=P-XIA98.par -calc=prediction
```

where -data=① provides the file name that contains the sequence information and the result will be in file simple-prediction.dat② which has the following form:

```
Main/Complementary alpha salt_concentration species_concentration
↪ temperature.measured temperature.adjusted temperature.predicted
GACGUC/CUGCAG 1 0 100 0 0 45.1147① 0 0 -54.71② 0 0 -153.6③ 0 0 0 0 0
```

where the melting temperature is $T_m = 45.1147\ ^\circ$ C ①, and $\Delta H_i^{\mathrm{tot}} = -54.71$ kcal/mol ② and $\Delta S_i^{\mathrm{tot}} = -153.6$ cal/(K·mol) ③.

Instead of a data file containing the sequences you may also simply provide the sequence ① info directly in the command line together with the total strand concentration $C_t = 100$ μM ②

```
vargibbs -calc=prediction -o=simple-prediction -par=P-XIA98.par \
-seq=AAGGUUGGAA① -cseq=UUCCAACCUU① -ct=100②
```

Note that this is only useful if you wish to calculate the melting temperatures for a single duplex and avoid setting up a new file for this.

Processing Time Less than a second per sequence on a laptop.

3.6.1 Result File of Type .dat

Here we will provide a more detailed description of the file with extension .dat. It shows all sequences, available information, and predicted quantities as in the following example:

```
Main/Complementary alpha① salt_concentration② species_concentration③
↪ temperature.measured④ ...
CCGG/GGCC 1① 1000② 100③ 27.2④ 27.2⑤ 25.2588⑥ 0 0 -33.81⑦ 0 0 -95⑧ 0 0 0
↪ 0 0
ACGCA/UGCGU 4① 1000② 200③ 29.4④ 29.4⑤ 28.9846⑥ 0 0 -36.31⑦ 0 0 -100.5⑧
↪ 0 0 0 0 0
```

The first line describes the content of each column (truncated here for claritya). The column alpha ① describes the detected self-complementarity and is the factor f in Eq. (4). Next comes the $[\mathrm{Na}^+]$ concentration ②, the total strand concentration C_t ③, the measured T_i ④, the salt adjusted T_i ⑤, the predicted T_i' ⑥, the predicted $\Delta H_i^{\mathrm{tot}}$ ⑦, and $\Delta S_i^{\mathrm{tot}}$⑧. All units are described in Subheading 3.3.1. The remaining columns are all zero and are not used by VarGibbs, but are kept for compatibility with other programs from our group.

3.6.2 Result File of Type `.ver`

The file with extension `.ver` contains the prediction quality parameters; it shows how the measured T_i compares to the predicted $T'_i(\{P\})$:

```
average diff_deviation sqr_diff sqrt_diff2 relative_sqr_diff
Tm 1.35812 ① 1.06513 324.709 ② 1.72597 0.258936
```

The first number ① is the average difference

$$\langle \Delta T \rangle = \frac{1}{N}\sum_{i=1}^{N} |\Delta T_i|, \quad \Delta T_i = T_i - T'_i(\{P\}) \qquad (11)$$

where N is the total number of sequences. $\langle \Delta T \rangle$ is one of the most commonly used quality parameters found in the literature. The other value of interest, especially for parameter optimization that we will discuss next, is χ^2 ②

$$\chi^2 = \sum_{i}^{N} \left[T_i - T'_i(\{P\})\right]^2 \qquad (12)$$

Note that χ^2 also depends on the number of sequences N, that is, it will be larger if more sequences are included.

3.7 Application: Parameter Optimization

While `VarGibbs` has a set of readily available parameters, its primary aim is to optimize new parameters $\{P\}$ such that the melting temperatures $T'_i(\{P\})$ predicted from Eq. (4) are as close as possible to the measured temperatures T_i. Mathematically, this corresponds to the minimization of the squared difference in Eq. (12), where `VarGibbs` changes the parameters $\{P\}$ continuously trying to get the smallest χ^2; this is algorithmically performed via the Nelder-Mead minimization method [34]. However, the ubiquitous difficulty in this type of minimization procedure is the occurrence of multiple minima and the possibility that the algorithm stops at a minimum which is not the smallest possible χ^2. In practical terms this means that depending on the initial set $\{P\}_0$, it may arrive at a less optimal χ^2. Therefore, the trick is to try many different initial parameters $\{P\}_0$ in the hope that one of them may arrive at the lowest χ^2. Here, we will refer to an attempt starting at a single $\{P\}_0$ as a *local minimization*. Attempting multiple $\{P\}_0$ will be called a *global minimization*, which is basically running many local minimization. `VarGibbs` has implemented both types of minimizations, global and local. You will rarely do local minimizations, but you need to understand the concept in order to perform global minimizations. A third type of minimization is used specifically to estimate the impact of the experimental uncertainty. Figure 1 shows an overview of how local and global minimizations relate to each other and to the control parameters in `VarGibbs`.

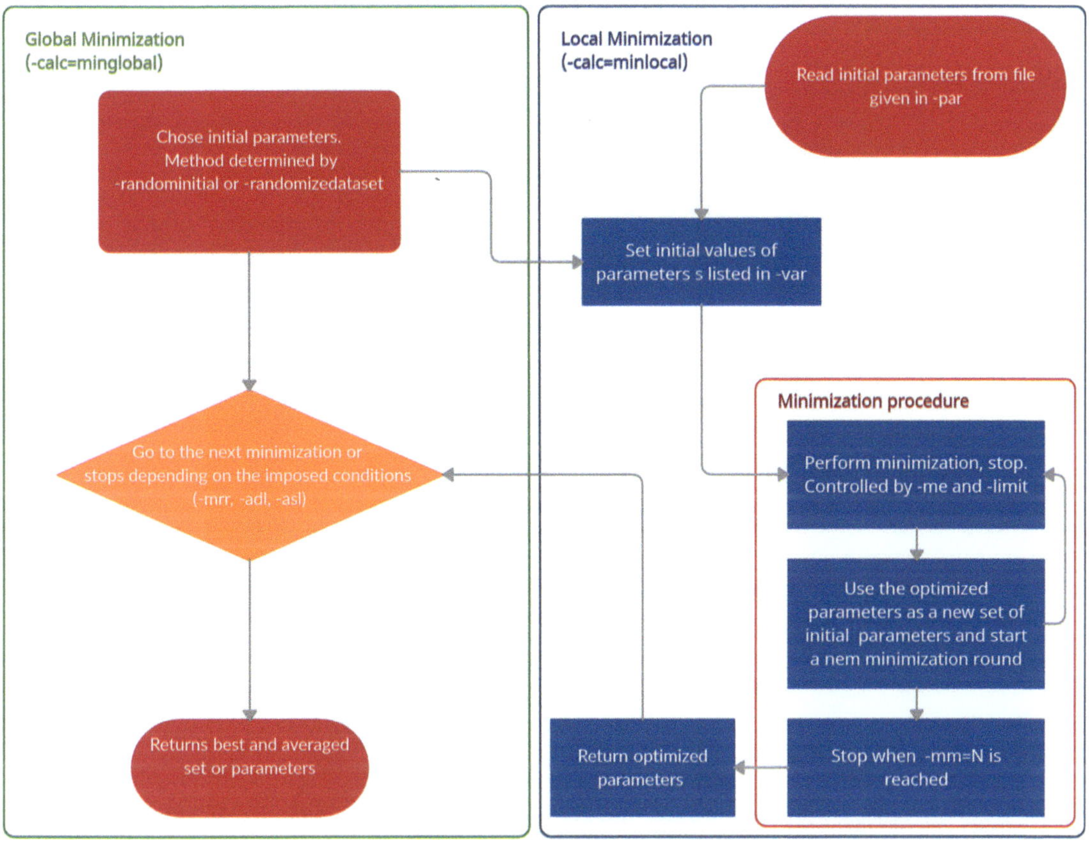

Fig. 1 Workflow and dependencies of global and local minimizations in `VarGibbs`

3.8 Local Minimizations `-calc=minlocal`

For local minimization we use the unmodified experimental melting temperatures T_i, provided via the `-data=` option together with a set of tentative initial parameters P_0 passed via the `-par=` option. The parameters that should be varied are given in a comma separated list through the `-var=` option. The parameters given in `-var=` will be varied continuously until the lowest χ^2 is obtained. The algorithm will end this process depending on a set of stop criteria. An example is shown here

```
vargibbs -var=AU_AU:entropy,AU_CG:entropy,AU_GC:entropy,AU_UA:entropy \
-data=experimental.dat -calc=minlocal -o=results -par=P-XIA98.par ①
```

where the parameter set `P-XIA98` ① was used as P_0; *see* Table 2.

The complete example script `XIA98-local.sh`, which took about 20 s to run on a laptop and uses 20 MiB of memory, is

```
vargibbs -cl=10① -me=3000② -mm=3③ -o=P-XIA98-local -par=P-XIA98.par \
-calc=minlocal -data=D-XIA98.dat \
④-var=AU_AU:entropy,AU_CG:entropy,AU_GC:entropy,AU_UA:entropy,CG_AU:entropy,...
```

where -cl=① is the characteristic length of the Nelder-Mead [34] minimization process which should be roughly of the order of magnitude as the enthalpy and entropy in their respective units. The option -me=② specifies how many evaluations of new parameters without improvement of χ^2 should be attempted before giving up, after which it restarts its cycle. The number of internal minimization cycles is given by -mm=③. Parameters that should be optimized are passed through the -var=④ option; note that we truncated the list for clarity.

3.8.1 Result Files of Type .par: All Parameters

Files with extension .par contain all parameters used to arrive at the lowest χ^2; they have the same format as input parameter files; *see* Subheading 3.4.3. If needed, they can be reused unmodified with the -par= option.

3.8.2 Result Files of Type .var: Optimized Parameters

It is similar to .par but only contains the optimized parameters that were given in the -var= option. If needed, they can be reused unmodified with the -par= option.

3.8.3 Result Files of Type .dat: Optimized Parameters

The file with .dat contains all optimized parameters in a single line

```
324.708① -6.34321 -17.4031 -11.4935 -29.8824 -9.65101 -24.4636 -10.0632
↪  -28.5737 -8.78905 -21.819 -14.2446 -35.2081 -10.7451 -26.8701 -13.2098
↪  -35.1058 -14.6937 -36.5463 -9.09548 -25.1389 -1.62576 2.35794 6.49579
```

where the first number is the lowest χ^2 ① that was achieved and the remaining columns are variable parameters, passed through -var=, in lexical order.

3.8.4 Result Files of Type .ver

Files with extension .ver are very large and contain numerous sections showing information before and after optimization. In the first part the parameters before optimization are shown:

```
local-XIA98
AU_AU:enthalpy -6.82
AU_AU:entropy -19
```

Further in the file you will find the initial quality parameters which have the same format as the `.ver` file of `-calc=prediction`

```
average diff_deviation sqr_diff sqrt_diff2 relative_sqr_diff
Tm 48.3798 11.6063 269809① 49.7525 109
```

where ① is the initial χ^2; *see* Subheading 3.6.2 for details of the other fields. Later in the file this information is repeated but now for the optimized parameter

```
average diff_deviation sqr_diff sqrt_diff2 relative_sqr_diff
Tm 1.35812 1.06512 324.708① 1.72597 0.258941
```

where we see that χ^2 ① has been reduced by several orders of magnitude. The rest of the file has the same format as `.ver` of `-calc=prediction`; *see* Subheading 3.6.1.

3.8.5 Result Files of Type `.log`: Minimization History

The file `.log` records the complete minimization history and therefore can be very large.

```
①*** Starting round 0
1② 402.212 -6.82 -19 -11.4 -29.5 -10.48 -27.1 -9.38 -26.7 -10.44 -26.9
↪  -13.39 -32.7
27② 402.212 -6.82 -19 -11.4 -29.5 -10.48 -27.1 -9.38 -26.7 -10.44 -26.9
↪  -13.39 -32.7
28② 402.212 -6.82 -19 -11.4 -29.5 -10.48 -27.1 -9.38 -26.7 -10.44 -26.9
↪  -13.39 -32.7
```

The first line ① indicates the internal minimization round controlled by `-mm=`. In the following lines the first number ② indicates the number of times that Eqs. (1) and (2) were evaluated; this number resets for each new round. The remainder of each line has the same format as the `.dat` file; *see* Subheading 3.8.3. Further in the file you will find the information between rounds

```
2488 325.43 -6.17678 -16.8771 -11.3391 -29.3951 -9.78153 -24.8599 -10.3885
↪  -29.5753 -8.74234
*** Round stopped with status GSL_SUCCESS
*** Starting round 2
1 325.43 -6.17678 -16.8771 -11.3391 -29.3951 -9.78153 -24.8599 -10.3885
↪  -29.5753 -8.74234
```

At the end of the file is the information of how VarGibbs ended the minimization process:

```
2321 324.708 -6.34321 -17.4031 -11.4935 -29.8824 -9.65101 -24.4636 -10.0632
↪  -28.5737 -8.78905 -
*** Round stopped with status GSL_SUCCESS
*** Local minimization stopped at round 3① (maximum rounds was 3①)
```

which was stopped as it reached the maximum -mm=3①. Note that you could increase -mm= to reach an even lower χ^2. You can either define a higher value for -mm= and restart the minimization from the beginning, or you can add the resulting `.par` to the `-par=` option in which case you will be starting from already optimized parameters, and repeat the process.

In case the `.log` files end in a message similar to

```
*** Round stopped after MaxEvaluations (10000)① without improvement reached,
↪  best chi2 is 4209.83
*** Difference between the last two rounds 0 below 0.1, stopping
↪  minimization altogether
*** Local minimization stopped at round 2 (maximum rounds was 100)
```

it means that the maximum value of `-me=10000` was reached. In this case increasing -mm= will make no difference, but you may try to increase `-me=` or adjust other options described in the package documentation.

As a general advice, make sure to always check endings of the `.log` to see how the minimization ended. You can use the `tail` command for this purpose.

3.8.6 Control File `.proceed`

The file with extension `.proceed` simply contains the number 1. If you wish to end the minimization process before its time you can change this to 0; the simplest way to do this is issuing an `echo` command

```
echo 0 > P-XIA98-local.proceed
```

`VarGibbs` will then end the minimization cleanly. Note that this is only of use for optimizations that are taking a very long time, say hours or even days.

3.9 Global Minimizations `-calc=minglobal`

A global minimization is, quite simply, to run many local minimizations controlled by `-mrr=`, but each time changing something. Figure 1 shows how global and local minimizations are related. `VarGibbs` implements two types of global minimizations; in one it starts many times from a different set of initial parameters $\{P^0\}$; this is controlled by the option `-randominitial`. The other method is to start always with the same initial parameters, but instead changing the experimental temperature T_i' by small random amounts, this is controlled by `-randomizedataset=1`. Before we explain how each method works, we will first show the type of files that will be generated as it is the same for both methods, and then briefly discuss parallelization.

Result Files for Global Minimizations Are the same as for local minimization, except that they have a numeric suffix of type .00001, .00002 for each local run. Therefore, a file named BOP-XIA98.log.00003 is the log file of local minimization number 00003. If you have set -mrr=10 you will get all files numbered from .00001 to .00010. In addition, there are files with extension .best, such as AOP-XIA98.par.best, which contains the best parameter set (lowest χ^2) of all runs. The file with extension .dat has the same format as for -calc=minlocal, except that each line contains one of the runs and can be imported easily into a spreadsheet program for further analysis of the results.

Serial Versus Parallel Execution Global minimization is the sequential processing of local minimizations, that is, setting -mrr=N will result in N consecutive minimizations running on one processor. If you need to run VarGibbs for very large N, this may take a long time, in which case it is advised to run multiple versions of VarGibbs concurrently. While VarGibbs has itself no parallel processing capabilities, it is relatively straightforward to set up a script that launches multiple versions of VarGibbs running over many processors.

If you run multiple instances of VarGibbs you need to ensure that each runs with a different seed for the random number generator, that is, each instance needs to run with a different argument for option -rs=. If you have access to a distributed computer system using SLURM [35], this can be automated by setting up a script that uses the -array option passed via the submission command

```
sbatch --array=1-10 ① submit.sh
```

which has the effect of running 10 instances of the script submit.sh, numbered 1 to 10 ①. For each instance an environment variable SLURM_ARRAY_TASK_ID takes one of these numbers. The numbering is arbitrary and you can choose whatever suits you. The script submit.sh is then set up by making use of SLURM_ARRAY_TASK_ID in the following way:

submit.sh

```
#!/bin/bash
srun vargibbs -calc=minglobal cl=10 data=D XIA98.dat -mc=3000 -mm=3 \
-mrr=1 ① -o=BOP-XIA98 -par=P-XIA98.par -randominitial=0.5
-var=AU_AU:entropy,AU_CG:entropy,AU_GC:entropy,AU_UA:entropy,CG_AU:entropy,... \
-rs=$RANDOM$SLURM_ARRAY_TASK_ID ② -o=BOP-XIA98-${SLURM_ARRAY_TASK_ID} ③
< /dev/null &> run-${SLURM_ARRAY_TASK_ID}.echo
```

where `-mrr` was set to one local minimization ①. The random number generator seed `-rs` ② is a mix of a random number `$RANDOM` from the `bash` environment and the `SLURM` array index `SLURM_ARRAY_TASK_ID`. It does not matter what number `-rs` is, as long as it is **different** for each instance. Here you also need to have different names for the output files; we accomplish this by also adding the array index to the `-o=` option ③. Therefore, the output files will have base names `BOP-XIA98-1` to `BOP-XIA98-10`.

3.9.1 Varying the Initial Parameters (-randominitial)

Selecting new sets of initial parameters manually is a tedious process, and `VarGibbs` automates this by using the concept of *seed parameters* $\{S\}$. Each time `VarGibbs` starts a new local minimization, new initial parameters will be randomly selected within a certain interval of seed parameters. Let's assume that the set of seed parameters is $\{S\} = \{s_1, s_2, \ldots\}$ and the *j*th attempt of initial parameters is $\{P^0\}_j = \{p_1^0, p_2^0, \ldots\}$. Each parameter p_i^0 in the tentative parameter set, $\{P^0\}_j$, is randomly varied in an interval

$$p_i \in [(1 - f)s_i, (1 + f)s_i] \tag{13}$$

That is, the value is sampled within a fraction $\pm f$ of a seed value for the parameter s_i. For example, if f is defined as 0.2, the parameter will be sampled in an interval of [0.8*f*, 1.2*f*], which means that all initial values will be modified by up to 20% every time a new local minimization starts. The parameter f is controlled by the option `-randominitial`, for example, for $f = 0.2$, we use `-randominitial=0.2`. The seed parameters $\{S\}$ are taken from the files that were passed through the `-par=` option.

In the following example we modified the local minimization script from Subheading 3.8:

```
vargibbs  -calc=minglobal ① -cl=10 -data=D-XIA98.dat -me=3000 -mm=3 \
-mrr=10 ② -o=BOP-XIA98 -par=P-XIA98.par ③ -randominitial=0.5 ④ \
-var=AU_AU:entropy,AU_CG:entropy,AU_GC:entropy,AU_UA:entropy,CG_AU:entropy,..
```

where we instructed to run a global minimization ① for 10 local minimizations ② and to modify the seed parameters taken from the file `P-XIA98.par` ③ by up to 50% ④. `VarGibbs` will carry on repeating the local minimization, every time for a new set of initial parameters $P_{0,l}$ until the maximum number of minimizations given by `-mrr` is reached. The complete example script is `XIA98-global.sh`, which took about 1 min to run on a laptop and used 21 MiB of memory. It is important to notice that this example runs for a small number minimization round, as defined by `-mm=3`, and also a small number of local minimization, defined by `-mrr=10`. The running time will increase considerably in a more realistic minimization procedure, such as the minimization dispersion shown in Fig. 2 which was repeated 3000 times and took about 15 h to run on a single processor.

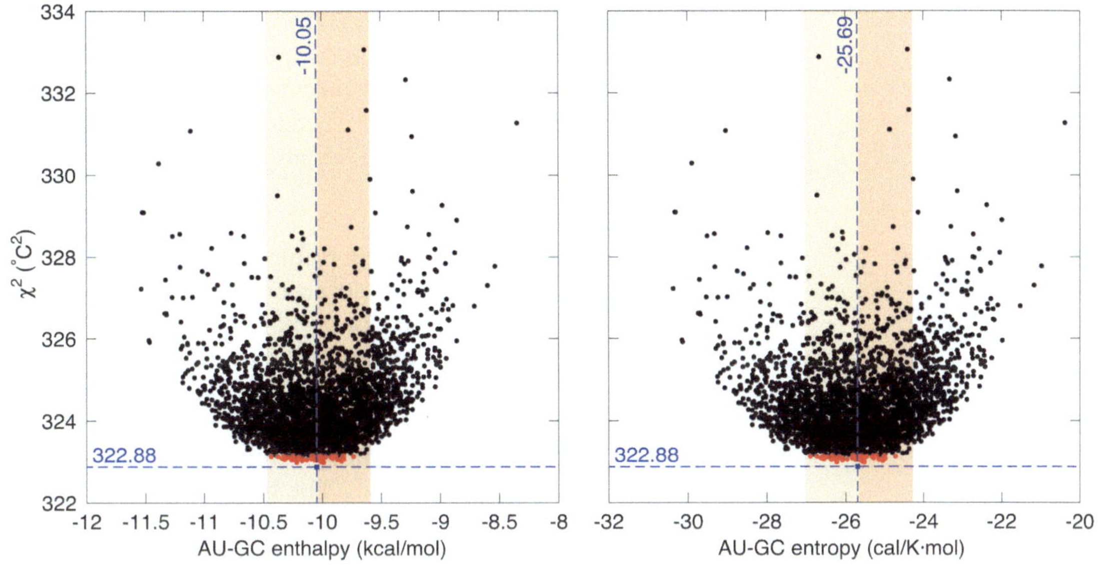

Fig. 2 χ^2 distribution of parameters `AU_GC:enthalpy` and `AU_GC:entropy` for `-mrr=3000` minimization runs. Each point represents a local minimization that started with different initial parameters. The red points show the 50 minimizations with lowest χ^2. The blue dashed horizontal line shows the absolute lowest χ^2 and the dashed vertical line its corresponding parameter which is the best overall parameter and presumably close to the global minima. The shaded area shows the interval for $\pm$ standard deviation; the lighter (darker) area is below (above) the calculated average

3.9.2 *Varying the Temperatures* - `randomizedataset`

`VarGibbs` also has implemented another type of global minimization that simulates the effect of the experimental uncertainty associated with temperature measurement in the final parameter set. That is, if the experimental uncertainty of the melting temperature set is declared as say 0.5 °C, each temperature measurement will be changed by small random amounts such that the standard deviation is close to the declared experimental uncertainty. In practical terms, we exchange the `-randominitial` with the `-randomizedataset=1` option and declare a temperature uncertainty `-ee` in °C. This type of minimization is often used by us as a last step to obtain error bars; therefore, we usually do not start this with generic parameters but with optimized ones from a previous `-randominitial` minimization [6, 36].

In the following example we take the best parameter set from `BOP-XIA98.par.best` ① and instruct to vary the temperatures ② by 0.3 °C ③

```
vargibbs  -calc=minglobal -cl=10 -data=D-XIA98.dat -me=3000 -mm=3 \
-mrr=10 -o=AOP-XIA98 -par=P-XIA98.par,BOP-XIA98.par.best ① \
-randomizedataset=1 ② -ee=0.3 ③
```

The standard deviation between the original set and the new artificial set will be close to the specified experimental error 0.3 °C ③. The uncertainty impact is then obtained from the standard deviation of the optimized parameters which needs to be done separately, for instance in a spreadsheet program, by importing the resulting `.dat` file.

3.9.3 Analyzing Your Global Minimization Results

The quality of the global minimization results depend on a number of factors, many of which require some trial and error testing. We recommend that you make use of graphical visualization software to analyze the behavior of your results, especially try to verify how your parameters are distributed. In Fig. 2 we show an example of parameter distribution as function of χ^2 for 3000 runs. Observe the characteristic parabolic "envelope" with its minima coinciding with the lowest χ^2. The best parameters in this case are $\Delta H_{\mathrm{AU\_CG}} = -10.05$ kcal/mol and $\Delta S_{\mathrm{AU\_CG}} = -25.69$ cal/(K·mol).

Unfortunately, one may not have always the luxury to run a very large number for `-mrr`, yet even in this case the results may not be poor as shown in Fig. 3 for only 100 runs. While in this case the parabolic envelope is not visible and the points appear to be randomly scattered, the lowest χ^2 has parameters $\Delta H_{\mathrm{AU\_CG}} = -10.06$ kcal/mol and $\Delta S_{\mathrm{AU\_CG}} = -25.75$ cal/(K·mol) which is still very close to the results for 3000 runs. Sometimes, one may not be lucky to have a low χ^2 within such a small number of runs, in which case one should also check the average. Usually the average result, which is the interface of the shaded areas in Figs. 2 and 3, is very close to the global minima. If the two do not coincide it is advisable to run further minimization steps until they do.

3.10 Work Flow Suggestion for Parameter Optimization

Here we will describe a step-by-step workflow for parameter optimization:

1. Retrieve the melting temperatures from a paper or format your own data according to Subheading 3.4.1. Make sure you have the correct values for total strand concentration C_t in μM. You can mix data with different strand concentrations, but not with different salt concentrations. If you need to optimize for various salt concentrations, do each one separately as in Ref. [6].
2. First check your data with `-calc=nncheck` command; *see* Subheading 3.4.3. This will allow you to catch typing mistakes as well as determine the type of parameters to optimize.
3. Define your initial parameters, `-par=`; ideally start with the most generic values possible, such as in file `data/gibbs-av.par`.
4. Define the variables that should be optimized, and add them to `-var=`, use the `-calc=nncheck` option for this; *see* Subheading 3.4.3.

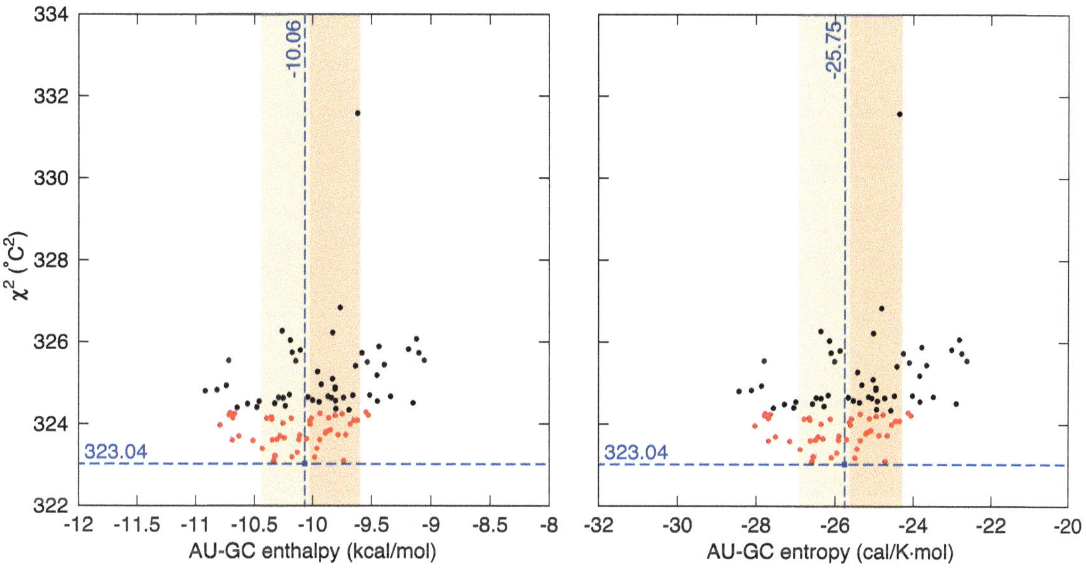

Fig. 3 Same as Fig. 2 for `-mrr=100` runs

5. Run global minimizations `-calc=minglobal` with varying initial parameters `-randominitial=`. For this step follow the recommendations in Subheading 3.9.1. Let it run for as many times as practically possible, but at least 100 times (0.5 h). It will take about 5 h if you let it run 1000 times. The processing time depends on the amount of sequences and the number of parameters that are varied. Use a distributed processing system if you have access to one; *see* Subheading 3.9.
6. Check your parameter dispersion as in Fig. 2; this can be achieved by a spreadsheet program.
7. If your distribution shows a clear minimum, then use the best parameters from files ending in `.best`, as described in Subheading 3.9.
8. If your distribution does look very scattered, as in Fig. 3, calculate the average of the parameters and use them as new initial parameters and run a new optimization again to see if you arrive at a clearer minimum.
9. If you require an estimate about experimental uncertainty, that is, error bars, then run a global minimization `-calc=minglobal` with `-randomizedataset=1`, as described in Subheading 3.9.2. Use the best parameters in previous steps as initial parameters. Your error bars will be the standard deviation of all rounds; import your `.dat` into a spreadsheet for this.
10. If your data has a clear minimum, use the best parameters found in `.best` files. Otherwise use the average parameters

and run a -calc=prediction for verification; *see* Subheading 3.6.

3.11 Salt Correction

VarGibbs comes with specific RNA parameters at various sodium concentrations; *see* Table 2. However, there may be cases where one might wish to use salt correction factors. These factors are empirically determined and are applied directly to the melting temperatures. There are several schemes which can be selected via the -saltscheme= option. It is possible to apply salt corrections in VarGibbs for both predictions and minimizations.

VarGibbs has basically two ways of applying salt corrections, directly on input melting temperatures such as those used for parameter optimizations or after performing predictions. This is controlled by the option -predsaltcorr=.

3.11.1 Adjust Temperatures After Prediction

There are cases where you performed a prediction with a specific file, like P-XIA98, which has parameters for [Na^+] 1000 mM but wish to obtain the melting temperatures at a lower concentration, say 100 mM. Let's adopt the example from Subheading 3.6:

```
vargibbs -data=D-XIA98.dat -o=simple-prediction -par=P-XIA98.par \
-calc=prediction -predsaltcorr=1 ①
-saltscheme=chen13eq22 ② -targetsalt=100 ③
```

-predsaltcorr=1 ① instructs VarGibbs to apply the salt correction **after** performing the prediction, using Eq. (22) from Ref. [37] ② at [Na^+] 100 mM ③. While there are other salt schemes from Ref. [37], using -saltscheme=chen13eq22 is recommended for RNA. For other correction factors please see the package documentation.

3.11.2 Adjusting Input Melting Temperatures

Let's assume that you wish to optimize parameters for a certain sodium concentration, say 50 mM, but your existing sequences are at 1000 mM as in the example of Subheading 3.8. Since we want to apply the salt correction *before* optimizing the parameters, we use -predsaltcorr=0. Instead of manually recalculating and editing the file, you can simply specify the salt correction at the command line. As an example, we will adopt the command of Subheading 3.8:

```
vargibbs -var=AU_AU:entropy,AU_CG:entropy,AU_GC:entropy,AU_UA:entropy \
-data=experimental.dat -calc=minlocal -o=results -par=P-XIA98.par \
 -predsaltcorr=0① -saltscheme=chen13eq22② -targetsalt=50 ③
```

-predsaltcorr=0 ① instructs VarGibbs to apply the salt correction **before** performing the optimization, using Eq. (22) from Ref. [37] ② at [Na^+] 50 mM ③.

4 Notes

1. The example scripts described in the previous sections were edited for clarity. In particular we omitted the file paths which are installation dependent. For a typical Linux installation, the file paths are those described in Subheading 3.1, which means that you will need to adapt the examples accordingly. For instance, take the script in Subheading 3.6; if the data path is /usr/share/VarGibbs/data, you will need to adapt the -data= option ①:

```
vargibbs -data=/usr/share/VarGibbs/data/D-XIA98.dat ① \
-o=simple-prediction -par=P-XIA98.par -calc=prediction
```

   Alternatively, you could copy the file P-XIA98.par to your working folder.

2. Specific scripts for RNA are included in the VarGibbs distribution package and should be found in /usr/share/VarGibbs/examples:
   - sequence-XIA98.sh calculates a simple prediction using P-XIA98.par parameters, *see* Subheading 3.6. Run time is negligible.
   - XIA98-local.sh performs a short local minimization starting from P-XIA98.par as initial parameters; *see* Subheading 3.8. It starts with $\chi^2 = 269,809\ ^\circ C^2$ and ends at $\chi^2 = 324.708\ ^\circ C^2$; run time is roughly 3 s.
   - XIA98-global.sh performs 10 runs for -calc=minglobal starting from P-XIA98.par as seed parameters as in Subheading 3.9.1, then another 10 minimization with -randomizedataset=1 as in Subheading 3.9.2. It ends at $\chi^2 = 319.87\ ^\circ C^2$; run time is roughly 80 s.

   Run any of these scripts by providing the path of your local data installation, as in the following example:

```
/usr/share/VarGibbs/examples/XIA98-global.sh\
usr/share/VarGibbs/data
```

This will create a folder with the same base name as the script, in this case XIA98-global where all results will be stored.

Funding Acknowledgment

This work is supported by Conselho Nacional de Desenvolvimento Científico e Tecnológico (CNPq, Brazil) and Coordenação de Aperfeiçoamento de Pessoal de Nível Superior (CAPES, Brazil).

References

1. Borer PN, Dengler B, Tinoco Jr I, Uhlenbeck OC (1974) Stability of ribonucleic acid double-stranded helices. J Mol Biol 86(4): 843–853. https://doi.org/10.1016/0022-2836(74)90357-X
2. Chamberlin M, Baldwin RL, Berg P (1963) An enzymically synthesized RNA of alternating base sequence: physical and chemical characterization. J Mol Biol 7(4):334–349, 1963. https://doi.org/10.1016/S0022-2836(63)80028-5
3. Freier SM, Petersheim M, Hickey DR, Turner DH (1984) Thermodynamic studies of RNA stability. J Biomol Struct Dyn 1(5): 1229–1242. https://doi.org/10.1080/07391102.1984.10507514
4. Freier SM, Kierzek R, Jaeger JA, Sugimoto N, Caruthers MH, Neilson T, Turner DH (1986) Improved free-energy parameters for predictions of RNA duplex stability. Proc Natl Acad Sci USA 83(24):9373–9377. https://doi.org/10.1073/pnas.83.24.9373
5. Xia T, SantaLucia Jr J, Burkard ME, Kierzek R, Schroeder SJ, Jiao X, Cox C, Turner DH (1998) Thermodynamic parameters for an expanded nearest-neighbor model for formation of RNA duplexes with Watson-Crick base pairs. Biochem 37:14719–14735. https://doi.org/10.1021/bi9809425
6. Ferreira I, Jolley EA, Znosko BM, Weber G (2019) Replacing salt correction factors with optimized RNA nearest-neighbour enthalpy and entropy parameters. Chem Phys 521:69–76. https://doi.org/10.1016/j.chemphys.2019.01.016. https://www.sciencedirect.com/science/article/abs/pii/S0301010418311200
7. Tinoco I, Uhlenbeck OC, Levine MD (1971) Estimation of secondary structure in ribonucleic acids. Nature 230(5293):362–367
8. Turner DH, Sugimoto N, Freier SM (1988) RNA structure prediction. Annu Rev Biophys Biophys Chem 17(1):67–192. https://doi.org/10.1146/annurev.bb.17.060188.001123
9. Mathews DH, Sabina J, Zuker M, Turner DH (1999) Expanded sequence dependence of thermodynamic parameters improves prediction of RNA secondary structure. J Mol Biol 288(5):911–940. https://doi.org/10.1006/jmbi.1999.2700
10. Lu ZJ, Turner DH, Mathews DH (2006) A set of nearest neighbor parameters for predicting the enthalpy change of RNA secondary structure formation. Nucleic Acids Res 34(17): 4912–4924. https://doi.org/10.1093/nar/gkl472
11. Gruber AR, Bernhart SH, Hofacker IL, Washietl S (2008) Strategies for measuring evolutionary conservation of RNA secondary structures. BMC Bioinformat 9(1):1–19. https://doi.org/10.1186/1471-2105-9-122
12. Spasic A, Berger KD, Chen JL, Seetin MG, Turner DH, Mathews DH (2018) Improving RNA nearest neighbor parameters for helices by going beyond the two-state model. Nucleic Acids Res 46(10):4883–4892. https://doi.org/10.1093/nar/gky270
13. Costa PR, Acencio ML, Lemke N (2013) Cooperative RNA polymerase molecules behavior on a stochastic sequence-dependent model for transcription elongation. PLoS One 8(2):e57328
14. Mathews DH, Burkard ME, Freier SM, Wyatt JR, Turner DH (1999) Predicting oligonucleotide affinity to nucleic acid targets. RNA 5(11):1458–1469
15. Joyeux M, Sahin B (2005) Dynamical model based on finite stacking enthalpies for homogeneous and inhomogeneous DNA thermal denaturation. Phys Rev E 72:051902. https://doi.org/10.1103/PhysRevE.72.051902
16. Weber G (2011) Finite enthalpy model parameters from DNA melting temperatures. Europhys Lett 96:68001. https://doi.org/10.1209/0295-5075/96/68001. http://iopscience.iop.org/0295-5075/96/6/68001
17. Gray DM (1997) Derivation of nearest-neighbor properties from data on nucleic acid oligomers. I. simple sets of independent sequences and the influence of absent nearest neighbors. Biopoly 42(7):783–793. https://doi.org/10.1002/(SICI)1097-0282(1

99712)42:7%3C783::AID-BIP4%3E3.0.CO;2-P
18. Gray DM (1997) Derivation of nearest-neighbor properties from data on nucleic acid oligomers. II. thermodynamic parameters of dna· RNA hybrids and dna duplexes. Biopoly 42(7):795–810. https://doi.org/10.1002/(SICI)1097-0282(199712)42:7%3C795::AID-BIP5%3E3.0.CO;2-O
19. Marky LA, Breslauer KJ (1987) Calculating thermodynamic data for transitions of any molecularity from equilibrium melting curves. Biopoly 26(9):1601–1620. https://doi.org/10.1002/bip.360260911
20. Breslauer KJ, Frank R, Blocker H, Marky LA (1986) Predicting DNA duplex stability from the base sequence. Proc Natl Acad Sci USA 83(11):3746–3750. https://doi.org/10.1073/pnas.83.11.3746
21. Weber G (2015) Optimization method for obtaining nearest-neighbour DNA entropies and enthalpies directly from melting temperatures. Bioinformatics 31(6):871–877. https://doi.org/10.1093/bioinformatics/btu751. http://bioinformatics.oxfordjournals.org/content/31/6/871
22. Chen H, Zhu G (1997) Computer program for calculating the melting temperature of degenerate oligonucleotides used in PCR or hybridization. Biotechniques 22(6):1158–1160. https://doi.org/10.2144/97226bc04
23. Schütz E, von Ahsen N (1999) Spreadsheet software for thermodynamic melting point prediction of oligonucleotide hybridization with and without mismatches. BioTechniques 27(6):1218–1224. https://doi.org/10.2144/99276bc04
24. Le Novere N (2001) Melting, computing the melting temperature of nucleic acid duplex. Bioinformatics 17(12):1226–1227
25. Panjkovich A, Melo F (2005) Comparison of different melting temperature calculation methods for short DNA sequences. Bioinformatics 21(6):711–722. https://doi.org/10.1093/bioinformatics/bti066. http://bioinformatics.oxfordjournals.org/cgi/content/abstract/21/6/711
26. Licinio P, Guerra JCO (2007) Irreducible representation for nucleotide sequence physical properties and self-consistency of nearest-neighbor dimer sets. Biophys J 92(6):2000–2006. https://doi.org/10.1529/biophysj.106.095059
27. SantaLucia Jr J, Allawi HT, Seneviratne PA (1996) Improved nearest-neighbour parameters for predicting DNA duplex stability. Biochem 35:3555–3562. https://doi.org/10.1021/bi951907q
28. Thompson A, Taylor BN (2008) Guide for the Use of the International System of Units (SI). National Institute of Standards and Technology Special Publication 811, Gaithersburg
29. Schreiber-Gosche S, Edwards S (2009) Thermodynamics of oligonucleotide duplex melting. J Chem Educ 86(5):644. https://doi.org/10.1021/ed086p644
30. SantaLucia Jr J (1998) A unified view of polymer, dumbbell, and oligonucleotide DNA nearest-neighbor thermodynamics. Proc Natl Acad Sci USA 95(4):1460–1465. https://doi.org/10.1073/pnas.95.4.1460. http://www.pnas.org/cgi/content/abstract/95/4/1460
31. de Oliveira Martins E, Weber G (2017) An asymmetric mesoscopic model for single bulges in RNA. J Chem Phys 147:155102. https://doi.org/10.1063/1.5006948
32. Miranda P, Oliveira LM, Weber G (2017) Mesoscopic modelling of Cy3 and Cy5 dyes attached to DNA duplexes. Biophys Chem 230C:62–67. ISSN 0301-4622. https://doi.org/10.1016/j.bpc.2017.08.007. http://www.sciencedirect.com/science/article/pii/S0301462217302831
33. Chen JL, Dishler AL, Kennedy SD, Yildirim I, Liu B, Turner DH, Serra MJ (2012) Testing the nearest neighbor model for canonical RNA base pairs: revision of GU parameters. Biochem 51(16):3508–3522. https://doi.org/10.1021/bi3002709
34. Nelder JA, Mead R (1965) A simplex method for function minimization. Comput J 7(4):308–313
35. SLURM Workload Manager. https://slurm.schedmd.com/
36. Barbosa VB, de Oliveira Martins E, Weber G (2019) Nearest-neighbour parameters optimized for melting temperature prediction of DNA/RNA hybrids at high and low salt concentrations. Biophys Chem 251C:106189. https://doi.org/10.1016/j.bpc.2019.106189
37. Chen Z, Znosko BM (2013) Effect of sodium ions on RNA duplex stability. Biochem 52(42):7477–7485. https://doi.org/10.1021/bi4008275

Chapter 3

RNA Secondary Structure Thermodynamics

Ronny Lorenz

Abstract

Several different ways to predict RNA secondary structures have been suggested in the literature. Statistical methods, such as those that utilize stochastic context-free grammars (SCFGs), or approaches based on machine learning aim to predict the best representative structure for the underlying ensemble of possible conformations. Their parameters have therefore been trained on larger subsets of well-curated, known secondary structures. Physics-based methods, on the other hand, usually refrain from using optimized parameters. They model secondary structures from loops as individual building blocks which have been assigned a physical property instead: the free energy of the respective loop. Such free energies are either derived from experiments or from mathematical modeling. This rigorous use of physical properties then allows for the application of statistical mechanics to describe the entire state space of RNA secondary structures in terms of equilibrium probabilities. On that basis, and by using efficient algorithms, many more descriptors of the conformational state space of RNA molecules can be derived to investigate and explain the many functions of RNA molecules. Moreover, compared to other methods, physics-based models allow for a much easier extension with other properties that can be measured experimentally. For instance, small molecules or proteins can bind to an RNA and their binding affinity can be assessed experimentally. Under certain conditions, existing RNA secondary structure prediction tools can be used to model this RNA-ligand binding and to eventually shed light on its impact on structure formation and function.

Key words Minimum free energy structure, Suboptimal structures, Equilibrium probabilities, Structure constraints, RNA-protein binding

1 Introduction

The foundations of algorithms for physics-based RNA secondary structure prediction today were laid back in the 1970s [1, 2]. The basic idea was to consider an RNA secondary structure to be composed of individual building blocks, the *loops* L, that each can be assigned a free energy contribution $E(L)$. The stability of any structure s in terms of its free energy $E(s)$ could then be approximated as

$$E(s) \approx \sum_{L \in s} E(L), \tag{1}$$

Ronny Lorenz (ed.), *RNA Folding: Methods and Protocols*, Methods in Molecular Biology, vol. 2726,
https://doi.org/10.1007/978-1-0716-3519-3_3,

i.e., the sum of the energies of the loops L that actually form s. Having reasonable parameters $E(L)$ at hand, the remaining task was to determine the most stable structure.

These early approaches were utterly inefficient, though, as they exhaustively enumerated and evaluated all possible structures. Later on, however, *efficient dynamic programming* algorithms emerged [3–7] that used recursive decomposition schemes (i) to count the number of possible secondary structures s for any particular RNA sequence, (ii) to determine the structure with the maximum number of base pairs, and (iii) to determine the structure that exhibits the minimum free energy (MFE). The development of an efficient algorithm to compute the latter, the *MFE structure*, was of particular importance. Borrowing the concepts of thermodynamics, the probability $p(s)$ of any structure s in thermodynamic equilibrium is proportional to its free energy $E(s)$, i.e.,

$$p(s) \propto exp(-E(s)/kT), \tag{2}$$

with Boltzmann constant k and thermodynamic temperature T. Therefore, in equilibrium, the MFE structure has the highest probability among all other conformations. Within this chapter we will use RT instead of kT, where R is the molar gas constant; *see* **Note 1**.

About a decade after these first attempts at physics-based RNA structure prediction, McCaskill [8] realized that the recursions previously used for MFE predictions can be transformed to compute the partition function

$$Z = \sum_s exp(-E(s)/RT). \tag{3}$$

This, in turn, opened up the application of a vast field of thermodynamics computations for RNA secondary structures. In particular, consider a subset of structures $\{s \mid F \in s\}$ that contain a certain feature F. Then, the equilibrium probability $p(F)$ can be computed as a fraction of the partition function of those structures that exhibit F

$$Z^F = \sum_{s|F\in s} exp(-E(s)/RT) \tag{4}$$

and the full ensemble captured in Z, i.e.,

$$p(F) = \frac{Z^F}{Z}. \tag{5}$$

The probability of a single secondary structure s becomes

$$p(s) = \frac{1}{Z} exp(-E(s)/RT). \tag{6}$$

In his original publication, McCaskill [8] further presented an efficient algorithm to compute base pairing probabilities

$$p_{ij} = \frac{1}{Z} \sum_{s|(i,j)\in s} exp(-E(s)/RT), \tag{7}$$

i.e., the probability to observe the base pair (i, j) within the equilibrium ensemble of structures.

To compute $p(F)$ for other features F proves tricky. In many cases, if not the majority, F may be a motif found frequently in the ensemble of structures. This renders an explicit enumeration of the subset of structures $\{s \mid F\in s\}$ virtually impossible. After all, the total number of secondary structures grows exponentially in the length of their sequence [4]. However, once an efficient method to compute Z^F is available, further quite useful applications arise.

Motif F may serve as an aptamer fold to bind a ligand, e.g., a small molecule or a protein. Then, having Z^F at hand opens up the possibility to compute many equilibrium properties for a mixed ensemble of structures with and without the bound ligand. The statistical mechanics view on the secondary structure space also allows for many more applications. Among them are strategies to find good representatives of the structure ensemble, to judge the reliability of predicted structures, and the investigation of the conformational diversity.

Within this chapter, I will present several showcase applications that exemplify the computations for a variety of the abovementioned RNA secondary structure equilibrium properties.

2 Materials

2.1 Hardware

- Preferably a Linux- or MacOS X-based computer

2.2 Software

- `ViennaRNA Package` (http://www.tbi.univie.ac.at/RNA).
- A working `Python` environment (optional).
- Text editor (**not** a word processor).
- A `Postscript` viewer (e.g., `gv`, `preview`, `gsview`).

3 Methods

3.1 General Remarks

In the following and wherever possible, the examples will be shown as both command line calls using the executable command line programs of the `ViennaRNA Package` and equivalent `Python` code that utilizes the corresponding *RNAlib* `C`-library. Depending on your installation and computer skills, you can use either of them to derive the same results. Instructions to execute command line programs are always preceded with a dollar sign ($), where the dollar sign is *not* part of the program call, e.g.:

```
$ echo "Hello world"
```

`Python` code will appear in boxes with line numbers on the left side, e.g.:

```
print("Hello world")
```

For the functions and methods used in the `Python` examples below, we also refer the reader to the `ViennaRNA Package` API reference manual at https://viennarna.readthedocs.io.

3.2 Installation and Usage

3.2.1 Bioconda

The most convenient way to install the `ViennaRNA Package` (executable programs as well as the `Python` interface) is to use the `bioconda` channel (https://bioconda.github.io/) [9] for the `conda` package manager (https://conda.io). This provides pre-compiled packages that can be easily installed using the following command:

```
$ conda install -c bioconda viennarna
```

3.2.2 Alternative Installation Instructions

The official `ViennaRNA Package` website (https://www.tbi.univie.ac.at/RNA) provides a series of pre-compiled binary packages for several `Linux`-based operating systems, as well as executable installers for `MacOS X` and `Microsoft Windows`™. Simply download the required files and follow the instructions given on the website for a system-wide installation of the `ViennaRNA Package`. Note, however, that the Windows installer is currently limited to the executable programs only. The `Python` interface for Windows can be obtained separately from https://pypi.org/project/ViennaRNA. Finally, the `ViennaRNA Package` can be compiled and installed from source code. For this purpose, follow the instructions at the official website https://viennarna.readthedocs.io/en/latest/install.html.

3.3 Equilibrium Probabilities

Let us begin with the following 48 nt long simple artificial example sequence `GGCUGUUGGUGCGGUCGUGCCCGUAAGGGUGUGGUCGCGUUGGUGGCC` and compute the partition function and the MFE structure with its corresponding equilibrium probability with the program `RNAfold`. Here, the `-p0` switch activates partition function computation and the `--noPS` switch omits the creation of a secondary structure PostScript output file:

```
$ echo "GGCUGUUGGUGCGGUCGUGCCCGUAAGGGUGUGGUCGCGUUGGUGGCC" | RNAfold \
  -p0 \
  --noPS
```

or equivalently in Python:

```python
import math
import RNA

RNAfold_output_simple = "{}\n{} ({:6.2f})\n" + \
                        " free energy of ensemble = {:6.2f} kcal/mol\n" + \
                        " frequency of mfe structure in ensemble {:g};"
sequence = "GGCUGUUGGUGCGGUCGUGCCCGUAAGGGUGUGGUCGCGUUGGUGGCC"
kT = RNA.exp_param().kT / 1000. # kT in kcal/mol

# create a fold_compound object to work with
fc = RNA.fold_compound(sequence)
# compute MFE and corresponding secondary structure
(ss, mfe) = fc.mfe()
# rescale Boltzmann factors to avoid numeric instabilities
fc.exp_params_rescale(mfe)
# compute partition function
(pp, dG) = fc.pf()
# compute probability of MFE structure in ensemble
p_mfe = math.exp((dG - mfe) / kT)
# print formatted output
print(RNAfold_output_simple.format(sequence, ss, mfe, dG, p_mfe))
```

Both variants (RNAfold and the Python code) result in the following output:

```
GGCUGUUGGUGCGGUCGUGCCCGUAAGGGUGUGGUCGCGUUGGUGGCC
((((((....))))))..((((....))))..((((((....)))))) (-21.10)
 free energy of ensemble = -22.41 kcal/mol
 frequency of mfe structure in ensemble 0.119414;
```

where the first line depicts the RNA sequence input. The second line shows the MFE structure in dot-bracket notation where base pairs are represented by matching parentheses and unpaired nucleotides by dots. At the end of the line follows the MFE of -21.10 kcal $\cdot$ mol^{-1}. The third line denotes the free energy of the entire ensemble of structures $\Delta G = -RT \ln Z$, and finally, the last line contains the equilibrium probability of the MFE structure, which in this case is about 12%. Note that instead of Z, it is more convenient to use ensemble free energy ΔG. Realistic actual values of the former are extremely large numbers that quickly exceed computer precision and thus usually introduce unnecessary errors into further computations, such as probabilities. The probability $p(s)$ of any structure s with energy $E(s)$ then becomes

$$p(s) = exp((\Delta G - E(s))/RT). \tag{8}$$

3.3.1 Base Pair Probabilities

Next, let us compute the base pairing probabilities

$$p_{ij} = \sum_{s|(i,j)\in s} p(s) \tag{9}$$

for any possible pair (i, j), i.e., the probability to observe base pair (i, j) within the ensemble of structures. However, instead of computing the above sum exhaustively, we make full use of the algorithms presented in McCaskill [8] and compute p_{ij} in cubic time. For this, we change the `-p0` parameter to `-p` such that the command line call becomes

```
$ echo "GGCUGUUGGUGCGGUCGUGCCCGUAAGGGUGUGGUCGCGUUGGUGGCC" | RNAfold -p
```

Note here that we also refrain from using the `--noPS` option to actually create the structure layout plot this time. The output now includes the following information:

```
GGCUGUUGGUGCGGUCGUGCCCGUAAGGGUGUGGUCGCGUUGGUGGCC
((((((....))))))..((((....))))..((((((....)))))) (-21.10)
(((((((({(((((((({((((....))))}))))))))}))))))))) [-22.41]
((((.(((((((((((((((((....))))))))))))))))).)))) {-20.80 d=7.83}
 frequency of mfe structure in ensemble 0.119414; ensemble diversity 11.28
```

Additionally, `RNAfold` creates two PostScript output files: (i) a structure layout plot named `rna.ps` and (ii) a dot-plot `dot.ps` denoting all base pair probabilities $p_{ij} \geq 10^{-5}$ (*see* also Fig. 1 and **Note 2**). The first dot-bracket string, again, represents the MFE structure. The third line almost looks like a dot-bracket string, but instead denotes the pairing propensity, i.e., pairing preference for each nucleotide, expressed by an extended dot-bracket alphabet (*see* **Note 3**). The pairing propensity is then followed by the ensemble free energy $\Delta G = -RT \ln Z = -22.41\ \text{kcal} \cdot \text{mol}^{-1}$. The fourth line shows the centroid, i.e., the structure with the least weighted base pair distance to all other structures in the ensemble (*see* the next subsection). In this example, the centroid only shares the innermost four base pairs with the MFE structure, forms an extended rod-like helix with a stability of $-20.8\ \text{kcal} \cdot \text{mol}^{-1}$, and has an average ensemble distance of 7.83 base pairs. In other words, for any structure s randomly picked from the ensemble according to its probability $p(s)$, on average about 8 base pairs have to be changed to transform s into the centroid structure. The last line, again, shows the probability of the MFE structure, but in addition reports the diversity of the ensemble expressed as the mean pairwise distance (*see* the next subsection).

For the two output files `rna.ps` and `dot.ps`, it is important to know that PostScript-formatted files are simple text files that specify a drawing or printer output in the PostScript language. To actually render the files, you need a viewer program, such as `gv` or `evince` in `Linux` environments. On MacOS X you can simply use the `preview` command to open the two files. The output should be the same as shown in Fig. 1a and b, respectively. The upper triangle

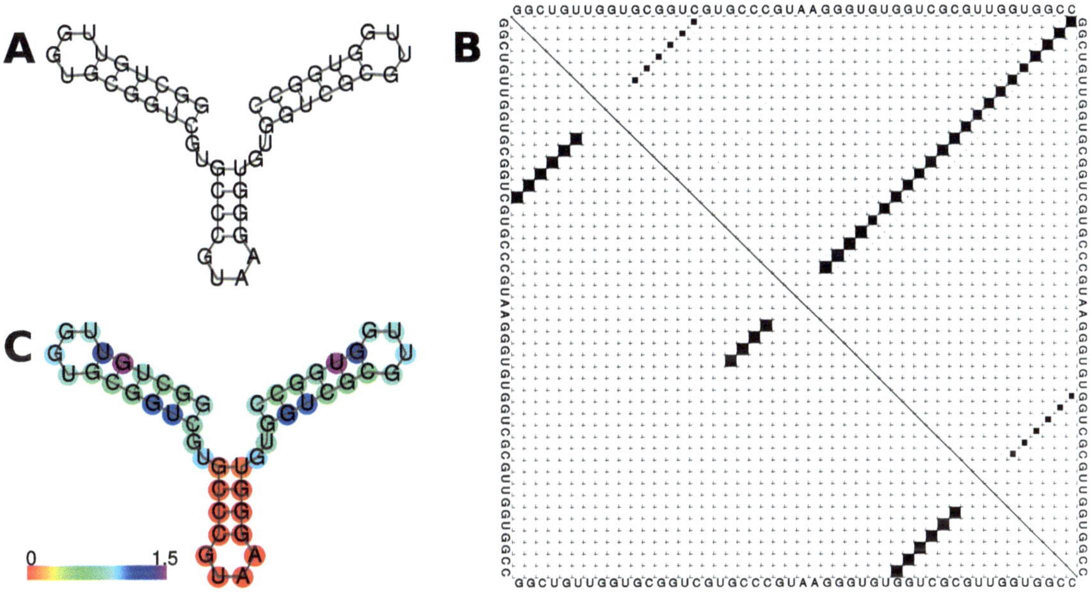

Fig. 1 Basic structure prediction for the artificial RNA sequence `GGCUGUUGGUGCGGUCGUGCCC-GUAAGGGUGUGGUCGCGUUGGUGGCC` . Shown are (**a**) the MFE structure and (**b**) the corresponding dot-plot that displays equilibrium base pair probabilities p_{ij} in the upper triangle. The area of the black squares scales with the corresponding base pair probability. The lower triangle denotes the base pairs present in the MFE structure. In (**c**), the MFE structure is shown with *positional entropy* $S(i)$ indicated as a color gradient. The middle helix (red part) can be considered reliable since it has low entropy, whereas the remaining parts of the structure have relatively high entropy. Therefore, for these parts, the MFE structure is not a good representative of the entire structure ensemble

of the dot-plot shows black squares where the area of the squares is proportional to the pairing probability it depicts. The lower triangle shows the base pairs of the MFE structure. We can immediately see a large number of consecutive base pairs with high probability that are not part of the MFE structure and likely refer to at least one other, alternative conformation that forms a long, rod-like helical structure.

To obtain the actual base pair probability values, the `dot.ps` file has to be parsed. As mentioned before, PostScript files are simple text files containing code to render the final image. The pairing probabilities in the `dot.ps` file are stored as lines with the following pattern:

```
i j p ubox
```

where `i` and `j` denote the 1-based nucleotide positions within the RNA sequence and `p` is the square root of the pairing probability $\sqrt{p_{ij}}$. Consequently, we have to square the values `p` to get the actual probabilities p_{ij}. Most `UNIX`-derived operating systems, such as `Linux` or `MacOS X`, provide the tool `awk`, which can be used to extract and convert the pairing probabilities with very little effort. We simply print the first, second, and the square of the third column if the fourth reads "`ubox`":

```
$ awk '$4=="ubox"{print $1,$2,$3*$3}' dot.ps
```

The resulting output should look like:

```
1 16 0.213678
1 47 5.58896e-05
1 48 0.765573
2 12 1.6887e-05
2 15 0.215397
2 16 0.000144763
2 30 2.23259e-05
...
```

In `Python` access to the base pair probability matrix is gained through the `bpp()` method that is attached to objects of type `fold_compound`. Hence, to print all probabilities p_{ij} that exceed the threshold of 10^{-5}, the following code can be used. Note, however, that base pair probabilities only become available *after* the partition function has been computed, i.e., after the call `fc.pf()`:

```
bppm = fc.bpp()
for i in range(1, len(sequence)):
    for j in range(i, len(sequence) + 1):
        if bppm[i][j] >= 1e-5:
            print("{:d} {:d} {:g}".format(i, j, bppm[i][j]))
```

For instructions to create dot-plot EPS files using the `Python` interface, please refer to **Note** 4.

3.4 Distances and Ensemble Diversity

In the previous example call to `RNAfold`, the last ensemble property in the output was the *ensemble diversity* with a value of 11.28. To understand the meaning of this measure, let us first define the base pair distance $d_{\mathrm{BP}}(s, t)$ between two structures s and t, which, in set-notation, is simply the total number of base pairs both structures do not have in common. In other words, this is the minimum number of base pairs to be added to or removed from s to transform s into t. The formal definition is

$$\begin{aligned} d_{\mathrm{BP}}(s, t) &= |s \cap t| - |s \cap t| \\ &= \sum_{i<j} \left(\delta_{ij}(s) + \delta_{ij}(t) - 2\delta_{ij}(s)\delta_{ij}(t) \right) \end{aligned} \tag{10}$$

where the second variant is an alternative derivation that makes use of the indicator function

$$\delta_{ij}(s) = \begin{cases} 1, & \text{if } (i,j) \in s \\ 0, & \text{otherwise} \end{cases} \tag{11}$$

(12)

which simply assumes value 1 if base pair (i, j) is present in s and 0 otherwise.

3.4.1 Ensemble Diversity

Let us now assume that we randomly draw two structures s and t from the Boltzmann-weighted ensemble according to their equilibrium probabilities $p(s)$ and $p(t)$. Then, a good *ensemble diversity* measure is the *expected* base pair distance $d_{\mathrm{BP}}(s, t)$, which can be computed as

$$\langle d_{\mathrm{BP}} \rangle = \sum_{s,t \in \Omega} p(s)p(t)d_{\mathrm{BP}}(s, t). \tag{13}$$

The higher this value, the more the ensemble is dominated by diverse structures with large individual probability mass. On the other hand, the lower this value, the more likely it is that the ensemble is dominated by a single structure or a set of structurally similar states. Note that the maximum value of $d_{\mathrm{BP}}(s, t)$ grows linearly with sequence length n. Hence, to compare values of $\langle d_{\mathrm{BP}} \rangle$ for sequences of different length, they can be normalized by dividing them by the corresponding sequence lengths. In `Python`, the ensemble diversity $\langle d_{\mathrm{BP}} \rangle$ can be computed using the `mean_bp_distance()` method of objects of type `fold_compound`.

3.4.2 Ensemble Defect

Another diversity measure related to the ensemble diversity is the so-called ensemble defect

$$\mathrm{ED}(t) = \frac{1}{n} \sum_{s \in \Omega} p(s) 2 d_{\mathrm{BP}}(t, s) \tag{14}$$

with respect to a target structure t. It quantifies the expected difference in the pairing pattern between structure t and the entire ensemble in a per-nucleotide fashion. Multiplied by n, this value specifies the expected number of nucleotides paired differently compared to t. This makes the ensemble defect a suitable quantity to assess the *reliability* of a structure as its value denotes how well a particular structure t represents the entire ensemble. Such quantities often play a role in the context of RNA sequence design [10].

The executable programs shipped with the `ViennaRNA package` do not provide access to the ensemble defect measure. However, an implementation exists in `RNAlib` which can be accessed from the `Python` interface by means of the `ensemble_defect()` method attached to objects of type `fold_compound`.

3.4.3 Positional Entropy

A more fine-grained reliability measure can be constructed from an information theory perspective. The idea is that for any structure within the ensemble, each nucleotide i either has a particular pairing partner j or remains unpaired. Moreover, in equilibrium, the resulting base pairs (i, j) are associated with the base pair probability p_{ij} and the probability that i remains unpaired is $q_i = 1 - \sum_j p_{ij}$. Now, the different pairing patterns of i can be considered an information and its diversity may be assessed by the inherent entropy [11], sometimes also referred to as *positional entropy*

$$S(i) = \quad -\sum_j p_{ij} \log(p_{ij}) - q_i \log(q_i). \tag{15}$$

Now, $S(i)$ measures the average uncertainty in the state of i, i.e., any position with high positional entropy cannot be represented by a single pairing pattern. Consequently, when given a particular structure s that is intended to represent the ensemble, for instance the MFE structure, only regions with low $S(i)$ can be considered reliable. To create a secondary structure plot with positional entropy annotation, we can use the `replot.pl` utility script that comes with the `ViennaRNA Package` installation. The only input required is two PostScript files: (i) a secondary structure plot, e.g., the one containing the MFE structure (`rna.ps`), and (ii) the dot-plot containing the base pair probabilities (`dot.ps`):

```
$ relplot.pl rna.ps dot.ps > rna_reliability.ps
```

Note here that the `relplot.pl` script prints the resulting annotated PostScript file directly into the terminal. We therefore use the shell redirection operator ">" to print all output into a new file `rna_reliability.ps`. As shown in Fig. 1c, the only region with low entropy is the innermost stem. The creation of secondary structure plots with positional entropy annotation from within `Python` is a little more elaborate and will not be shown in this chapter. However, the computation of $S(i)$ can be done using the `positional_entropy()` method attached to objects of type `fold_compound`. The list returned by this method is 1-based, i.e., $S(1)$ is stored as the second element of the list, $S(2)$ as the third, etc. The first (0th) element of the list is the length n of the input sequence.

3.5 Representative Structures

As mentioned above, sometimes the MFE structure does not adequately represent the entire structure ensemble. In fact, for most biologically relevant sequences there are almost always structural alternatives s with probabilities $p(s)$ comparable to that of the MFE structure. Therefore, depending on the research question, it is reasonable to consider a different representative or entire sets of representative structures.

3.5.1 The Centroid Structure

Besides the MFE structure, which is the most probable in thermodynamic equilibrium, we have already encountered the *centroid* structure as part of the `RNAfold -p` output. Similar to clustering methods, the centroid s_c is the structure with the shortest distance to all other structures in the ensemble (or cluster) Ding et al. [12], i.e.,

$$s_c = \underset{s}{argmin} \sum_t p(s)p(t)d_{\mathrm{BP}}(s,t). \tag{16}$$

Benchmarking experiments by Hamada et al. [13] on sets of curated reference sequence/structure pairs have shown that the predicted centroid structure often contains fewer false positive (FP) base pairs or more true positives (TP). In their work, the authors proposed the concept of the γ-centroid together with a dynamic programming algorithm to compute it from base pair probabilities p_{ij}. Here, γ is a variable that can be used to adjust the prediction of the centroid: low γ values yield better positive predictive values (PPV), i.e., fewer FP base pairs, while higher γ values yield better sensitivities, i.e., more true positives. In its simplest form, $\gamma = 1$, the γ-centroid is identical to that of Eq. 16 and can be computed without any need for an additional dynamic programming algorithm. In fact, it turns out that the centroid structure s_c simply consists of all base pairs with a probability larger than 0.5, i.e.:

$$s_c = \left\{(i,j) \mid p_{ij} > 0.5\right\} \tag{17}$$

To compute the centroid structure with the help of the `Python` interface, the `centroid()` method attached to objects of type `fold_compound` can be used. The following `Python` code computes and prints the centroid structure in dot-bracket notation, its corresponding free energy in kcal · mol^{-1}, and the average distance to the ensemble.

```
# compute centroid structure and distance to ensemble
s_c, d_c = fc.centroid()
# compute energy of the centroid structure
e_c = fc.eval_structure(s_c)
# print everything
print("{} {{{:6.2f} d={:.2f}}}".format(s_c, e_c, d_c))
```

Here we print all information as a single line exactly as in the fourth line of the output of a corresponding `RNAfold -p` command line call, resulting in the following output:

```
((((.(((((((((((((((((....))))))))))))))))).)))) {-20.80 d=7.83}
```

Note, however, that the centroid structure s_c does not always adequately describe the ensemble of secondary structures. In general, it tends to consist of only a small number of base pairs due to its construction; *see* Eq. 17. This may, in extreme cases where the ensemble is dominated by two or more structures that do not share any or just a very small number of base pairs, lead to a centroid that is devoid of any base pairs. Nevertheless, combined with another structure, e.g., the MFE structure, the centroid already gives a good idea of how well this other structure represents the ensemble, since any base pair shared between the two can be considered reliable.

3.5.2 Maximum Expected Accuracy

Finally, a third structure representative, the *maximum expected accuracy* (MEA) structure s_{MEA}, can be found in the literature [14]. Here, the idea is that the ensemble is well characterized by the base pairing probabilities p_{ij} which already provide a reasonable measure of correctness of a base pair (i, j). Now, the expected accuracy $\text{EA}(s)$ of a given structure s can be computed from the probabilities of the base pairs it is composed of and, for each unpaired nucleotide i, the corresponding probability $q_i = 1 - \sum_j p_{ij}$ that it is unpaired:

$$\text{EA}(s) = \sum_{(i,j)\in s} 2\gamma p_{ij} + \sum_{i\notin s} q_i. \tag{18}$$

The scaling factor γ, again, is usually set to 1 and can be used to balance the importance of base pairs against that of unpaired positions. The question now is to find the structure s with the largest amount of correctly predicted base pairs, i.e., the one that maximizes $EA(s)$

$$s_{\text{MEA}} = \underset{s}{argmax}\,\text{EA}(s). \tag{19}$$

This immediately leads to an optimization problem that can be solved with a Nussinov-like dynamic programming algorithm in cubic time.

To compute the MEA structure with `RNAfold`, simply add the `--MEA` command line flag

```
$ echo "GGCUGUUGGUGCGGUCGUGCCCGUAAGGGUGUGGUCGCGUUGGUGGCC" | RNAfold -p --MEA
```

The structure prediction output then contains an additional (5th) row:

```
GGCUGUUGGUGCGGUCGUGCCCGUAAGGGUGUGGUCGCGUUGGUGGCC
((((((....))))))..((((....))))..((((((....)))))) (-21.10)
(((((((({(((((((({((((....))))}))))))))}))))))))) [-22.41]
((((.(((((((((((((((((....))))))))))))))))).)))) {-20.80 d=7.83}
((((((((((((((((((((((....)))))))))))))))))))))) {-21.10 MEA=37.45}
 frequency of mfe structure in ensemble 0.119414; ensemble diversity 11.28
```

In our example, the MEA structure, just like the centroid, consists of an extended rod-like helix with expected accuracy of 37.45 and a free energy of 21.1 kcal $\cdot$ mol^{-1}, the same as that of the MFE structure (*see* second row). Such cases of multiple (degenerate) MFE structures also occur for natural RNA sequences, albeit not that frequently. Still, this finding underlines the importance to not just take a *single* structure representative for granted, but to always consider alternative conformations.

Similar to the centroid structure above, the `Python` interface can be used to compute the MEA structure. For this purpose, we may call the `MEA()` method of a `fold_compound` object, evaluate the energy of the returned structure, and print all the results in the same manner as `RNAfold`:

```
# compute MEA and corresponding structure
s_mea, mea = fc.MEA()
# compute energy of the MEA structure
e_mea = fc.eval_structure(s_mea)
# print everything
print("{} {{{:6.2f} MEA={:.2f}}}".format(s_mea, e_mea, mea))
```

3.6 Suboptimal Structures

Whenever structure prediction yields quite different MFE, centroid, and/or MEA structures, or if any other diversity measure indicates a structural alternative, *a set of structures* should be considered to best represent the most likely conformations of an RNA. This, in principle, can be done by simply using the three representatives known so far. A far more sophisticated approach, on the other hand, is to investigate the space of suboptimal structures.

3.6.1 Suboptimal Structures Sensu Zuker

Back in the late 1980s, Zuker [15] presented an algorithm to compute a set of suboptimal structures. For sequences of length n that can form at most n^2 different base pairs (i, j), this set consists of the energetically best structures containing any (i, j). Thus, the set of suboptimal structures proposed by Zuker [15] has a size of at most n^2. To compute the *Zuker* suboptimals, use the `RNAsubopt` program in conjunction with the `-z` option

```
$ echo "GGCUGUUGGUGCGGUCGUGCCCGUAAGGGUGUGGUCGCGUUGGUGGCC" | RNAsubopt -z
```

or the equivalent Python code snippet

```
print(sequence)
for s in fc.zubopt_zuker():
    print("{} [{:6.2f}]".format(s.structure, s.energy))
```

The resulting output for both is a list of all suboptimal structures sensu Zuker, printed according to their stability in descending order. The first five suboptimals are shown below:

```
GGCUGUUGGUGCGGUCGUGCCCGUAAGGGUGUGGUCGCGUUGGUGGCC
((((((....))))))..((((....))))..((((((....)))))) [-21.10]
((((((((((((((((((((((....)))))))))))))))))))))) [-21.10]
((((((((((((((.(.(((((....)))))..))))))))))))))) [-17.60]
((((..(.(.((((((((((((....))))))))))))..).).)))) [-17.40]
((((((....))))))..((((.....)).))((((((....)))))) [-17.30]
...
```

Here, we can also see that the example sequence can adopt two dominating conformations that simultaneously resemble the MFE. The majority of structural alternatives with lower stability (higher free energies) are variations of the two dominant states with slightly shifted base pairs. Note that the mfold program Zuker [15] by default prints all Zuker suboptimals within a 5% energy range around the MFE instead of just returning a single structure representative. Therefore, for this example sequence, its output would consist of both the MFE structures.

3.6.2 All Suboptimal Structures

Even though this seminal work by Michael Zuker bore the title "On finding all suboptimal foldings of an RNA molecule," the notion of suboptimality can also be defined in a much broader sense. Instead, *all* secondary structures could be exhaustively enumerated by energy, beginning with the most stable, the MFE structure, up to a specified energy threshold $\text{MFE} + \delta$. This approach was first presented by Wuchty et al. [16] and its implementation is also available from the RNAsubopt program as well as from within the Python interface of the ViennaRNA Package.

Let us keep the default $5\% \approx 1.1\ \text{kcal} \cdot \text{mol}^{-1}$ energy range around the MFE from mfold which in our example amounts to energies up to about $-20\,\text{kcal} \cdot \text{mol}^{-1} \approx -21.1\,\text{kcal} \cdot \text{mol}^{-1} + 5\%$. To *exhaustively enumerate all secondary structures* of a sequence, run the following command:

```
$ echo "GGCUGUUGGUGCGGUCGUGCCCGUAAGGGUGUGGUCGCGUUGGUGGCC" | RNAsubopt \
  -e 1.1 \
  -s
```

where we specify an energy range $\delta = 1.1\ \text{kcal} \cdot \text{mol}^{-1}$ around the MFE (-e 1.1) and instruct RNAsubopt to return the structures sorted by energy (`-s`). For the equivalent `Python` commands, we make use of the `subopt()` method:

```
for s in fc.subopt(110):
    print("{} {:6.2f}".format(s.structure, s.energy))
```

Here, the value of 110 is the energy range in units of *decacalories* per mole (1 dacal/mol = 10 cal/mol = 10^{-2} kcal · mol^{-1}). This is a common unit in many of the implementations of the `ViennaRNA Package`. The resulting output then lists *all* structures that exhibit an energy within the specified energy range:

```
GGCUGUUGGUGCGGUCGUGCCCGUAAGGGUGUGGUCGCGUUGGUGGCC -21.10   1.10
((((((((((((((((((((((....)))))))))))))))))))))) -21.10
((((((....))))))..((((....))))..((((((....)))))) -21.10
(((((((((((((((((.((((....)))).))))))))))))))))) -20.80
((((.(((((((((((((((((....))))))))))))))))).)))) -20.80
((((((....)))))).(((((....))))).((((((....)))))) -20.60
((((.((((((((((((.((((....)))).)))))))))))).)))) -20.50
(((((((.((((((((((((((....)))))))))))))).))))))) -20.00
(((((.((((((((((((((((....)))))))))))))))).))))) -20.00
```

Again, one can see both of the MFE structures and some other structures with similar stability but fewer base pairs. Note, however, that increasing the energy range δ quickly results in an enormous number of structures, as the secondary structure space grows exponentially in the sequence length. Thus, it is advisable to keep δ small and/or filter and cluster the resulting structures in any appropriate manner. Such filter or cluster strategies are currently not available from the `RNAsubopt` program, but may be implemented with relatively little effort in `Python`; *see* **Note 5**.

3.6.3 Stochastic Backtracking

Finally, suboptimals can also be gathered from randomly drawing structures s from the ensemble according to their equilibrium probabilities $p(s)$ [17]. This approach is commonly known as *Boltzmann sampling* and can be considered a *stochastic backtracking* in the dynamic programming matrices of the partition function algorithm. In contrast to the suboptimal structure generating approaches above, this method usually generates a redundant set of secondary structures, since each backtracking step is performed independently. However, the `ViennaRNA Package` also provides an implementation for *nonredundant* Boltzmann sampling Michálik et al. [18] (*see* **Note 7**).

To generate a set of ten random structures, we may use the `RNAsubopt` program in conjunction with the `-p` option:

```
$ echo "GGCUGUUGGUGCGGUCGUGCCCGUAAGGGUGUGGUCGCGUUGGUGGCC" | RNAsubopt -p 10
```

or the equivalent `Python` method `pbacktrack()` bound to objects of type `fold_compound`. However, for the latter, a few more steps are required (i) to initialize the random number generator and (ii) to properly instantiate the `fc` object; *see* **Note 6**.

```
# initialize random number generator
RNA.init_rand()

# create model details object
md = RNA.md()
# activate unique multi-branch loop decomposition
md.uniq_ML = 1
# create fold_compound with updated model settings
fc = RNA.fold_compound(sequence, md)
# compute MFE
ss, mfe = fc.mfe()
# rescale Boltzmann factors
fc.exp_params_rescale(mfe)
# compute partition function
pp, dG = fc.pf()

# generate Boltzmann samples
print(sequence)
for s in fc.pbacktrack(10):
    print(s)
```

Now, both the above command line call of `RNAsubopt` and the `Python` code produce an output similar to

```
GGCUGUUGGUGCGGUCGUGCCCGUAAGGGUGUGGUCGCGUUGGUGGCC
((((.((((((((.((.(((((....))))).)).)))))))).))))
((((.(.(((((((((..((((....))))..))))))))).).))))
((((((....))))))..((((....))))..((((((....))))))
((((((((((((((((((((((....))))))))))))))))))))))
((((((((((((((((((((((....))))))))))))))))))))))
((((.((((((((((((.((((....)))).)))))))))))).))))
((((.((((((((((((.((((....)))).)))))))))))).))))
((((((....))))))..((((....))))..((((((....))))))
((((.(((((((((((((((((....))))))))))))))))).))))
((((((....))))))((((((....))))))((((((....))))))
```

where we can spot both the MFE structures and a few other structures with relatively high probability. Of course, due to its randomness, each run of the above commands results in a different outcome.

Using larger sets of randomly generated structures can help to assess the equilibrium probability of structural features, such as particular helices or certain unpaired regions. For that purpose, simply count how often these features F appear among the set of Boltzmann samples to derive corresponding motif frequencies $f(F)$. In the limit of exhausting the entire structure space, i.e., extremely large sample sizes, these frequencies approach the actual equilibrium probabilities $p(F)$. Furthermore, similar to the suggestions for exhaustive structure enumeration above, the resulting set of sampled structures should be subjected to a filtering and/or clustering step to handle the usually large amount of data and derive more information on the relevant research questions. For instance, Giegerich et al. [19] used a sampling and clustering strategy to predict RNA switches. Similarly, Ding et al. [12] used centroid structures derived from Boltzmann sampling to identify relevant structural representatives of the ensemble.

3.7 Structure Constraints

All prediction methods presented above always considered the entire secondary structure space, i.e., all structures compatible with the input sequence that consist of nested Watson-Crick (GC, AU) or wobble pairs (GU). However, in some situations, one would rather investigate a particular restricted sub-ensemble. For instance, an RNA sequence might be chemically modified such that a particular nucleotide is unable to form canonical Watson-Crick base pairs. While this does occur in nature in the form of post-transcriptional modifications, artificial modifications are found in the literature as well, where researchers attempt to resolve and model RNA structures by means of experiments [20]. Sometimes such modifications are even photo-reactive and can be removed during the experiment [21, 22], for instance to identify the functional importance of particular base pairing partners.

To understand the effect of such constraining modifications on the RNA secondary structure space, one requires the means of constraints in the prediction method. In the following we will explore the versatile toolbox of *structure constraints* within the `ViennaRNA Package` [23].

3.7.1 Nucleotide-Wise Constraints

The simplest form of a constraint would be to exclude one or more nucleotides from potential base pairing. Most prediction programs treat the sequence letter "N" as an unknown base that cannot form pairs. Therefore, we can replace the modified nucleotides of the input sequence by "N" and proceed with prediction as usual. The `RNAfold` program offers two more possibilities to provide structure constraints: (i) by passing a dot-bracket-like string with a special alphabet that provides per-nucleotide constraints and (ii) by using a dedicated *command file* (*see* also **Note** 8). In the following, we will focus on the latter, as it provides the most flexible way to pass structure constraints. Therefore, let us create a `ViennaRNA` command file first that simply prevents the fourth nucleotide of our input sequence from forming a base pair. For this purpose, open a text editor and create a text file named "simple.const" with the content

```
P 4 0 1
```

This command essentially translates to: prohibit (P) 1 nucleotide starting at position 4 from pairing with any (0) other nucleotide (see also the man page of `RNAfold` and **Note** 8). Now, run `RNA-`

`fold` as before, but use the command file to restrict the secondary structure conformation space to those structures where the 4th base stays unpaired

```
$ echo "GGCUGUUGGUGCGGUCGUGCCCGUAAGGGUGUGGUCGCGUUGGUGGCC" | RNAfold \
  -p \
  --MEA \
  --commands simple.const
```

which yields

```
GGCUGUUGGUGCGGUCGUGCCCGUAAGGGUGUGGUCGCGUUGGUGGCC
(((..(((((((((((((((((....)))))))))))))))))..))) (-18.40)
(((..{((((((((((({((((....))))}))))))))))))}..))) [-19.72]
(((..(((((((((((((((((....)))))))))))))))))..))) {-18.40 d=2.98}
(((..(((((((((((((((((....)))))))))))))))))..))) {-18.40 MEA=42.60}
 frequency of mfe structure in ensemble 0.118227; ensemble diversity 4.57
```

When comparing the prediction to the unconstrained predictions before, it becomes apparent that all three representatives, the MFE structure, the centroid, and the MEA structure, are the same. At the same time, the average distance of the centroid dropped from 7.83 to 2.98, the MEA increased from 37.45 to 42.6, and the ensemble diversity decreased from 11.28 to just 4.57, which all indicates that the ensemble is now mostly dominated by structures similar to what we see in the prediction output. This observation can also be confirmed by a call to `RNAsubopt -z`

```
$ echo "GGCUGUUGGUGCGGUCGUGCCCGUAAGGGUGUGGUCGCGUUGGUGGCC" | RNAsubopt \
  -z \
  --commands simple.const
```

where the next best alternative conformation with three helices is about 1.3 $\text{kcal} \cdot \text{mol}^{-1}$ less stable than the MFE:

```
GGCUGUUGGUGCGGUCGUGCCCGUAAGGGUGUGGUCGCGUUGGUGGCC
(((..(((((((((((((((((....)))))))))))))))))..))) [-18.40]
(((.((((((((((((((((((....)))))))))))))))))).))) [-18.10]
(((.((....)).)))..((((....))))..((((((....)))))) [-17.10]
(((...(.(.((((((((((((....))))))))))))..).)..))) [-15.30]
...
```

To reproduce the same predictions from within a `Python` script, either use the command file *simple.const* that has been created above

```
fc.commands_apply("simple.const")
```

or directly use the *hard constraints* method `hc_add_up()` that is attached to objects of `fold_compound` to force the 4th base to stay unpaired:

```
fc.hc_add_up(4)
```

After either of these commands, proceed with the predictions of MFE, partition function, centroid, MEA, suboptimals, etc. as before.

To remove all constraints from `fc`, call the `fc.hc_init()` method and all subsequent predictions will be performed without any further constraints again.

3.7.2 Constraints on Base Pairs

Similarly, one can also constrain individual base pairs either to be prohibited (P) due to some reason or to forcefully (F) appear in all the structures of the ensemble. The latter can be of interest when a base pair is chemically cross-linked. Let us now assume that in an experiment, we somehow introduced a covalent bond on the closing pair of the first stem of the three-helix conformation, i.e., base pair $(1, 16)$ between the first and the 16th nucleotide. The respective command for the command file is

```
F 1 16 1
```

which we store in a file "crosslink.const". When we now perform the structure prediction:

```
$ echo "GGCUGUUGGUGCGGUCGUGCCCGUAAGGGUGUGGUCGCGUUGGUGGCC" | RNAfold \
  -p \
  --MEA \
  --commands crosslink.const
```

we see that this constraint drastically affects the structure ensemble:

```
GGCUGUUGGUGCGGUCGUGCCCGUAAGGGUGUGGUCGCGUUGGUGGCC
((((((....))))))..((((....))))..((((((....)))))) (-21.10)
((((((....))))))..((((....))))..((((((....)))))) [-21.46]
((((((....))))))..((((....))))..((((((....)))))) {-21.10 d=0.51}
((((((....))))))..((((....))))..((((((....)))))) {-21.10 MEA=46.99}
 frequency of mfe structure in ensemble 0.558851; ensemble diversity 0.80
```

The MFE structure now has almost 56% probability, and the diversity of the ensemble has dropped substantially to just 0.8.

3.7.3 Motif Probabilities

Apart from simply looking at the predicted structures and the reduced diversity within the constrained ensemble of structures above, such constrained predictions provide a rich source for many more applications. For instance, in the previous section, Boltzmann sampling with fairly large sample sizes was suggested to estimate the equilibrium probability $p(F)$ of structural features or motifs F. Now recall that the exact value of $p(F)$ is given by the relation of the corresponding partition functions

$$p(F) = \frac{Z^F}{Z} \quad \text{with} \quad Z^F = \sum_{s|F \in s} exp(-E(s)/RT), \tag{20}$$

where Z^F is the partition function over the sub-ensemble of structures that exhibit F. Whenever the subset $\{s \mid F \in s\}$ can be expressed in terms of a set of constraints, the exact value of $p(F)$ can be computed from two predictions, one with and a second without constraints, instead of simply relying on an approximation from Boltzmann sampling. For instance, consider the probability $p(H_1)$ of the 5′ helix (H_1) with closing pair $(1, 16)$ that is part of one of the two MFE structures. It consists of six consecutive base pairs and four unpaired bases at positions 7–10. Create a command file "H1.const" with the corresponding set of constraints

```
F 1 16 6
P 7 0 4
```

and re-run `RNAfold` to obtain $\Delta G^{H_1} = -RT \ln Z^{H_1}$:

```
$ echo "GGCUGUUGGUGCGGUCGUGCCCGUAAGGGUGUGGUCGCGUUGGUGGCC" | RNAfold \
  -p \
  --commands H1.const
```

This should produce the output

```
GGCUGUUGGUGCGGUCGUGCCCGUAAGGGUGUGGUCGCGUUGGUGGCC
((((((....))))))..((((....))))..(((((((....))))))) (-21.10)
((((((....))))))..((((....))))..(((((((....))))))) [-21.44]
((((((....))))))..((((....))))..(((((((....))))))) {-21.10 d=0.48}
 frequency of mfe structure in ensemble 0.57932; ensemble diversity 0.73
```

with $\Delta G^{H_1} = -21.44$ kcal · mol^{-1}. In `Python`, we can (i) again simply use the `commands_apply()` method or (ii) use the `fc.hc_add_up()` and `fc.hc_add_bp()` methods

```
# for base pairs (1,16), (2,15), etc. to form
for i in range(1, 7):
    fc.hc_add_bp(i, 16 - i + 1)

# force position 7-10 to stay unpaired
for i in range(7, 11):
    fc.hc_add_up(i)
```

to apply the constraints. After that, recompute the MFE to properly rescale the Boltzmann factors with `exp_params_rescale(mfe)` again, and compute ΔG^{H_1} with the `fc.pf()` method.

Finally, with $RT \approx 0.616321$ kcal · mol^{-1} at default temperature $T = 37$ °C $= 310.15$ K, the exact probability of the 5′ helix can be computed similar to Eq. 8, as

$$\begin{aligned} p(H_1) &= \frac{Z^{H_1}}{Z} \\ &= exp((\Delta G - \Delta G^{H1})/RT) \\ &= exp((-22.41 + 21.44)/RT) \\ &\approx 0.20725. \end{aligned} \quad (21)$$

Hence, the probability of observing the first helix in a structure randomly drawn from the ensemble is about 20.7%.

Note that an approximation from Boltzmann sampling in this case would most certainly have resulted in an equally accurate approximation of $p(H_1)$. After all, stem H_1 has a relatively large equilibrium probability. However, generally speaking, motif probabilities tend to be very small, and, especially for longer RNA molecules with a fairly large structure ensemble, good estimates of $p(F)$ from frequencies $f(F)$ require vast sample sizes. Consequently, their computation from constrained partition functions is much more reliable and, in many cases, easier to perform.

3.8 Aptamer Pockets and Bound Ligands

Riboswitches, whether artificially designed or naturally occurring, usually require a bound ligand to adopt their alternative structure conformation. Such ligands usually bind selectively to short sequence/structure motifs, so-called aptamer domains, where the ligand binding induces the stabilization of the corresponding structure. Many naturally occurring ligand binding riboswitches are known today, most of them belonging to the realm of bacterial genomes [24]. However, artificially designed riboswitches can be obtained from SELEX experiments [25, 26], too. This allows for a wide range of applications as biosensors in synthetic biology and other related subjects [27, 28].

3.8.1 The Theophylline Aptamer

Some chemical compounds that are known to act as ligands have been well studied and characterized. One of them is theophylline [29], a naturally occurring compound found in tea or cocoa, which exhibits chemical structure similar to caffeine. A corresponding RNA aptamer has already been identified using SELEX in 1994 Jenison et al. [30], and a truncated yet functional form of the aptamer has been determined to result in a dissociation constant $K_d = 0.1$ μM. Its sequence/structure motif is shown in Fig. 2.

To model the binding of theophylline to the RNA, first determine the binding free energy ΔG_{bind} that corresponds to the $K_d = 0.1$ μM value measured experimentally. At a concentration of 1 M, this is simply

$$\begin{aligned} \Delta G_{\text{bind}} &= RT \cdot \ln K_d \\ &\approx -9.93 \text{ kcal} \cdot \text{mol}^{-1} \end{aligned} \quad (22)$$

Hence, at very high theophylline concentrations ($c_{\text{theo}} = 1$ M), binding the molecule to the RNA has a considerable stabilizing

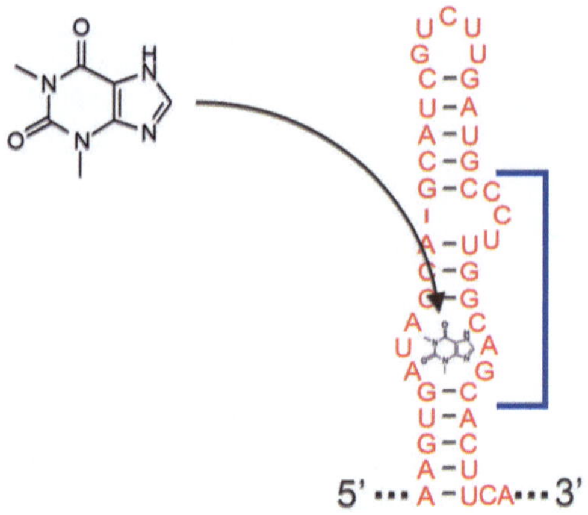

Fig. 2 Theophylline aptamer conformation. The theophylline molecule specifically binds to an RNA sequence/structure aptamer motif (indicated by the blue section) with high affinity. The dissociation constant has been determined as $K_d = 0.1\ \mu\text{M}$ in Jenison et al. [30]. The structure motif itself is enclosed by two GC base pairs, forming a 3/3 internal loop configuration followed by three more base pairs (CG,CG,AU), and a final bulge of three nucleotides on the 3′ side

effect of about $-9.93\ \text{kcal} \cdot \text{mol}^{-1}$. Let us now create the following FASTA file "theo.fa" that contains an RNA sequence which is capable of forming the aptamer motif:

```
>theo_bind
GGUGAUACCAGAUUUCGCGAAAAAUCCCUUGGCAGCACCUCGCACAUCUUGUUGUCUGAU
UAUUGAUUUUUCGCGAAACCAUUUGAUCAUAUGACAAGAUUGAG
```

To include the binding potential into the secondary structure prediction, we can now resort to an approach similar to the constraints example in the previous section. In fact, if we compute the constrained partition function Z^{theo} that captures all conformations that actually do form the aptamer fold, the binding free energy can be easily incorporated to yield the partition function $Z^{\text{theo}}_{\text{bound}}$ consisting of all conformations with bound ligand:

$$\begin{aligned} Z^{\text{theo}}_{\text{bound}} &= Z^{\text{theo}} \cdot exp(-\Delta G_{\text{bind}}/RT) \\ &= Z^{\text{theo}} \cdot \frac{1}{K_d}. \end{aligned} \tag{23}$$

Finally, the corrected partition function $\bar{Z} = Z + Z^{theo}_{\text{bound}}$ including both the bound and unbound states can be computed by simply adding the corresponding partition functions. This allows for computing a variety of equilibrium properties, for instance, the probability that theophylline is bound to an aptamer is

$$p^{\text{theo}}_{\text{bound}} = Z^{\text{theo}}_{\text{bound}} / \bar{Z} \tag{24}$$

3.8.2 Modeling Ligand Binding with Soft Constraints

To simplify matters, the `ViennaRNA Package` already implements a convenience function that enables such aptamer motifs to be included without having to go to the considerable effort of computing multiple (hard constrained) partition functions. This high-level convenience function is available in both `RNAfold` and the `Python` interface. The underlying implementation requires the structure motif of the aptamer to be delimited by 1 or 2 enclosing base pairs and the corresponding sequence to be well determined. In such cases, the bound state can be included in a *single* partition function (or MFE prediction) run as a so-called soft constraint [23] that simply adjusts the loop energy contribution for the loop that delimits the aptamer motif accordingly. If the aptamer motif is found multiple times in the input sequence, all possible combinations of bound and unbound states are considered while computing $\bar{Z}$. For `RNAfold`, the essential command line parameter to include such ligand binding is `--motif` and the `Python` interface provides the `sc_add_hi_motif()` method attached to `fold_compound` objects.

With the theophylline binding example considered here, the requirements for easily including it via the mentioned convenience functions are fulfilled: it is delimited by two base pairs (*see* also Fig. 2) and all three, the sequence motif, the structure motif, and the binding free energy, are well characterized. The corresponding command line call to `RNAfold` is now

```
$ RNAfold \
  -p \
  --motif="GAUACCAG&CCCUUGGCAGC,(...((((&)...)))...),-9.93" \
  theo.fa
```

Note here that the $5'$ and $3'$ parts of the aptamer motif are combined with the letter "`&`" in both the sequence and structure. This indicates that the motif is delimited by two enclosing pairs. The full aptamer motif is specified in dot-bracket format as the second parameter to the `--motif` option, followed by the binding free energy ΔG in units of $\mathrm{kcal}\cdot\mathrm{mol}^{-1}$. The resulting output is shown in Fig. 3, where the MFE structure and the centroid structure form the theophylline aptamer fold. Also recall that the `--motif` option prepares the computations of `RNAfold` to consider both the bound and the unbound states. Thus, the ensemble free energy as shown in the fourth line of Fig. 3a is already $\Delta\bar{G} = -RT\ln\bar{Z}$, i.e., it includes all bound as well as unbound states. To compute the probability $p^{\mathrm{theo}}_{\mathrm{bound}}$ as in Eq. 24, run `RNAfold` once again without passing the `--motif` option to obtain ΔG. The probability that theophylline is bound to the RNA at excess concentrations can then be computed as

A

```
>theo_bind
GGUGAUACCAGAUUUCGCGAAAAAUCCCUUGGCAGCACCUCGCACAUCUUGUUGUCUGAUUAUUGAUUUUUCGCGAAACCAUUUGAUCAUAUGACAAGAUUGAG
((((...((((((((......)))))...)))...))))...((.((((((((((.((((((.((.((((....)))).))..)))))).))))))))))))..  (-34.53)
((((...((((((((......)))))...)))...))))...,,.((((((((((.((((((.((.((((....)))).))..)))))).))))))))))))..  [-35.85]
((((...((((((((......)))))...)))...))))......((((((((((.((((((.((.((((....)))).))..)))))).))))))))))))....  {-24.20 d=4.35}
 frequency of mfe structure in ensemble 0.116952; ensemble diversity 6.71
```

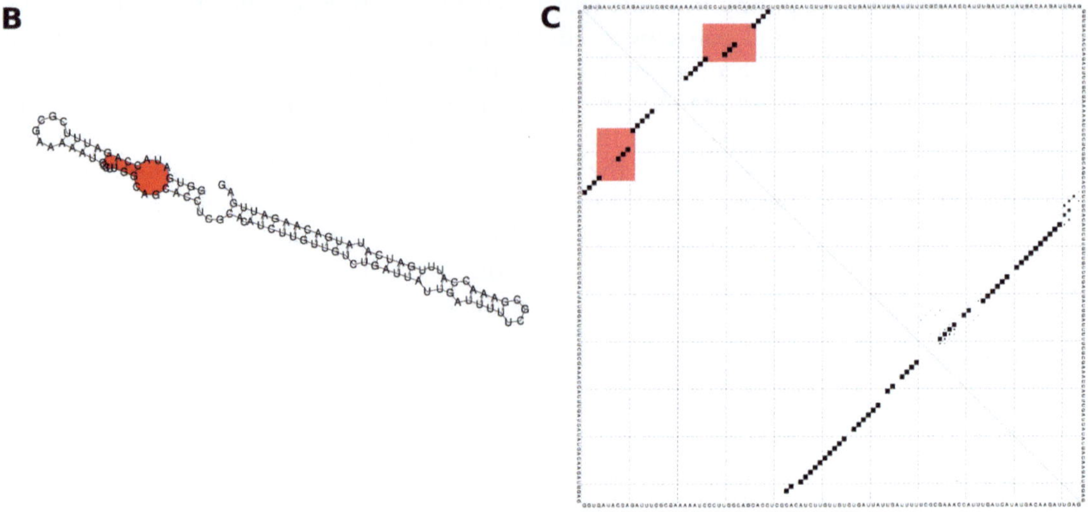

Fig. 3 Structure prediction with theophylline binding. (**a**) shows the output of the `RNAfold` program when the theophylline aptamer is provided via the `--motif` parameter and the ligand is present in excess (1 M). The two PostScript output files produced by `RNAfold` (*theo_bind_ss.ps* and *theo_bind_dp.ps*) depict the MFE structure (**b**) and the base pair probabilities in the upper triangle of (**c**). The aptamer motif is highlighted in both plots by red areas

$$
\begin{aligned}
p^{\text{theo}}_{\text{bound}} &= Z^{\text{theo}}_{\text{bound}}/\bar{Z} \\
&= 1 - Z/\bar{Z} \\
&= 1 - exp((\Delta\bar{G} - \Delta G)/RT) \\
&= 1 - exp((-35.85 + 28.18)/RT) \\
&\approx 0.999996,
\end{aligned}
\tag{25}
$$

i.e., there is an almost 100% chance that theophylline is bound.

A more elaborate example can be sketched using the `Python` interface. Assume this time that ligand is present at lower concentrations c. By adjusting the binding free energy accordingly $\Delta G_{\text{bind}}(c) = RT \ln \frac{K_d}{c}$, we can then compute $p^{\text{theo}}_{\text{bound}}(c)$ in a concentration-dependent manner. The following code fragment computes the partition function for the previous example (i) without taking the ligand into account and (ii) at a range of different concentrations c. For each of the concentrations, it also computes the probability $p^{\text{theo}}_{\text{bound}}(c)$ and prints the corresponding MFE structure and MFE.

```
import math
import RNA

# kT in kcal/mol
kT   = RNA.exp_param().kT / 1000.

theo_kd           = 0.1 * 1e-6 # 0.1 mu mol/l taken from Jenison et al. 1994
theo_motif_seq    = "GAUACCAG&CCCUUGGCAGC"
theo_motif_struct = "(...((((&)...)))...)"
theo_mu_0         = kT * math.log(theo_kd)

seq = "GGUGAUACCAGAUUUCGCGAAAAAUCCCUUGGCAGCACCUCGCACAUCUUGUUGUCUGAU" + \
      "UAUUGAUUUUUCGCGAAACCAUUUGAUCAUAUGACAAGAUUGAG"

fc = RNA.fold_compound(seq)

# Perform computations without ligand
ss, mfe = fc.mfe()
fc.exp_params_rescale(mfe)
pp, dG = fc.pf()

# print MFE (without ligand)
print("{}\n{} ({:-6.2f})".format(seq, ss, mfe))

for conc in [1e-6, 1e-5, 1e-4, 1e-3]:
    dG_bind = theo_mu_0 - kT * math.log(conc)
    fc.sc_add_hi_motif(theo_motif_seq, theo_motif_struct, dG_bind)
    ss, mfe = fc.mfe()
    fc.exp_params_rescale(mfe)
    pp_b, dG_b = fc.pf()
    p_theo = 1 - math.exp((dG_b - dG) / kT)
    print("p_theo({}) = {}".format(conc, p_theo))
    print("{}\n{} ({:-6.2f})".format(seq, ss, mfe))
```

After executing the code, we can see that at low concentrations such as $c = 10^{-6} M$, the probability that any structure taken from the ensemble at random is bound by the ligand in about 20% of cases. Increasing the concentration 10-fold to $c = 10^{-5} M$ already increases the probability to almost 72%, and with $c = 1$ mM theophylline concentration $p_{\text{bound}}^{\text{theo}}(c)$ increases further to more than 99%.

3.9 Single-Strand Binding (SSB) Proteins

Similar to the ligand binding example above, the binding of a protein to a single-stranded RNA motif can be easily modeled by utilizing structure constraints. RNAs in living cells are known to be covered by hundreds of different kinds of proteins [31]. Many of such proteins regulate gene expression at different levels, e.g., by controlling splicing or increasing RNA stability. This makes them interesting targets for inclusion in RNA structure prediction.

Again, as long as the binding affinity of the protein in question is known, the additional states that need to be considered can be incorporated into the thermodynamic ensemble by computing two partition functions: one for all unbound states Z^{u} and a second Z^{b} for all bound states. For the latter, only structures where the binding motif is accessible have to be considered, and their energy contribution needs to be adjusted such that it includes the potentially stabilizing effect of the bound protein, similar to Eq. 23.

However, in many cases an SSB protein may bind to multiple locations along the RNA sequence or, even worse, does not require a highly specific sequence motif to bind to. This immediately makes the entire computation much more complicated. Even for a single type of SSB protein and non-overlapping motifs along the RNA sequence, k partition functions Z_k^b have to be adjusted for all the subset of those structures exposing k accessible binding motifs, i.e.:

$$\bar{Z} \quad = Z + Z_1^b \cdot \frac{c}{K_d} + Z_2^b \cdot \left(\frac{c}{K_d}\right)^2 + \ldots \tag{26}$$

When multiple different proteins are involved or binding sites overlap, the entire approach quickly becomes infeasible even for relatively small k. A solution to this problem has been proposed in Forties and Bundschuh [32], and the `ViennaRNA Package` follows these lines and implements an efficient extension to the decomposition scheme for the prediction algorithms [23]. This extension is known as the ***unstructured domains*** feature and allows an ***arbitrary*** number of different SSB proteins to be included that potentially bind to multiple locations.

3.9.1 A Simple Example

Let us start with a very simple example of an artificial SSB protein with a footprint of three nucleotides that specifically binds to `GGU` motifs. Let us further assume that the protein binds to the RNA with $\Delta G_{\text{bind}} = -3$ kcal · mol^{-1}. In this example, we will refrain from any concentration dependency and assume that the protein is available in excess. To include these properties of protein-RNA binding into RNA structure prediction, create a `ViennaRNA` command file "simple_SSB.const" with the content

```
UD GGU -3
```

and run `RNAfold` with our example sequence from before as follows:

```
$ echo "GGCUGUUGGUGCGGUCGUGCCCGUAAGGGUGUGGUCGCGUUGGUGGCC" | RNAfold \
  -p \
  --commands simple_ssb.const
```

to receive the following output:

```
GGCUGUUGGUGCGGUCGUGCCCGUAAGGGUGUGGUCGCGUUGGUGGCC
((((......((((((((((((....))))))))))))......)))) (-24.50)
((((......(({|||{{((((....))))}}}}}}))......)))) [-25.45]
((((......((((((((((((....))))))))))))......)))) {-24.50 d=8.32}
 frequency of mfe structure in ensemble 0.215258; ensemble diversity 11.32
```

Recall that in previous predictions for this sequence without any further constraints, the sequence formed two equally stable MFE structures: one with three helices and another with an extended rod-like helix spanning the entire sequence. Now, with the SSB protein, the MFE structure has two longer stretches of

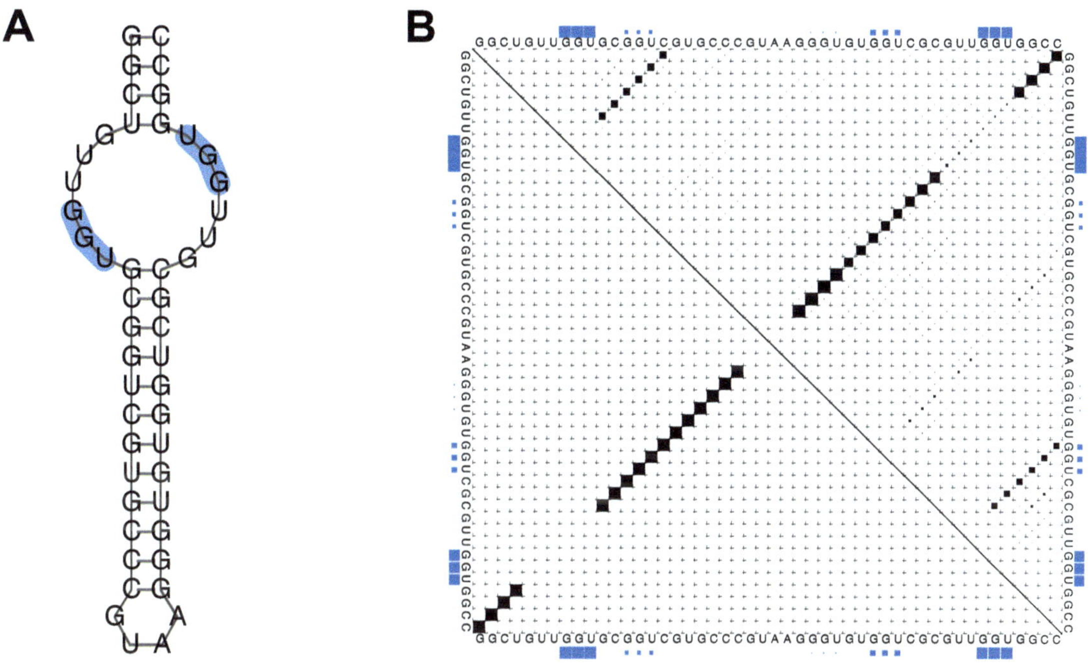

Fig. 4 Structure prediction with an artificial SSB protein that binds to GGU sequence motifs. The SSB protein interaction with a binding free energy $\Delta G_{\text{bind}} = -3\ \text{kcal} \cdot \text{mol}^{-1}$ has been incorporated into `RNAfold` prediction via the *unstructured domains* feature. The left panel (**a**) shows the MFE structure with two motifs bound by the protein highlighted in light blue. The right panel (**b**) is the corresponding dot-plot with base pair probabilities as black squares in the upper triangle and additional protein binding probability annotation in light blue at the four sides. Like base pair probabilities, the area of the light-blue squares at the side scales with their corresponding probability. Besides the two motifs present in the MFE structure, three further motifs are visible in the dot-plot, albeit with (much) lower binding probability $p_b(i)$

6 unpaired bases: one starting at position 5 and the other at the end of the sequence at position 39. A closer inspection of the sequence in this regions reveals that both unpaired segments contain the GGU motif and thus are actually bound by the SSB protein. This is also visible from the secondary structure and dot-plots created by the `RNAfold` (*see* Fig. 4). In fact, the RNA sequence contains in total five potential binding sites, of which three have much lower probabilities to be bound, measured in probabilities

$$p_b(i) \quad = \hat{Z}_i / \bar{Z} \tag{27}$$

that sequence position i is bound (occupied) by a protein. In the predicted MFE and centroid structures, those positions are sequestered by regular base pairs. Note that computing the partition function $\hat{Z}_i$ requires all cases where a protein binds to the RNA such that position i is covered while, at the same time, other nucleotides may be bound by further proteins. This introduces a large number of combinations, and for simplicity's sake, we omit a detailed definition here. The algorithms implemented in the

ViennaRNA Package efficiently compute $p_b(i)$ similar to the base pair probabilities p_{ij}.

The actual binding probabilities can be extracted from the "dot.ps" PostScript file using the awk tool again, knowing that they are stored as array data in lines of the form

```
[ i p 0.6000 0.8000 0.9500]
```

where i is the sequence position, p the square-root $\sqrt{p_b(i)}$ of the probability that the protein is bound to that sequence position, and the remaining three numbers simply specify the color of the corresponding square for rendering the PostScript file. The following command line can therefore be used to obtain the list of binding probabilities:

```
$ awk -e '$0 ~ /0.6000 0.8000 0.9500/ {print $2, $3*$3}' dot.ps \
  | sort -u -k1,1n
```

Note here that we let awk match any line ($0) against the three numbers that specify the color code. Since by default, lines read by awk are split at any whitespace, the data we are interested in are in columns 2 and 3. For the output, we already square the third value to obtain the actual probability $p_b(i)$. Next, since the probabilities are redundantly stored for each of the four sides of the dot-plot, the sort command is used with -u option to make the output unique, i.e., to omit redundant data. The -k1,1n option simply instructs sort to sort numerically by only using the first column of data (the sequence position). The resulting output is

```
8 0.939701
9 0.939701
10 0.939701
13 0.0865337
14 0.0865337
15 0.0865337
28 0.0054715
29 0.0054715
30 0.0054715
33 0.189336
34 0.189336
35 0.189336
42 0.764945
43 0.764945
44 0.764945
```

where we can see that the first (pos. 8–10) and last sequence motif (pos. 42–44) show quite high binding probabilities of 94% and 76.5%, respectively. The remaining three binding sites (pos. 13–15, 28–39, and 33–35) have much lower probabilities to be bound.

To reproduce the same prediction from within Python, we can use the ud_add_motif() method attached to objects of type fold_compound. This method is given three parameters, where the first is the sequence motif, the second is the binding free energy in units of kcal · mol^{-1}, and the third is a name of the motif that serves as an identifier:

```
fc.ud_add_motif("GGU", -3, "simple SSB")
```

After adding the motif, again compute MFE and partition function with the `fc.mfe()` and `fc.pf()` methods. The latter also computes the binding probabilities but does not provide a dedicated output for $p_b(i)$. Instead, we need to create a list of *all* equilibrium probabilities computed in the partition function run using the `fc.plist_from_probs()` method. Then, manually filter out those entries that correspond to the unstructured domains computation. A complete working code fragment that produces the same output as filtering of the *dot.ps* file using `awk` and `sort` above may therefore look like this:

```
import RNA

# create fold_compound with input sequence
fc = RNA.fold_compound("GGCUGUUGGUGCGGUCGUGCCCGUAAGGGUGUGGUCGCGUUGGUGGCC")
# add unstructured domain motif, i.e. SSB protein
fc.ud_add_motif("GGU", -3, "simple SSB")
# compute MFE
ss, mfe = fc.mfe()
# rescale Boltzmann factors
fc.exp_params_rescale(mfe)
# compute partition function
pp, dG = fc.pf()
# obtain list of equilibrium probabilities larger than 1e-6
pl = fc.plist_from_probs(1e-6)

# go through list, extract protein binding probabilities
# and print results
for p in pl:
    if p.type == RNA.PLIST_TYPE_UD_MOTIF:
        for i in range(p.i, p.j + 1):
            print("{} {}".format(i, p.p))
```

3.9.2 Nonspecific SSB Proteins with Fixed Footprint

In their seminal work, Forties and Bundschuh [32] presented an example of a hypothetical SSB protein with a 6-nucleotide footprint that binds nonspecifically to a DNA sequence. The binding affinity of the protein in their example was set such that it exhibits a chemical potential for binding $\mu_0^P(i) = 10$ kcal · mol^{-1} at 1 M concentration. The authors then predicted probabilities $p_b(i)$ for each nucleotide position i and a wide range of protein concentrations (*see* Figure 3a in Forties and Bundschuh [32]). This allowed them to identify regions within the DNA that form regular base pairs at low protein concentrations and only open up to bind the protein at higher concentration.

Since such an in silico experiment is quite tedious using the `RNAfold` program and requires a lot of post-processing of the prediction output, we will only provide `Python` code to reproduce

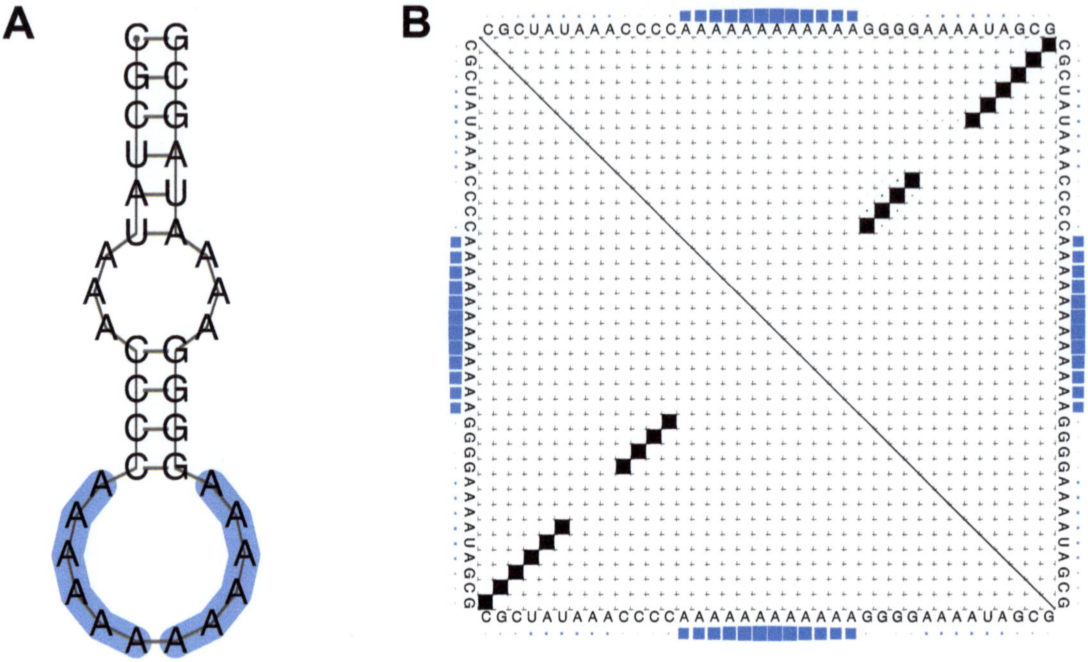

Fig. 5 Structure prediction with SSB protein binding. Shown here is a prediction for the nonspecifically binding SSB protein with a footprint of 6 nt at concentration $c = 20$ nMc with `RNAfold` and corresponding "`UD NNNNNN -0.93`" command file content ($\Delta G_{\text{bind}}(20 \text{ nM}) = -0.93 \text{ kcal} \cdot \text{mol}^{-1}$, *see* Eq. 28). The MFE structure (**a**) is occupied by two proteins at its central loop (indicated by the light-blue annotation). Other parts of the DNA sequence form regular base pairs. In (**b**), the dot-plot with corresponding equilibrium base pair probabilities is shown. Probabilities $p_b(i)$ that sequence positions i are occupied by the SSB protein are denoted as light-blue squares at the four sides of the plot

the experiment shown in Figure 3a of Forties and Bundschuh [32]. First, the means to compute binding free energy $\Delta G_{\text{bind}}(c)$ at any concentration c is required. This turns out fairly easy, since

$$\begin{aligned} \Delta G_{\text{bind}}(c) &= -\mu^P(c, i) \quad \text{with} \\ \mu^P(c, i) &= \mu_0^P(i) + RT \ln c. \end{aligned} \tag{28}$$

The protein is assumed to bind nonspecifically to the nucleic acid sequence with a footprint of six nucleotides. Since the unstructured domains feature of the `ViennaRNA Package` recognizes *IUPAC* notation for the sequence motif, we can use the `NNNNNN` motif to indicate that each of the six positions may be *any* nucleotide. To get an idea of the secondary structure formed by the nucleic acid sequence, Fig. 5 shows the predicted MFE structure and corresponding dot-plot at low protein concentrations of $c = 20$ nM. We can see that the DNA forms a helical structure with a loop region in the center enclosed by four base pairs. The ends of the sequence form a helix of six nucleotides that connects to the central part of the structure via an internal loop with four unpaired bases on either sides. At the low concentration considered in this prediction, two SSB proteins are bound to the MFE structure at the central loop. The remaining parts of the sequence form a relatively stable secondary structure.

Now, to proceed with the computations of $p_b(c, i)$, we can use the `binding_probs(s, mu_0, motif, steps = 50, stepsize = 2e-8)` function below. It receives the sequence `s`, the chemical potential at 1 M concentration `mu_0`, and the binding sequence motif `motif`. It will compute the position-wise binding probabilities $p_b(c, i)$ for a range of concentrations c of 20 nM to 1000 nM in steps of 20 nM, as specified by the default values of `stepsize` and `steps`. The function finally returns the result in the form of a matrix-like list.

```python
def binding_probs(s, mu_0, motif, steps = 50, stepsize = 2e-8):
    """
    Create position-wise protein binding probabilities for sequence 's'
    and SSB protein with chemical potential 'mu_0' at 1M concentration
    and binding motif 'motif'. Default settings computes probabilities
    for concentrations from 20nM to 1000nM in steps of 20nM, i.e. 50
    different data points
    """

    # Load DNA parameters (comment next line to treat sequence as RNA)
    RNA.params_load_DNA_Mathews2004()
    # kT in kcal/mol
    kT = RNA.exp_param().kT / 1000.
    # initialize binding probabilities
    prob = [[0 for i in range(0, steps + 1)] for j in range(0, len(s) + 1)]

    for i in range(1, steps + 1):
        # current concentration
        c = i * stepsize
        # compute current binding free energy
        dG_bind = -(mu_0 + kT * math.log(c))
        # prepare fold_compound
        fc = RNA.fold_compound(s)
        fc.ud_add_motif(motif, dG_bind, "Protein")
        # compute MFE and partition function
        (ss, mfe) = fc.mfe()
        fc.exp_params_rescale(mfe)
        (pp, dG) = fc.pf()
        # go through probabilities above 1e-14
        for p in fc.plist_from_probs(1e-14):
            if p.type == RNA.PLIST_TYPE_UD_MOTIF:
                for j in range(p.i, p.j + 1):
                    prob[j][i] += p.p

    return prob
```

Execute the `binding_probs()` function and store its result in variable `probabilities`:

```python
# example sequence from Forties and Bundschuh 2010
sequence = "CGCTATAAACCCCAAAAAAAAAAAAAGGGGAAAATAGCG"

# compute binding probabilities
probabilities = binding_probs(sequence, 10., "NNNNNN")
```

The probabilities $p_b(c, i)$ for $c \in [20\ \text{nM}, 40\ \text{nM}, \ldots, 1000\ \text{nM}]$ and each sequence position i are now accessible as `probabilities [i][cc]`, where `cc = 1` for $c = 20$ nM, `cc = 2` for $c = 40$ nM, etc.

Lastly, the following function uses `matplotlib` and the probabilities we have just computed to create a figure similar to Figure 3a in Forties and Bundschuh [32].

```
def create_figure(seq, probs, filename, steps = 50, stepsize = 2e-8):
    """
    This function produces a heatmap like plot similar
    to Figure 3a of Forties and Bundschuh 2010
    """
    import matplotlib.pyplot as plt
    import numpy as np

    conc_ticks  = [ i for i in range(0, steps + 1, 10) ]
    conc_labels = [ i * stepsize * 1e9 for i in range(0, steps + 1, 10) ]
    seq_ticks   = [ i for i in range(1, len(seq) + 1) ]
    seq_labels  = [ c for c in seq ]

    probs_arr = np.array(probs)
    f = plt.figure(figsize = (12, 6))

    plt.imshow(probs_arr,
               cmap='gist_yarg',
               vmin=0,
               vmax=1,
               interpolation=None,
               origin = 'lower')

    plt.ylim(0.5, len(seq) + 0.5)
    plt.xlim(1, steps)

    plt.xticks(ticks = conc_ticks,
               labels = conc_labels,
               fontsize=16)
    plt.xlabel("[Protein] (nM)", fontsize=20)

    plt.yticks(ticks = seq_ticks,
               labels = seq_labels,
               rotation='vertical',
               fontsize=12)
    plt.ylabel("Sequence", fontsize=20)
    plt.tick_params(left = False)

    cb = plt.colorbar(shrink = 0.8)
    cb.ax.tick_params(labelsize=16)
    cb.set_label(label = "Binding Probability", size=18)

    f.savefig(filename, bbox_inches='tight')
```

Executing the function with

```
# create figure
create_figure(sequence, probabilities, "fig3a_forties2010.pdf")
```

yields the plot "*fig3a_forties2010.pdf*" shown in Fig. 6a. We can see that the hypothetical protein binds to the loop region in the center of the DNA sequence at low concentrations. Upon increasing the concentration, the protein starts to bind to the ends of the sequence breaking up the energetically weak outermost stem. Only at very high concentrations is the strong binding of the

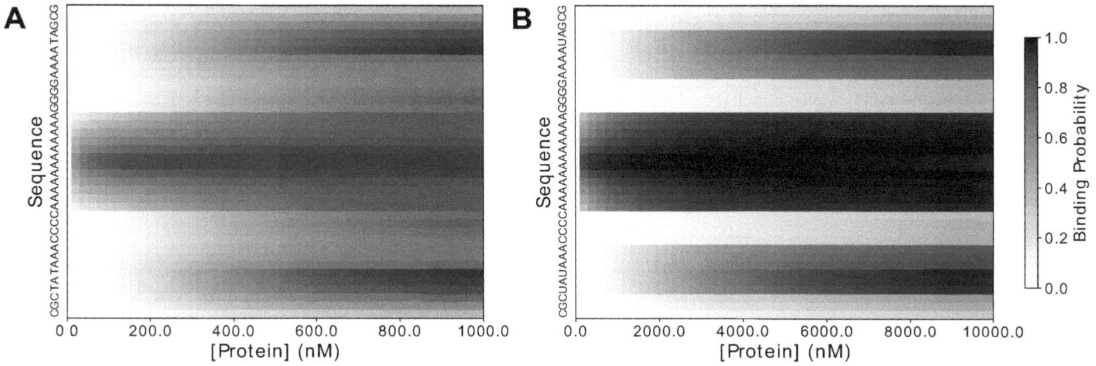

Fig. 6 Concentration-dependent binding of a SSB protein to a nucleic acid. Shown here is the position-wise probability $p_b(c, i)$ that particular positions i on the nucleic acid are occupied by the protein (Y-axis) at concentration c. The X-axis denotes the different protein concentrations. In (**a**), the sequence investigated is DNA where at low concentrations the protein binds to the loop region in the center first. Increasing the concentration results in the replacement of the outermost base pairs by regions occupied with the protein. Only at very high concentrations are the innermost base pairs broken up to give space for the protein to bind. Predictions for an RNA with the same sequence (**b**) reveal that even at very low concentrations, the center loop region is very likely to be occupied by the protein. However, breaking any further base pairs to allow for the protein to bind requires much higher concentrations compared to the DNA sequence. Moreover, the innermost base pairs open up and are covered by the protein only at very high concentrations ($c \gg 50$ μM, data not shown)

innermost base pairs broken up, allowing the protein to bind to these sequence parts. This tallies closely with the experiment in Forties and Bundschuh [32].

When the above experiment is reproduced with an RNA sequence instead, as shown in Fig. 6b, the picture changes quite drastically. At low concentrations, the binding behavior is very similar to the DNA case, the SSB protein being likely to bind to the central loop. However, for the remainder of the RNA structure, much higher concentrations are required to replace the base pairs with bound proteins. The nucleotides forming the innermost base pairs closing the central loop even require a concentration about 100-fold of the equivalent DNA example to become mostly occupied by the SSB protein.

4 Notes

1. Many secondary structure prediction programs operate on free energy parameter $E(s)$ scaled to amounts of substance, i.e., the unit of $E(s)$ is $\text{kcal} \cdot \text{mol}^{-1}$. Knowing that and since R is the molar equivalent of the Boltzmann constant k_B (or simply k),

RT should be used instead of *kT* to properly cancel out the units in all formulas used throughout this chapter.

2. The executable programs of the `ViennaRNA Package` that compute probabilities, such as `RNAfold`, limit their output to those values that exceed a certain threshold. However, in most cases, this threshold is user-adjustable. `RNAfold`, for instance, provides the command line parameter `--bppmThreshold x` that can be used to set the threshold to any value `x`. Consequently, to include *all* probabilities, use a threshold of $x = 0$, i.e.,

```
$ RNAfold --bppmThreshold 0
```

3. After computing partition function and base pair probabilities, the `RNAfold` program prints for each nucleotide i the corresponding pairing propensity, i.e., the preference in the pairing pattern of i in the form of a dot-bracket-like string of length n. Each character in the string addresses the corresponding base of the RNA sequence. The following characters are used: "." the base is mostly unpaired ($\sum_j p_{ij} < \frac{1}{3}$), "," the base is weakly paired, "|" the base is strongly paired ($\sum_j p_{ij} > \frac{2}{3}$) but without preferred pairing direction, "{" the base pairs weakly downstream, "}" the base pairs weakly upstream, "(" the base pairs strongly downstream ($\sum_{i<j} p_{ij} > \frac{2}{3}$), and ")" the base pairs strongly upstream ($\sum_{j<i} p_{ij} > \frac{2}{3}$).
4. The `Python` interface of the `ViennaRNA Package` also allows for the creation of the `PostScript` dot-plot produced by `RNAfold` and shown in Fig. 1. However, the base pair probabilities as well as the MFE structure have to be converted to a suitable list format first, as shown below:

```
# create a pair list from base pair probabilities
list_pr_bp  = fc.plist_from_probs(1e-5)
# create a pair list from dot-parenthesis structure stored in ss
# and use 0.95 as pairing probability
list_pr_mfe = RNA.plist(ss, 0.95)
# save dot plot as file named dot.ps
RNA.PS_dot_plot_list(sequence, "dot.ps", list_pr_bp, list_pr_mfe, "")
```

5. Suboptimals, as discussed in Subheading 3.6.2, are usually huge sets of structures which can hardly be inspected itemwise. A reasonable approach to limit the enormous data they generate is to filter or cluster them and only inspect a coarse-grained result. Such clustering can, for instance, be done through the use of abstract shapes [33]. In its most abstract form (shape level 5), base pairs that constitute helices are

collapsed to a single pair of matching brackets ([]) and unpaired bases are neglected entirely. Thus, only the branching pattern of the helices of a structure is left. Now, by only keeping the energetically most stable *shape representative* (shrep), any large set of structures can be reduced to a reasonable amount of shreps. The `Python` code below presents an example that uses a *callback* function `store_shrep()` to store all shreps from a set of suboptimal structures generated by `fc.subopt_cb()`. In contrast to `fc.subopt()`, this function automatically executes a callback function for each suboptimal structure it generates. In each call, it passes the corresponding dot-bracket notation, free energy in $\text{kcal} \cdot \text{mol}^{-1}$, and a further user-defined data structure.

```
import RNA

def store_shrep(structure, energy, data):
    """
    Callback function for subopt_cb() method that
    maps each structure into abstract shapes level 5
    according to Giegerich 2004 and only stores the
    energetically best representative for each shape.
    It also counts how many structures in total have
    been mapped to each shape
    """
    if structure != None:
        # convert to shape
        shape = RNA.abstract_shapes(structure)
        # increase counter
        count = data[shape]["N"] + 1 if shape in data else 1
        # store structure if we discovered a new shape or
        # if the current structure is energetically better
        if shape not in data or energy < data[shape]["en"]:
            data[shape] = { "en": energy,
                            "ss": structure,
                            "N": count
                          }
        else:
            # otherwise simply update counter
            data[shape]["N"] = count

seq = "GGCUGUUGGUGCGGUCGUGCCCGUAAGGGUGUGGUCGCGUUGGUGGCC"

# create fold_compound
fc = RNA.fold_compound(sequence)

# create an empty dictionary to store results in
data = dict()

# compute suboptimals up to 15kcal/mol above the MFE and process
# each generated structure with the 'store_shrep' function
fc.subopt_cb(1500, store_shrep, data)

# print total number of suboptimal structures generated
print("N_total = {:d}".format(sum([d["N"] for d in data.values()])))

# print abstract shape representatives by energy in ascending order
for k, v in sorted(data.items(), key=lambda item: item[1]["en"]):
    print("{} {:6.2f} {} {}".format(v["ss"], v["en"], k, v["N"]))
```

which yields

```
N_total = 300976
((((((((((((((((((((((....)))))))))))))))))))))) -21.10 [] 255582
((((((....))))))..((((....))))..((((((....)))))) -21.10 [][][] 23251
((((((....))))))................((((((....)))))) -15.00 [][] 16222
((((((....))))))(.((((....))))...(((((....)))))) -13.50 [][[][]] 683
((((((....)))))(((((((....)))))))(((((....)))))) -12.60 [[][][]] 1276
((((((....)))))...((((....)))).)((((((....)))))) -12.00 [[][]][] 361
((((......((((((((((((....))))))))))))(....))))) -11.90 [[][]] 2943
.(((...)))((....))((((....))))..((((((....)))))) -11.80 [][][][] 635
((((((....))))))((((((....))))(....)((....)).)).  -8.40 [][[][][]] 10
.((.((....))(....)((((....))))))((((((....))))))  -7.50 [[][][]][] 5
((((((....))......((((....))))(....)((....))))))  -6.80 [[][][][]] 4
.(((...)))(((....)((((....))))))((((((....))))))  -6.60 [][[][]][] 4
```

We can see that of the more than 300,000 suboptimals generated, the majority form a rod-like helical shape (`[]`). The next most frequent shape encountered is a three-helix conformation (`[][][]`). Both of these shapes have a representative with MFE. All other shapes are less frequent and less stable than the two dominating conformations.

6. The default mode for partition function and equilibrium probability computation is to keep the number of dynamic programming matrices as small as necessary to save memory. However, for Boltzmann sampling, an additional matrix is required to uniquely backtrack multi-branch loops in an efficient way. To allow for access to this additional matrix in any `fc.pbacktrack()` run, we therefore have to *inform* the partition function algorithm to create and fill the additional matrix when the `fc.pf()` method is executed. This is done by setting the `md.uniq_ML = 1` flag in the model details `RNA.md()`, which are an optional argument for the `RNA.fold_compound()` constructor.

7. Boltzmann sampling in the `ViennaRNA Package` implements a nonredundant sampling mode similar to that presented in Michálik et al. [18]. To activate it, pass the `RNA.PBACKTRACK_NON_REDUNDANT` flag as last argument to the `fc.pbacktrack()` method or use the `-N` parameter flag for `RNAsubopt`. Both will result in a set of structures s randomly generated according to their probability $p(s)$, but every structure s will appear at most once. In other words, this mode of sampling follows a sampling scheme that draws samples from an urn without replacement.

8. Most programs of the `ViennaRNA Package` such as `RNAfold` allow for two different ways to provide structure constraints: (i) through an *extended dot-bracket notation* in the input and (ii) *command files.* For the former, the input must be followed by a dot-bracket-like structure string where each character represents a constraint on the corresponding nucleotide. This extended dot-bracket string may consist of the following

characters: (i) "x" to indicate that this base must not pair, (ii) matching parentheses "()" to indicate that the respective nucleotides are to form a base pair, (iii) "|" to indicate the respective base should pair with any other base, (iv) "<" to force a base to only pair downstream, (v) ">" to force a base to only pair upstream, and finally (vi) "." to indicate that the corresponding base is unconstrained.

The usage of *command files* allows for a wider range of constraints to be passed to the prediction algorithm. These files are simple text files that generalize and extend the constraint syntax used in the `mfold` program Zuker [15]. Each line in the file specifies one constraint starting with a one- or two-letter command followed by some variables while obeying one of the following grammar rules:

```
F i 0 k    [TYPE] [ORIENTATION] # Force nucleotides i...i+k-1 to pair
F i j k    [TYPE] # Force helix of size k starting with (i,j) to form
P i 0 k    [TYPE] # Prohibit nucleotides i...i+k-1 to be paired
P i j k    [TYPE] # Prohibit pairs (i,j),...,(i+k-1,j-k+1)
P i-j k-l  [TYPE] # Prohibit pairing between two ranges
C i 0 k    [TYPE] # Nucleotides i,...,i+k-1 must appear in context TYPE
C i j k           # Remove pairs conflicting with (i,j),...,(i+k-1,j-k+1)
E i 0 k e         # Add pseudo-energy e to nucleotides i...i+k-1
E i j k e         # Add pseudo-energy e to pairs (i,j),...,(i+k-1,j-k+1)
UD m e     [LOOP] # Add ligand binding to unstructured domains with
                  # motif m and binding free energy e
```

where `i`, `j`, `k`, and `l` assume positive integer numbers, `e` assumes a floating point energy value in kcal · mol^{-1}, and `m` is a sequence string in uppercase IUPAC notation. The optional `LOOP`, `TYPE`, and `ORIENTATION` flags may take on the following values:

```
[LOOP]        = { E, H, I, M, A }
[TYPE]        = [LOOP] + { i, m }
[ORIENTATION] = { U, D }
```

to indicate the closing pairs of an external loop (`E`), hairpin loop (`H`), internal loop (`I`), multi-branch loop (`M`), or any loop (`A`). Base pairs enclosed by another pair can be addressed, too, in particular those enclosed in an internal loop (`i`) or multi-branch loop (`m`). Pairing directions are indicated by `U` for upstream and `D` for downstream pairing partners.

References

1. Tinoco I, Uhlenbeck OC, Levine MD (1971) Estimation of secondary structure in ribonucleic acids. Nature 230(5293):362–367. https://doi.org/10.1038/230362a0
2. Tinoco I, Borer PN, Dengler B, Levine MD, Uhlenbeck OC, Crothers DM, Gralla J (1973) Improved estimation of secondary structure in ribonucleic acids. Nat New Biol 246(150): 40–41. https://doi.org/10.1038/newbio246040a0
3. Waterman MS (1978) Secondary structure of single—stranded nucleic acids. Stud Found

Comb Adv Math Suppl Stud Academic Press NY 1:167–212. https://doi.org/10.1.1.15.4425
4. Waterman MS, Smith TF (1978) RNA secondary structure: a complete mathematical analysis. Math Biosci 42:257–266. https://doi.org/10.1016/0025-5564(78)90099-8
5. Nussinov R, Pieczenik G, Griggs JR, Kleitman DJ (1978) Algorithms for loop matchings. SIAM J Appl Math 35(1):68–82. https://doi.org/10.1137/0135006
6. Nussinov R, Jacobson AB (1980) Fast algorithm for predicting the secondary structure of single-stranded RNA. Proc Natl Acad Sci 77(11):6309–6313. https://doi.org/10.1073/pnas.77.11.6309
7. Zuker M, Stiegler P (1981) Optimal computer folding of large RNA sequences using thermodynamics and auxiliary information. Nucleic Acids Res 9(1):133–148. https://doi.org/10.1093/nar/9.1.133
8. McCaskill JS (1990) The equilibrium partition function and base pair binding probabilities for RNA secondary structure. Biopolymers 29(6–7):1105–1119. https://doi.org/10.1002/bip.360290621
9. Grüning B, Dale R, Sjödin A, Chapman BA, Rowe J, Tomkins-Tinch CH, Valieris R, Köster J Bioconda: sustainable and comprehensive software distribution for the life sciences. Nat Methods 15(7):475–476. https://doi.org/10.1038/s41592-018-0046-7
10. Zadeh JN, Wolfe BR, Pierce NA (2011) Nucleic acid sequence design via efficient ensemble defect optimization. J Comput Chem 32(3):439–452. https://doi.org/10.1002/jcc.21633
11. Shannon CE (1948) A mathematical theory of communication. Bell Syst Tech J 27(3):379–423. https://doi.org/10.1002/J.1538-7305.1948.TB01338.X
12. Ding YE, Chan CY, Lawrence CE (2005) RNA secondary structure prediction by centroids in a Boltzmann weighted ensemble. RNA 11(8):1157–1166. https://doi.org/10.1093/nar/gkg938
13. Hamada M, Kiryu H, Sato K, Mituyama T, Asai K (2009) Prediction of RNA secondary structure using generalized centroid estimators. Bioinformatics 25(4):465–473. https://doi.org/10.1093/bioinformatics/btn601
14. Do CB, Woods DA, Batzoglou S (2006) CONTRAfold: RNA secondary structure prediction without physics-based models. Bioinformatics 22(14):e90–e98. https://doi.org/10.1093/bioinformatics/btl246
15. Zuker M (1989) On finding all suboptimal foldings of an RNA molecule. Science 244(4900):48–52. https://doi.org/10.1126/science.2468181
16. Wuchty S, Fontana W, Hofacker IL, Schuster P (1999) Complete suboptimal folding of RNA and the stability of secondary structures. Biopolymers: Original Res Biomol 49(2):145–165. https://doi.org/10.1002/(SICI)1097-0282(199902)49:2<145::AID-BIP4>3.0.CO;2-G
17. Ding Y, Lawrence CE (2003) A statistical sampling algorithm for RNA secondary structure prediction. Nucleic Acids Res 31(24):7280–7301
18. Michálik J, Touzet H, Ponty Y (2017) Efficient approximations of RNA kinetics landscape using non-redundant sampling. Bioinformatics 33(14):i283–i292. https://doi.org/10.1093/bioinformatics/btx269
19. Giegerich R, Haase D, Rehmsmeier M (1999) Prediction and visualization of structural switches in RNA. In: Biocomputing'99. World Scientific, Singapore, pp 126–137. https://doi.org/10.1142/9789814447300_0013
20. Peracchi A, Matulic-Adamic JA, Wang S, Beigelman L, Herschlag D (1998) Structure-function relationships in the hammerhead ribozyme probed by base rescue. RNA 4(11):1332–1346. https://doi.org/10.1017/s1355838298980979
21. Chaulk SG, MacMillan AM (1998) Caged RNA: photo-control of a ribozyme reaction. Nucleic Acids Res 26(13):3173–3178. https://doi.org/10.1093/nar/26.13.3173
22. Chaulk SG, MacMillan AM (2007) Synthesis of oligo-RNAs with photocaged adenosine 2-'-hydroxyls. Nat Protoc 2(5):1052–1058. https://doi.org/10.1038/nprot.2007.154
23. Lorenz R, Hofacker IL, Stadler PF (2016) RNA folding with hard and soft constraints. Algorithms Mol Biol 11(1):1–13. https://doi.org/10.1186/s13015-016-0070-z
24. Breaker RR (2018) Riboswitches and translation control. Cold Spring Harbor Perspect Biol 10(11):a032797. https://doi.org/10.1101/cshperspect.a032797
25. Klug SJ, Famulok M (1994) All you wanted to know about SELEX. Mol Biol Rep 20(2):97–107. https://doi.org/10.1007/BF00996358
26. Darmostuk M, Rimpelova S, Gbelcova H, Ruml T (2015) Current approaches in SELEX: An update to aptamer selection technology. Biotechnol Adv 33(6):1141–1161. https://doi.org/10.1016/j.biotechadv.2015.02.008
27. Wachsmuth M, Domin G, Lorenz R, Serfling R, Findeiß S, Stadler PF, Mörl M

(2015) Design criteria for synthetic riboswitches acting on transcription. RNA Biol 12(2):221–231. https://doi.org/10.1080/15476286.2015.1017235

28. Ge H, Marchisio MA (2021) Aptamers, riboswitches and ribozymes in S. cerevisiae synthetic biology. Life 11(3):248. https://doi.org/10.3390/life11030248
29. Wrist A, Sun W, Summers RM (2020) The theophylline aptamer: 25 years as an important tool in cellular engineering research. ACS Synth Biol 9(4):682–697. https://doi.org/10.1021/acssynbio.9b00475
30. Jenison RD, Gill SC, Pardi A, Polisky B (1994) High-resolution molecular discrimination by RNA. Science 263(5152):1425–1429. https://doi.org/10.1126/science.7510417
31. Van Nostrand EL, Freese P, Pratt GA, Wang X, Wei X, Xiao R, Blue SM, Chen J-Y, Cody NAL, Dominguez D et al (2020) A large-scale binding and functional map of human RNA-binding proteins. Nature 583(7818):711–719. https://doi.org/10.1038/s41586-020-2077-3
32. Forties RA, Bundschuh R (2010) Modeling the interplay of single-stranded binding proteins and nucleic acid secondary structure. Bioinformatics 26(1):61–67. https://doi.org/10.1093/bioinformatics/btp627
33. Giegerich R, Voß B, Rehmsmeier M (2004) Abstract shapes of RNA. Nucleic Acids Res 32(16):4843–4851. https://doi.org/10.1093/nar/gkh779

Chapter 4

RNA Secondary Structure Modeling Following the IPANEMAP Workflow

Delphine Allouche, Grégoire De Bisschop, Afaf Saaidi, Pierre Hardouin, Francois-Xavier Lyonnet du Moutier, Yann Ponty, and Sargueil Bruno

Abstract

The structure of RNA molecules and their complexes are crucial for understanding biology at the molecular level. Resolving these structures holds the key to understanding their manifold structure-mediated functions ranging from regulating gene expression to catalyzing biochemical processes. Predicting RNA secondary structure is a prerequisite and a key step to accurately model their three dimensional structure. Although dedicated modelling software are making fast and significant progresses, predicting an accurate secondary structure from the sequence remains a challenge. Their performance can be significantly improved by the incorporation of experimental RNA structure probing data. Many different chemical and enzymatic probes have been developed; however, only one set of quantitative data can be incorporated as constraints for computer-assisted modelling. IPANEMAP is a recent workflow based on RNAfold that can take into account several quantitative or qualitative data sets to model RNA secondary structure. This chapter details the methods for popular chemical probing (DMS, CMCT, SHAPE-CE, and SHAPE-Map) and the subsequent analysis and structure prediction using IPANEMAP.

Key words RNA structure prediction, Chemical probing, RNAfold, Secondary structure, IPANEMAP

1 Introduction

RNA secondary structure modelling has been a challenge since the early days of molecular biology. Although algorithms for RNA structure modelling are more and more efficient and accurate, they significantly benefit from the integration of experimental structure probing data. RNA structure probing consists in submitting an RNA to enzymes or small molecules that specifically react with individual nucleotides according to their pairing status. Most enzymes used are single-strand-specific RNAses (RNAses T1, U2, nuclease S1 . . .), with the notable exception of the double-strand-

Ronny Lorenz (ed.), *RNA Folding: Methods and Protocols*, Methods in Molecular Biology, vol. 2726,
https://doi.org/10.1007/978-1-0716-3519-3_4,

specific RNAse V1. Although they are low molecular weight proteins, they are too bulky to access some nucleotides of a folded RNA. Small molecules can essentially reach any nucleotide, and most of them are also single-strand-specific, although psoralen has recently been successfully used a double-strand probe [17]. For the longest time, RNA probing experiments remained tedious and rather qualitative than quantitative.

RNA structure probing recently reached the medium, and then high, throughput. Pioneered and mostly developed within the Weeks lab, the SHAPE technology uses small molecules that react with flexible ribose, thus essentially reporting single-stranded nucleotides with some subtleties [10, 26]. A medium throughput version of the SHAPE protocol was first developed based on capillary electrophoresis, later to be transformed into a high throughput method using next-generation sequencing. The same workflows can be applied to more traditional probes such as DiMethyl Sulfate (DMS) and N-Cyclohexyl-N′-(2-morpholinoethyl)carbodiimide metho-p-toluenesulfonate (CMCT), which reveal unpaired A,C and G,U, respectively.

It appeared that using dissimilar probes provides complementary information that further improves RNA structure prediction. We, therefore, developed IPANEMAP, an experimental and computational workflow that models RNA secondary structure from different sets of RNA structure probing performed with different probes, and/or in different conditions, and/or on mutants [23]. This workflow relies on medium- or high-throughput structure probing and combines statistical sampling, clustering [7], and pseudo-potentials [6]. The method was shown to produce accurate and stable, even when a single reactivity profile is available, while the availability of multiple reactivities was shown to increase robustness and, to a lesser extent, accuracy of the modelling [23].

Below, we detail a whole IPANEMAP workflow, starting with experimental probing with DMS and/or CMCT and/or SHAPE reagent. Such probing can be carried out in various relevant conditions (varying temperature, Mg^{2+} concentration, introducing point mutations in the RNA to be modelled, etc) [5, 23]. Two versions of the experimental procedure (medium throughput and high throughput) are proposed; DMS and CMCT probing were adapted from [1, 8, 13, 29], while the SHAPE probing is described in K. Weeks team publications [14, 16, 20, 21, 24, 30, 31]. We then detail instructions for executing the IPANEMAP algorithm to obtain the RNA secondary structure model.

2 Materials

2.1 Medium Throughput RNA Structure Probing with DMS, CMCT, or a SHAPE Reagent

1. Disposable labware.
2. Capillary electrophoresis (ABI or Ceq8000 AB Sciex).
3. Thermocycler.
4. Dry bath.
5. Refrigerated centrifuge.
6. Fume hood.
7. Ammonium acetate 5 M. Store at −20 °C.
8. 70% and 100% cold ethanol.
9. Dideoxynucleotide 10 mM, either ddTTP, ddATP or ddCTP. Store at −20 °C.
10. MMLV Reverse Transcriptase RNAse H (−) 200 u/μL (e.g., Promega M1705). Store at −20 °C.
11. MMLV RT buffer 5X (e.g., Promega M5313). Store at −20 °C.
12. Deoxynucleotide mix dATP, dTTP, dCTP and dITP, 10 mM each. Store at −20 °C.
13. Two differently labelled fluorescent primers at 2 μM in water, for example, D2-PA and D4-PA (Sigma-Aldrich), if running the sequences on a CEQ8000 capillary sequencer (*see* **Note 1**). Store at −20 °C.
14. RT Stop buffer 5X: 1.2 M sodium acetate pH 5, 40 mM EDTA, 4 mg/mL glycogen. Store at −20 °C.
15. Sample loading solution (deionised formamide). Store at −20 °C.
16. Glycogen 20 mg/mL. Store at −20 °C.
17. Yeast tRNA 10 mg/mL. Store at −20 °C.
18. DMS (diluted in ethanol 1:12) CAS # 77-78-1.
19. DMS buffer (10X): 400 mM HEPES pH 7.5, 1 M KCl, 50 mM $MgCl_2$.
20. DMS stop reaction: 400 mM Tris pH 7.5.
21. 500 mL of a solution saturated with sodium hydroxide.
22. CMCT (42 mg/mL) CAS # 2491-17-0.
23. CMCT buffer (10X): 500 mM potassium borate pH 8, 1 M KCl, 50 mM $MgCl_2$.
24. 10% acetic acid.

25. SHAPE reagent 10X in anhydrous DMSO: 40 mM 1 M7 (CAS # 4693-01-0) or 40 mM NMIA (CAS # 10328-92-4) or 400 mM BzCN (CAS # 613-90-1). Store at −20 °C.
26. SHAPE buffer (10X): 400 mM HEPES pH 7.5, 1 M KCl, 50 mM $MgCl_2$.

2.2 High Throughput SHAPE Probing: The SHAPE-MaP Protocol

1. Disposable labware.
2. Mini column (Biorad).
3. Ampure XP Beads (Beckman). Store at 4 °C.
4. Micro-volume spectrophotometer.
5. Agilent Bioanalyser—Chip and Reagents for DNA analysis (usually available on sequencing platforms). Store at 4 °C.
6. Thermocycler.
7. RNA folding buffer (3X): 300 mM HEPES pH 8; 1.2 M KOAc and 0–30 mM $MgCl_2$.
8. Fragmentation buffer (2X): 300 mM Tris-HCl pH 8.3, 450 mM KCl, 18 mM $MgCl_2$.
9. RNA denaturing buffer (10X): 500 mM HEPES and 40 mM EDTA.
10. SHAPE-MaP pre-buffer (10X): 500 mM Tris-HCl pH 8, 750 mM KCl and 100 mM DTT.
11. Anhydrous DMSO.
12. SHAPE reagents: 1 M7 (CAS # 4693-01-0) 100 mM or NMIA (CAS# 10328-92-4) 100 mM or BzCN mM (CAS # 613-90-1) dissolved at 100 mM in anhydrous DMSO. Store at −20 °C.
13. $MnCl_2$ 30 mM. Store at −20 °C.
14. Specific primers at 2 μM or Random 9-mer primers 200 ng/μL (NEB S1254). Store at −20 °C.
15. dNTP stock 5 mM. Store at −20 °C.
16. SuperScript III (Invitrogen). Store at −20 °C.
17. 70%, 80% and 100% cold ethanol. Store at −20 °C.
18. NEBNext Ultra II RNA Library Prep Kit for Illumina (NEB E7770). Store at −20 °C.
19. Multiplex Oligo (NEB E7335/E7500/E7710/E7730). Store at −20 °C.
20. G50 sephadex (GE Healthcare). Store at 4 °C.

3 Methods

3.1 Medium Throughput RNA Structure Probing with DMS, CMCT, or a SHAPE Reagent

3.1.1 RNA Refolding

RNA purification protocols from cells or from in vitro transcription often include a denaturation step. In addition, RNA may also go through freeze and thaw cycles which also alter the native folding. In such cases, a denaturation-renaturation step helps to homogenize RNA folding. In case a nondenaturing protocol has been used to prepare the RNA, this step may be skipped. Numerous refolding protocols have been proposed; their relative relevance may depend on the RNA considered [4, 11, 12, 28]. We propose below a "classical" protocol:

1. Prepare two tubes with 6 pmol of RNA in 36 μL of water. Incubate for 5 min at 85 °C.
2. Add 4 μL of prewarmed appropriate buffer (DMS, CMCT or SHAPE), and let it cool down for 10 min at room temperature. Incubate for an additional 15 min at the probing temperature (30–37 °C). (*see* **Note 2**).

3.1.2 RNA Modifications

With DMS

1. Add 4 μL of DMS (or 4 μL ethanol for the mock reaction). Incubate 5 min at 37 °C.
2. Stop the reaction by addition of 400 mM of Tris pH 7.5 (final) and immediately place on ice.

With CMCT

1. Add 10 μL of CMCT (or 10 μL of H_2O for the mock reaction). Incubate for 10 min at 37 °C.
2. Stop the reaction by adding 1 μL of tRNA (10 μg/μL). Put on ice.

With SHAPE Reagents

1. Add 5 μL of SHAPE reagent 10X (or 5 μL DMSO for the mock reaction) (*see* **Note 3**).
2. Incubate for five reagent half-lives at 37 °C (30 min for NMIA, 2 min for 1M7 or few seconds for BzCN), after which the shape compound is hydrolyzed (no inactivation step required) (*see* **Notes 4, 5, 6**).

3.1.3 Ethanol Precipitation

1. Add 1 μL of glycogen (20 mg/mL) or tRNA (10 mg/mL), 10% of ammonium acetate 5 M, and 400 μL of 100% cold ethanol, and incubate for at least 30 minutes at −20 °C.
2. Centrifuge for 30 min at 16000 g at 4 °C. Discard the supernatant.
3. Wash with 400 μL of 70% cold ethanol.

4. Centrifuge for 5 min at 16000 g at 4 °C. Discard the supernatant. Repeat this step.
5. Air dry the pellet, and resuspend it in 10 μL of water.

STOP POINT: Probed RNA can be stored at −20 °C for at least 6 months.

3.1.4 Reverse Transcription

Nucleotides that have reacted with DMS, CMCT, or the SHAPE reagent are modified and induce premature reverse transcription stops. The reactivity is assessed for each nucleotide by the difference of intensity of the stop in the modification reaction and in the mock reaction. The nucleotide is mapped by comparison with the sequencing reaction

1. Transfer the resuspended RNAs to PCR tubes and prepare two PCR tubes with 6 pmol of RNA in 10 μL of water for the sequencing reactions.
2. Add 1 μL of DMSO to the samples (modified RNA, mock reaction and sequencing reactions), denature for 3 min at 95 °C, then flash cool on ice.
3. Add 3 μL of 2 μM D2-labelled primer to the sequencing reaction and 3 μL of D4-labelled primer to the modified/mock RNA. Incubate for 5 minutes at 65 °C, then anneal for 10 min at 35 °C, and store on ice.
4. Add 4 μL of 5X RT buffer, 1 μL of the 10 mM deoxynucleotides mix and 1 μL of MMLV (H-) reverse transcriptase. For the sequencing reactions, also add 1 μL of 10 mM dideoxynucleotide. Incubate for 2 minutes at 35 °C, 30 min at 42 °C, and 5 min at 55 °C, then store on ice (*see* **Note 7**).
5. Add 5 μL of 5X RT Stop buffer.
6. Combine one elongation from the modified RNA or the mock RNA with one sequencing reaction.
7. Add 2.5 volumes of 100% ethanol, incubate for at least 30 min at −20 °C and precipitate the DNA as in Subheading 3.1.3, (**steps 2–4**).
8. Air dry the pellet, and resuspend in 40 μL of sample loading solution (*see* **Note 8**).

STOP POINT: Elongated cDNA can be stored at −20 °C.

3.1.5 Sequencing by Capillary Electrophoresis

1. Load side-by-side 20 μL of the elongated cDNA corresponding to the modified RNA and to the mock RNA.
2. Run the capillary electrophoresis.
3. Export the raw data as text files.

3.1.6 Data Analysis with QuSHAPE

Different software such as QuSHAPE [14], RiboCAT [3], CAFA [19], FAST [22], and HiTRACE [15, 32] have been developed to analyze electropherograms.

We mostly use QuSHAPE [14], please note that the experimental setup might slightly differ when using another analysis software.

QuSHAPE can treat any type of chemical probing experiment after capillary electrophoresis separation of the RT products. Output files from ABi sequencer can be directly fed to QuSHAPE while those from CEQ8000 (AB Sciex) contains a header that needs to be removed beforehand (open the file with Excel and delete the header).

For a description of QuSHAPE and its use, please *see* [14] and the "help" of the software.

As a result, QuSHAPE exports a tab delimited text file that contains the areas of the (+) ('RX') and (−) ('BG') reactions and their difference ("RX-BG") for each nucleotide, row-by-row. QuSHAPE also offers the possibility to normalize the results, however we have developed our own script for normalization including some specific treatment for DMS and CMCT, it also allows for the averaging of the reactivities issued from replicates (*see* next chapter). If using directly QuSHAPE output format the result file as described in Subheading 3.3.2.

3.1.7 Normalization and Averaging of the Replicates

QuSHAPE data analysis yields raw reactivity values ("RX-BG") corresponding to the differences of the cognate area for the (+) and the (−) reactions for each nucleotide. Values corresponding to intrinsic reverse transcriptase stops need to be eliminated; the data are then normalized to allow the comparison between different experiments. To this end, two different approaches have been proposed in [16] and should be used depending on the RNA length.

Quantile Normalization (RNA Shorter Than 300 bps)

The 2% of the most reactive peaks of all reactivities (modified and unmodified condition) are removed since they empirically correspond to intrinsic stops (outliers). Then the next 8% of the most reactive peaks are averaged (M_8), and finally, all reactivities are normalized by the average (M_8).

Parametric Outlier Removal (RNAs Beyond 300 bps)

In this approach, peaks whose value is higher than the 75th percentile by at least 1.5 times the interquartile range are considered as outliers and removed. The 10% of the highest remaining reactivities (outliers removed) are averaged. All reactivities (outliers included) are normalized by this average value.

Typically, the probing of an RNA in a given condition is performed in triplicate before further analysis.

To automate the normalization process described above [16], we have developed a python script that allows the normalization and yields for each nucleotide the average value of the replicates and the standard deviation (code is available at https://github.com/Sargueil-CiTCoM/ShapeCE-Normalize).

To use the script, pass as argument Qushape.txt output file followed by –output [Normalized data folder], (the folder where to store normalized files) (*see* **Note 9**). The script is highly customizable by command line arguments. List is accessible using. / *normalize_reactivity –help*.

- *Parameters for normalization:*
  - *Reactive_nucleotides*: SHAPE reagents hit any type of nucleotides, but reagents such as DMS and CMCT hit only two types of nucleotides. We, therefore, define the outlier on all nucleotides, but normalize the values considering the targeted nucleotides only. To configure this, enter the nature of the nucleotides targeted by the chemical agent (A,C,G,U for SHAPE reagents; A,C for DMS; G,U for CMCT) (*see* **Note 10**).
  - *Norm_methods*: Two normalization methods are available, as defined in [16]–the first is a simple normalization method using the mean of the 8% most reactive nucleotides as normalization factor after having removed the 2% outliers, whereas the second uses the 1.5 interquartile range method. By default, the results of both methods are present in the output file. Users can select the method by indicating *simple* or *interquartile* as argument.
  - *Stop percentile*: When the background is too high, data cannot be interpreted and must be discarded. This parameter indicates the percentile above which nucleotides are discarded and considered as a stop. The value by default is 90.
  - *Low_norm_reactivy_threshold*: Reactivities are not supposed to be negative; notwithstanding, it happens that even after the scale factor correction, some values are still negative. The negative threshold for which the reactivity will be set to zero can be configured. Default value = −0.3.
  - *Plot*: Users have the possibility to plot the normalized reactivity by setting an output filename (−output [filename], *see* below).

Once the parameters set, run the script by typing the following command line from the Qu-shape-output folder:

```
> python normalize_reactivity.py [INPUTS] –output [OUTPUT_FOLDER]
```

The script generates two output files for each input file:

- A .shape file, which can be passed to IPANEMAP (*see* below).
- A .tsv file containing all the normalized information.
- *Parameters considered in the final analysis:*

  The averaged reactivity for each nucleotide is calculated from the replicates in an excel file.

  Rules for the analysis:

  - If at least two values in the set are different from −10, compute the mean over all non-10 values. Otherwise, the average is set to −10.
  - *Threshold*: This parameter has been added to characterize nucleotides whose reactivity is variable and ensure the exclusive use of solid data. It is a maximal difference to the mean accepted for one of the replicate. Default value = 0.3. If the difference <0.3, for all the replicates, the average is calculated; in contrast, if the difference >0.3, in one of the replicates, the reactivity for the nucleotide is set to −20.
  - *Desactive_threshold*: It is the highest reactivity value for which the "threshold" correction will be applied (Default value = 0.6).

3.2 High-Throughput SHAPE Probing: The SHAPE-MaP Protocol

Different protocols have been established for high-throughput SHAPE probing; the *SHAPEseq* on premature stops to detect the modified nucleotides [18, 27], while the SHAPE-MaP relies on the property of the reverse transcriptase to make errors when copying a modified nucleotide in the presence of manganese ions [24]. Although coupled to NGS, such techniques do not allow to probe a full transcriptome because each nucleotide should be read 5000–20,000 times to obtain a significant reactivity. To overcome this limitation, *ic*SHAPE selects for modified RNA fragments before the reverse transcription steps [9, 25].

Below is described the SHAPE-MaP protocol that we routinely use; note that it can be readily adapted to DMS [33] or to any other probe.

3.2.1 RNA Modifications with SHAPE Reagents

Modification Condition and Negative Control

1. In PCR tubes, add 10 pmol of RNA in 12 μL of water. Incubate for 3 min at 95 °C, then immediately put on ice for 3 min (*see* **Note 11**).
2. Flash spin before adding 6 μL of RNA folding buffer (3X), and incubate for 20 min at 37 °C (*see* **Note 12**).
3. In separate PCR-tubes, add 1 μL of the SHAPE-probe at 100 mM (modification reaction) or 1 μL of DMSO (negative control).

4. Add 9 μL of RNA folded directly on the SHAPE-probe and mix well by pipetting. Repeat this step for the negative control.
5. Incubate for 6 min at 37 °C for 1 M7 and about 30 min for NMIA (approximately 5 times the half-life). Store the PCR-tubes on ice.

Denaturing Condition

1. In a PCR-tube, add 5 pmol of RNA in 3 μL of water. Then add 1 μL of RNA denaturing buffer (10X) and 0.5 volume of formamide.
2. Incubate for 1 min at 95 °C (*see* **Note 13**).
3. In another tube add 1 μL of SHAPE-probe at 100 mM. Make a short spin, then quickly add the denatured RNA, mix well by pipetting, and incubate for another minute at 95 °C. Place the PCR-tube on ice.

3.2.2 RNA Purification

Bring the final volume of each sample to 50 μL with water, and place the tubes on ice time while preparing the G50 Sephadex mini-column.

1. Place a mini column in a 1.5 mL Eppendorf tube, and add 800 μL of G50 slurry. Spin for 30 s at 1000 g. Discard the "eluate."
2. Complete with G50 slurry and centrifuge for 30 s at 1000 g. Discard the eluate.
3. Repeat the centrifugation step twice. The first for 30 s, the second for 1 min.
4. Place the mini-column in a new 1.5 mL Eppendorf tube. Load RNA on the sephadex, and spin for 2 min at 1000 g. Recover the eluate.

STOP POINT: RNA probed and denatured can be stored at −20 °C for at least 6 months.

RNA Fragmentation (Optional: Only for RNA over 300 nt)

1. Add 50 μL of fragmentation buffer (2X).
2. Incubate at 94 °C for 4 min.
3. RNA purification (repeat Subheading 3.2.2).

3.2.3 cDNA Synthesis by Reverse Transcription and Purification

Prepare 5X SHAPE-MaP pre-buffer by combining an equal volume of SHAPE-MaP pre-buffer 10X and of dNTP stock at 5 mM. Then add an equal volume of 30 mM of $MnCl_2$ to the SHAPE-MaP pre-buffer to obtain 2.5X SHAPE-MaP buffer (*see* **Note 14**).

The reverse transcription can be performed with specific or random primers. We recommend specific primers for RNA up to 300–400 nt long.

Reverse Transcription with Specific Primers (RNA up to 400 nt)

1. In separate PCR tubes add 10 μL (corresponding to roughly 1 pmol) of the modified/unmodified/denatured RNA. Incubate for 3 min at 95 °C then place on ice for at least 1 min.
2. Spin briefly, add 1 μL of 2 μM specific primers, incubate for 5 min at 65 °C, and cool down on ice.
3. Add 8 μL of 2.5X SHAPE-MaP buffer, mix well by pipetting, and incubate for 2 min at 50 °C.
4. Place the tube on ice and add 1 μL of SuperScript II (or III) reverse transcriptase. Flick the tube and spin briefly before incubating for 3 h at 50 °C.
5. Inactivate the SuperScript II (or III) by heating samples at 70 °C for 15 min. Place the samples on ice.

Reverse Transcription with Random Primers

1. In separate PCR tubes, add 10 μL modified/unmodified/denatured RNA. Incubate for 3 min at 95 °C, then place on ice for at least 1 min.
2. Spin briefly and add 200 ng/μL of random nonamer primers, incubate for 5 min at 65 °C, and cool down on ice (*see* **Note 15**).
3. Add 8 μL of 2.5X SHAPE-MaP buffer, mix well by pipetting, and incubate for 2 min at 25 °C.
4. Place the tube on ice and add 1 μL of SuperScript II (or III) reverse transcriptase. Flick the tube and spin briefly before incubating for 10 min at 25 °C, then for 3 h at 50 °C (*see* **Note 16**).
5. Inactivate the SuperScript II (or III) by heating samples at 70 °C for 15 min. Then place samples on ice.

3.2.4 cDNA Purification

Proceed as in Subheading 3.1.3, but do not bring the final volume to 50 μL.

STOP POINT: cDNA can be stored at −20 °C for at least 1 year.

The steps described below are designed for Illumina sequencing. If using another type of NG sequencing, the library preparation must be adapted. Also note that the ShapeMapper script will also have to be adapted to accept output files other than Illumina's. At least three different barcodes (libraries) need to be use, one for each condition: modified (BC1)/unmodified (BC2) and denatured (BC3). In each library, several RNA can be mixed if their sequence can be unambiguously distinguished. For RNA with identical or close sequences, use different barcodes for each condition.

3.2.5 Illumina Library Preparation and Quality Control

Prepare the samples on ice, and follow the manufacturer instructions (*see* **Note 17**).

Second Strand Synthesis (Orange Cap, E7770)

dsDNA Purification with Ampure XP Beads

1. Leave the beads on the bench for at least 30 minutes until they reach room temperature (*see* **Note 18**).
2. In a 1.5 mL Eppendorf tube, add 144 μL of the beads slurry on the "second strand synthesis" reaction, mix well by pipetting, and incubate for 5 min at room temperature.
3. Place the 1.5 mL Eppendorf tube on a magnetic rack, and wait for 2 more min. Discard the liquid.
4. Wash the beads with 200 μL of cold ethanol (80%) (*see* **Note 19**).
5. Discard the liquid and repeat the **step 4**. Spin briefly to discard all the liquid.
6. Air dry the beads for 5–10 min (*see* **Note 20**).
7. Add 53 μL of 0.1X TE, mix well and incubate for 2 min on magnetic rack.
8. Remove 50 μL of purified DNA and place it in PCR tubes.

STOP POINT: dsDNA can be stored at −20 °C for at least 1 year.

End-it Repair and Adaptors Ligation (Green and Red Cap, E7770)

Samples are prepared of ice following the manufacturer instructions.

Ligated DNA Purification with Ampure XP Beads

Proceed as for "Subheading 3.3.2 dsDNA purification," but add 85 μL of beads and elute with 15 μL of 0.1X TE.

STOP POINT: dsDNA can be stored at −20 °C for at least 1 year.

Library Enrichment (Bleu Cap from E7770 and Barcode from E7335/E7500/E7710/E7730)

Follow the manufacturer's indications (*see* **Note 21**).

Library Purification with Ampure XP Beads

Proceed as for "Subheading 3.3.2 dsDNA purification," but add only 45 μL of beads and elute with 25 μL of 0.1X TE.

Quantify the samples by OD_{260nm}.

STOP POINT: dsDNA library can be stored at −20 °C for at least 1 year.

Library Verification with a Bioanalyzer (Agilent DNA 1000 Chip)

Incubate the gel dye-mix at least 30 min at room temperature before using.

Follow manufacturer's indications (*see* **Note 22**).

3.2.6 Illumina Sequencing

See **Note 23**.

We have also adapted the ShapeMapper script for Ion Torrent sequencing (available at https://github.com/afafbioinfo/ReadsMap), but probably due to a higher error rate of this sequencing technique, we found the results obtained less accurate.

3.2.7 Data Analysis Is Carried Out Using ShapeMapper2 as Described in [2]

ShapeMapper2 generates four different output files:

1. Condition.map: It contains a table composed of four columns: The first is the sequence position, the second contains SHAPE reactivities, with the standard errors in the third column for each nucleotide (fourth column).
2. Condition.shape: It is a tab delimited text file with the position of the nucleotide and the associate SHAPE reactivities. This file can be used as input for IPANEMAP software (*see* **Note 24**).
3. Condition_histogram.pdf: It is a graphical summary of the sequencing. Those histograms represent the mutation rate for each condition, the read depth and the reactivities analysis. In a successful experiment, the mutation rate in the modified condition is higher than the untreated condition; the majority of nucleotides have read above 5000 and finally reactivities are positive with low standard errors.
4. Condition_profiles.pdf: It contains three graphs. The first graph represents SHAPE reactivity value for each nucleotide. The second graph corresponds to the mutation rate for each condition (modified, unmodified and denatured) and for each nucleotide. The final graph shows the coverage for each condition. These graphs allow the experiment validation or are helpful for troubleshooting the problems.

3.3 Predicting Structural Models from One or Several Probing Experiments Using IPANEMAP

Once reactivity profiles have been produced in one or several conditions, IPANEMAP can jointly integrate them and produce one or several secondary structure models that are well-supported by the experimental data.

SHAPE reactivities can be interpreted as follows:

1. **R** = −20: Nucleotides with variable values (when multiple values available).
2. **R** = −10: Removed outliers (*see* **Note 25**).
3. Lowervalue < **R** < 0.4: Nucleotides with low reactivity.
4. 0.4 < **R** < 0.7: Nucleotides with medium reactivity.
5. 0.7 < **R**: Nucleotides with high reactivity.

3.3.1 Installing IPANEMAP

IPANEMAP can indifferently be installed on recent versions of Mac, Windows, or Linux operating systems. Using Windows, some dependencies have to be installed manually.

Install Miniconda

https://docs.conda.io/en/latest/miniconda.html

See **Note 26**.

Download IPANEMAP

The software can be downloaded from:

https://gitlab.inria.fr/amibio/ipanemap/-/tags/ipanemap-1.1.0-python3

Download the ZIP archive for the most recent version, and unzip it in the directory of your choice.

Prepare Conda Environnement and Install IPANEMAP

In a miniconda terminal, inside the ipanemap folder, create a virtual environment:

```
conda env create –name ipanemap –file conda-env.yaml -c conda-forge -c bioconda.
```

Install IPANEMAP:

```
pip install -e.
```

See **Note 27**.

[Windows Only] Install the ViennaRNA Package, Version Posterior to 2.0, Downloadable from the TBI Vienna

https://www.tbi.univie.ac.at/RNA/#download

ViennaRNA is installed automatically with the conda environment when using mac or linux.

See **Note 28**.

Install VARNA

```
> conda install varna -c conda-forge -c bioconda -c sargueil-citcom
```

3.3.2 Setting Up the Analysis

Throughout the following, we assume that the name of the analyzed RNA is "myRNA" and that the conditions are named "Cond1," "Cond2"...

1. Copy the folder structure of the *example* folder in a folder in which the data will be treated.
2. In this folder prepare a FASTA file named "myRNA.fa," whose first line contains the FASTA header (e.g., "> myRNA") and the second line consists of the RNA sequence (e.g., "CCCAAAGGG").

   In the *Constraints_Probing* subfolder, prepare reactivity files "myRNACond1.shape," "myRNACond2.shape"... (for instance, "HIVgRNADMS.shape," "HIVgRNANMIA.shape" ...) These txt files should be formatted to contain one line per nucleotide, each consisting of: (i) the nucleotide number and (ii) the reactivity score, both separated by space/tabulations (*see* **Note 29**).

3.3.3 Running the Analysis and Retrieving the Results

1. Open a terminal and activate conda environment.

```
> conda activate ipanemap
```

2. Move to your data folder and execute the command:

```
> ipanemap --RNA myRNA.fa --cond Cond1 Cond2 > Best.txt.
```

See **Notes 30–32**.

4 Results

4.1 Dominant Conformations

The main results are now located in the file "Best.txt", featuring one or several dominant secondary structure models, along with additional information.

For instance:

Structure	dG	#SupportingConditions	BoltzmannProbability
(((...)))...(((...)))	−4.3	1	0.546

Indicates a single dominant conformer, associated with a secondary structure in dot-bracket notation. The Turner free-energy, number of conditions supporting the conformer, and the accumulated pseudo-Boltzmann probability across conditions are also reported for each conformer.

4.2 Complete Conformers List

The complete list of the main conformations across the pseudo-Boltzmann ensemble can be found in a file "output/Centroids.fa." The file is simply a list of sequence/structures, and includes conformers that are dominated, that is, either less stable or less supported by experiments than the ones reported in the primary output.

4.3 Visual Representation

Using default options, IPANEMAP produces the files "output/img/Optimal-1.svg," "output/img/Optimal-2.svg," and "output/img/Centroid-1.svg". These files are VARNA-rendered drawings of the conformers present in both the above-described lists. Individual nucleotides are color-coded with the reactivity of the first condition ("Cond1") (Fig. 1). While running successive analyses, one should be aware that the output file(s) will have always the same name and will be overwritten. In case you would like to preserve the results, you need to save them under another name or simply move them out of the output folder (*see* **Note 33**).

The IPANEMAP workflow was extensively tested using different benchmark RNAs and multiple probing conditions [23]. It appeared from that study that IPANEMAP provides more accurate and stable results than other existing workflows, even when using a single set of probing conditions. It is safer to use two conditions

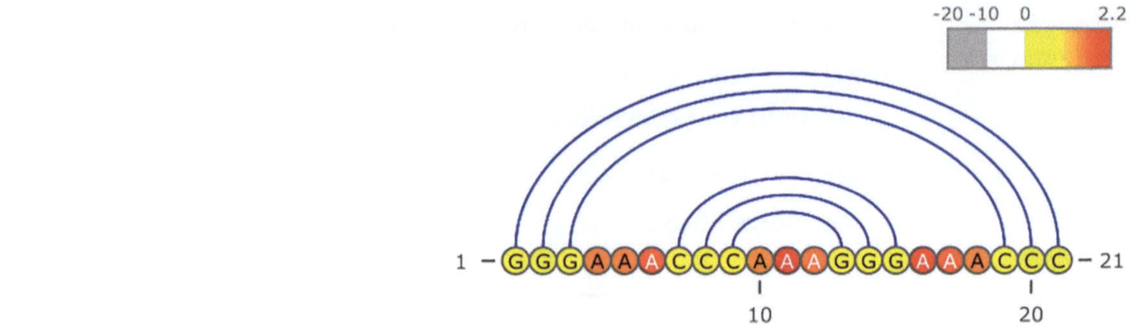

Fig. 1 Typical VARNA-rendered output of IPANEMAP, color-coded using the reactivity of the first condition

because if one of them is artefactual, the second one may correct, or at least mitigate, the modelling error. However, it is important to note that the predictions obtained from two conditions do not seem to improve over the results obtained from the best of the two conditions. Modelling accuracy is generally improved when using at least three different conditions [23].

5 Notes

1. Primer sequence should be chosen at least 50–75 nucleotides 3′ (downstream) to the region of interest.
2. Reaction Buffers can be adapted to each situation, notably monovalent and divalent cations nature, and concentrations may be changed to mimic at the best a physiological situation. $MgCl_2$ can even be left out to destabilize tertiary structure. However, do not use Tris buffer with DMS because it reacts with amine groups. Use potassium borate pH8 with CMCT and increase the buffer (any) concentration when using BzCN (SHAPE reagent) to compensate for the reagent acidification.
3. In order to rapidly mix the solution, be careful to add the RNA solution onto the SHAPE reagent rather than the opposite.
4. The solution will turn yellowish upon reaction with 1M7 or cloudy with BzCN.
5. All these reagents are genotoxic and/or skin and/or eye irritant. They should be handled with care under a fume hood wearing nitrile gloves and safety glasses. Waste tips and dispose soiled materials in a saturated sodium hydroxide solution or a 10% acetic acid solution for DMS and CMCT respectively. SHAPE reagents get hydrolyzed relatively rapidly in water.
6. As the modification of one nucleotide may influence the rest of the structure, multiple modifications per RNA molecule should be avoided. As a rule of thumb, we consider that such

conditions are met when the amount of the full length product in at least 70% of that mock reaction. The modification conditions described above are pretty standard and work well for RNA between 200 and 500 nucleotides. They can be adjusted by modifying the incubation time and/or reagent concentration to get more or less modification (NB: concentration and/or time should be increased for shorter RNAs). This applies for DMS and CMCT, but not for SHAPE reagents, which do not produce more than one modification per RNA under the above-described conditions.

7. For the sequencing reaction pick a nucleotide that is well represented all along the RNA sequence.
8. Resuspension will take time at room temperature. It can be accelerated by incubating the samples for 10 min at 50 °C.
9. The file title should be as follows: RNA_type-of-probe_replicate-number.txt (ex: TtLSU_DMS_2.txt).
10. Do not use any space after commas.
11. In this protocol, it is particularly important to use DNA-free RNA.
12. Other denaturation/renaturation protocol can be applied according to the RNA to be modified or any other specificity of the system studied. The Probing buffer can also be adapted.
13. For RNA with strong secondary structure may be incubated up to 5 min at 95 °C.
14. It is important to prepare this buffer extemporaneously because manganese oxidizes over time.
15. Nonamers ensure a better sequence coverage than shorter or longer randomers.
16. It is important to use a reverse transcriptase devoid of RNAse H activity to allow for the DNA second strand synthesis.
17. To improve the yield of DNA second strand synthesis, incubate for 2 h instead of 1 h, as indicated in manufacturer's instructions.
18. This step significantly increased the purification yield.
19. Rotate the tube a quarter turn every 30 s to mix the beads and improve the wash.
20. Do not over dry the beads. The beads become mat when they are dry.
21. Typically, 8–10 PCR cycles are enough and avoid the formation of adaptor dimers.
22. For a clear image load 5 ng sample in a final volume of 6 μL per well. If in the elution profile you observed a thin peak at ≈ 85 bp, it is due to the presence of primers. It is suitable to

perform another purification with Ampure XP beads to improve the sequencing quality. If in your elution profile you observed a thin peak at ≈ 120 bp, it is due to the presence of adaptor-dimer. To reduce this peak, dilute your sample 10 times with 0.1X TE, verify the library quality on a new chip, and use the diluted sample for sequencing.

23. Separate projects or a project including multiple RNA can be loaded on the same Illumina sequencing. If loading separate projects with RNA of the same or similar sequences, be careful to use different bar codes for each project (*see* above). As for the number of projects that can be loaded on the same run, keep in mind that each nucleotide should be covered about 20,000 times (the significance threshold being 5000 reads).
24. A SHAPE reactivity equal to −999 corresponding to a nucleotide which isn't mapped 5000 times.
25. Negative values are not interpreted by IPANEMAP.
26. It is highly recommended to install miniconda only for one user (as opposed to system wide installation).
27. Windows users have to open conda-env.yaml and remove the line containing viennarna because this package is not yet available in conda for windows.
28. *For Windows users:* The Vienna package installation does not automatically modify the PATH variable. It has to be set manually; many tutorials are available on the web. The variable to be added is the address of the Vienna package programs e.g. "C:\Program Files (x86)\ViennaRNA Package".
29. All names/paths can be changed to any other name (alphanumerical characters only), but such modifications should be reflected in the file names. The README.md file also describes how to set up many additional options.
30. Many additional options, including the possibility to specify user-defined hard constraints, are available and documented in the "README.md" file found at the root of the installation.
31. Unless you modify the PATH variable IPANEMAP should be run from the IPANEMAP master folder.
32. If myRNA.fa is in the folder fasta_files, refer to it as "fasta_files \myRNA.fa".
33. SVG files can be visualized within any modern browser and can be converted to a large list of formats using software like Inkscape or Adobe Illustrator.

Acknowledgments

Research in B.S laboratory is funded by the CNRS, the Université Paris Cité, a grant from "la Fondation pour la Recherche Médicale" (FRM DBI20141423337), ANR INSANNE (ANR 21-CE45-0034-03), ANR PARNASSUS (ANR 19-CE45-0023-02), and ANR DECRYPTED (ANR 19-CE30-0021-03), awarded to BS and YP laboratories. AS and DA were recipient of a PhD fellowship from FRM (FRM DBI20141423337); GdB was recipient of a fellowship from the French Ministry for Education and Research (MESR); PH was recipient of a fellowship from ANR PRE-RIBO60S (ANR 20-CE12-0026-02).

References

1. Brunel C, Romby P (2000) [1] Probing RNA structure and RNA-ligand complexes with chemical probes. In: Methods in enzymology. Academic, United States, pp 3–21
2. Busan S, Weeks KM (2018) Accurate detection of chemical modifications in RNA by mutational profiling (MaP) with ShapeMapper 2. RNA 24:143–148
3. Cantara WA, Hatterschide J, Wu W, Musier-Forsyth K (2017) RiboCAT: a new capillary electrophoresis data analysis tool for nucleic acid probing. RNA 23:240–249
4. Chillón I, Marcia M, Legiewicz M, Liu F, Somarowthu S, Pyle AM (2015) Chapter One - Native purification and analysis of long RNAs. In: Woodson SA, Allain FHT (eds) Methods in enzymology. Academic, United States, pp 3–37
5. Deforges J, de Breyne S, Ameur M, Ulryck N, Chamond N, Saaidi A, Ponty Y, Ohlmann T, Sargueil B (2017) Two ribosome recruitment sites direct multiple translation events within HIV1 Gag open reading frame. Nucleic Acids Res 45:7382–7400
6. Deigan KE, Li TW, Mathews DH, Weeks KM (2009) Accurate SHAPE-directed RNA structure determination. PNAS 106:97–102
7. Ding Y, Lawrence CE (2003) A statistical sampling algorithm for RNA secondary structure prediction. Nucleic Acids Res 31:7280–7301
8. Ehresmann C, Baudin F, Mougel M, Romby P, Ebel JP, Ehresmann B (1987) Probing the structure of RNAs in solution. Nucleic Acids Res 15:9109–9128
9. Flynn RA, Zhang QC, Spitale RC, Lee B, Mumbach MR, Chang HY (2016) Transcriptome-wide interrogation of RNA secondary structure in living cells with icSHAPE. Nat Protoc 11:273–290
10. Frezza E, Courban A, Allouche D, Sargueil B, Pasquali S (2019) The interplay between molecular flexibility and RNA chemical probing reactivities analyzed at the nucleotide level via an extensive molecular dynamics study. Methods 162–163:108–127
11. Herschlag D (1995) RNA chaperones and the RNA folding problem. J Biol Chem 270: 20871–20874
12. Jaeger L, Westhof E, Michel F (1993) Monitoring of the cooperative unfolding of the sunY group I intron of bacteriophage T4: the active form of the sunY ribozyme is stabilized by multiple interactions with 3′ terminal intron components. J Mol Biol 234:331–346
13. James L, Sargueil B (2008) RNA secondary structure of the feline immunodeficiency virus 5′UTR and Gag coding region. Nucleic Acids Res 36:4653–4666
14. Karabiber F, McGinnis JL, Favorov OV, Weeks KM (2013) QuShape: rapid, accurate, and best-practices quantification of nucleic acid probing information, resolved by capillary electrophoresis. RNA 19:63–73
15. Kim H, Cordero P, Das R, Yoon S (2013) HiTRACE-Web: an online tool for robust analysis of high-throughput capillary electrophoresis. Nucleic Acids Res 41:W492–W498
16. Low JT, Weeks KM (2010) SHAPE-directed RNA secondary structure prediction. Methods 52:150–158
17. Lu Z, Chang HY (2016) Decoding the RNA structurome. Curr Opin Struct Biol 36:142–148
18. Lucks JB, Mortimer SA, Trapnell C, Luo S, Aviran S, Schroth GP, Pachter L, Doudna JA, Arkin AP (2011) Multiplexed RNA structure characterization with selective 2′-hydroxyl acylation analyzed by primer extension

sequencing (SHAPE-Seq). Proc Natl Acad Sci U S A 108:11063–11068

19. Mitra S, Shcherbakova IV, Altman RB, Brenowitz M, Laederach A (2008) High-throughput single-nucleotide structural mapping by capillary automated footprinting analysis. Nucleic Acids Res 36:e63–e63
20. Mortimer SA, Weeks KM (2007) A fast-acting reagent for accurate analysis of RNA secondary and tertiary structure by SHAPE chemistry. J Am Chem Soc 129:4144–4145
21. Mortimer SA, Weeks KM (2009) Time-resolved RNA SHAPE chemistry: quantitative RNA structure analysis in one-second snapshots and at single-nucleotide resolution. Nat Protoc 4:1413–1421
22. Pang PS, Elazar M, Pham EA, Glenn JS (2011) Simplified RNA secondary structure mapping by automation of SHAPE data analysis. Nucleic Acids Res 39:e151
23. Saaidi A, Allouche D, Regnier M, Sargueil B, Ponty Y (2020) IPANEMAP: integrative probing analysis of nucleic acids empowered by multiple accessibility profiles. Nucleic Acids Res 48:8276–8289
24. Smola MJ, Rice GM, Busan S, Siegfried NA, Weeks KM (2015) Selective 2′-hydroxyl acylation analyzed by primer extension and mutational profiling (SHAPE-MaP) for direct, versatile and accurate RNA structure analysis. Nat Protoc 10:1643
25. Spitale RC, Flynn RA, Zhang QC, Crisalli P, Lee B, Jung J-W, Kuchelmeister HY, Batista PJ, Torre EA, Kool ET et al (2015) Structural imprints *in vivo* decode RNA regulatory mechanisms. Nature 519:486–490
26. Steen K-A, Rice GM, Weeks KM (2012) Fingerprinting non-canonical and tertiary RNA structures by differential SHAPE reactivity. J Am Chem Soc 134:13160–13163
27. Strobel EJ, Yu AM, Lucks JB (2018) High-throughput determination of RNA structures. Nat Rev Genet 19:615
28. Uhlenbeck OC (1995) Keeping RNA happy. RNA 1:4–6
29. Weill L, Louis D, Sargueil B (2004) Selection and evolution of NTP-specific aptamers. Nucleic Acids Res 32:5045–5058
30. Wilkinson KA, Merino EJ, Weeks KM (2006) Selective 2′-hydroxyl acylation analyzed by primer extension (SHAPE): quantitative RNA structure analysis at single nucleotide resolution. Nat Protoc 1:1610–1616
31. Wilkinson KA, Gorelick RJ, Vasa SM, Guex N, Rein A, Mathews DH, Giddings MC, Weeks KM (2008) High-throughput SHAPE analysis reveals structures in HIV-1 genomic RNA strongly conserved across distinct biological states. PLoS Biol 6:e96
32. Yoon S, Kim J, Hum J, Kim H, Park S, Kladwang W, Das R (2011) HiTRACE: high-throughput robust analysis for capillary electrophoresis. Bioinformatics 27:1798–1805
33. Zubradt M, Gupta P, Persad S, Lambowitz AM, Weissman JS, Rouskin S (2017) DMS-MaPseq for genome-wide or targeted RNA structure probing *in vivo*. Nat Methods 14:75–82

Chapter 5

How Parameters Influence SHAPE-Directed Predictions

Torin Greenwood and Christine E. Heitsch

Abstract

The structure of an RNA sequence encodes information about its biological function. Dynamic programming algorithms are often used to predict the conformation of an RNA molecule from its sequence alone, and adding experimental data as auxiliary information improves prediction accuracy. This auxiliary data is typically incorporated into the nearest neighbor thermodynamic model22 by converting the data into pseudoenergies. Here, we look at how much of the space of possible structures auxiliary data allows prediction methods to explore. We find that for a large class of RNA sequences, auxiliary data shifts the predictions significantly. Additionally, we find that predictions are highly sensitive to the parameters which define the auxiliary data pseudoenergies. In fact, the parameter space can typically be partitioned into regions where different structural predictions predominate.

Key words RNA secondary structure, Thermodynamic optimization with auxiliary data, Nearest neighbor thermodynamic model, SHAPE data

1 Introduction

RNA sequences fold into three-dimensional structures that encode functional information about the molecules. However, it is difficult to determine the base pairings of an RNA sequence experimentally. Instead, discrete optimization methods are used to predict the most probable folding for an RNA sequence. Because three-dimensional structures contain complicated bonding relationships between nucleotides, optimization methods instead search for two-dimensional approximations, called secondary structures, that still contain important functional information. To do so, the algorithms use the nearest neighbor thermodynamic model (NNTM) from [1] and [2] to assign energies to every potential structure for a sequence. Then, dynamic programming algorithms (many of which are based on the Zuker algorithm [3]) find the minimum free energy (MFE) structure.

Unfortunately, since the NNTM is only a model, these prediction methods seldom have perfect accuracy on their own. However,

Ronny Lorenz (ed.), *RNA Folding: Methods and Protocols*, Methods in Molecular Biology, vol. 2726,
https://doi.org/10.1007/978-1-0716-3519-3_5,

incorporating auxiliary experimental data often improves prediction accuracy. Through chemical footprinting experiments, every nucleotide in a folded RNA sequence is assigned a reactivity score. Ideally, the reactivity would be 0 for paired nucleotides and a positive constant A for unpaired nucleotides, but for a variety of experimental reasons, the reactivity signal is less clear. Nonetheless, these reactivity scores are used to shift the assigned energies for all potential structures corresponding to a sequence. Several competing methods of incorporating this noisy data into the thermodynamic prediction model have been proposed, described below. When auxiliary data is included, the prediction accuracy of these methods is often high enough to recover important structural information about wide classes of RNA sequences.

Even when auxiliary data is incorporated into prediction methods, sometimes the correct structure is still not predicted. One possible problem is that the MFE structure and the native structures have energies that are so far apart that auxiliary data cannot overcome this difference. In this chapter, we will investigate how much of the distribution of structures is accessible to the prediction methods when incorporating auxiliary data, and we ask the following question:

Question 1 How close do the energies of the MFE structure and the native structure need to be in order for auxiliary data to overcome the difference?

For example, consider a tRNA sequence from the organism, *Tropheryma whipplei*. The NNTM assigns energies to every possible structure the tRNA sequence could form, as shown in Fig. 1. Incorporating auxiliary data into the models shifts the energies of all structures, allowing the prediction model to consider all structures within a band of energies above the MFE structure. Sometimes, this band is too narrow to contain the native structure. For this *Tropheryma whipplei* sequence, the native structure is within the region recoverable by incorporating auxiliary data, and indeed, the data-directed prediction recovers the native structure.

To analyze this more broadly, we use a test set of 50 tRNA sequences and 50 5S rRNA sequences and compare their MFE and native structures. We first find a mathematical relation showing that the more structural differences there are between the MFE and native structures, the more auxiliary data guides predictions towards the native structure. As a result, when there are large structural differences and small energy differences between the MFE and native structures, auxiliary data is likely to help. Then, for our test set of 100 sequences, we confirm that auxiliary data generally shifts predictions enough to favor the native structure over the MFE structure. For more details, *see* Subheading 4.

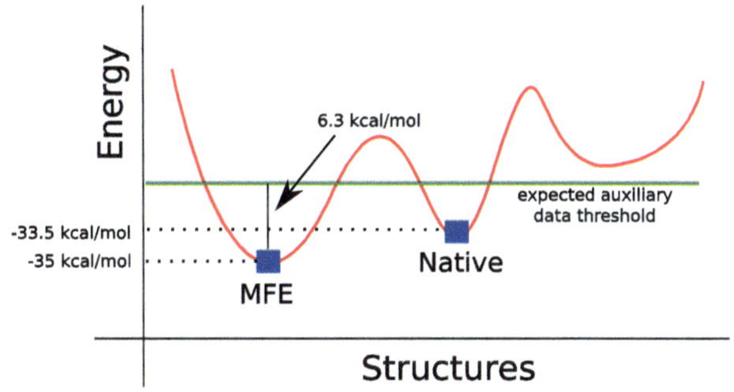

Fig. 1 Each point on the horizontal axis represents a possible structure for the tRNA sequence from *Tropheryma whipplei*. The vertical axis corresponds to the associated structure's energy, assigned by the NNTM. If the native structure is below the auxiliary data threshold, it could potentially be recovered by incorporating auxiliary data into a prediction model. In this case, the native structure is below the line, and it is also successfully recovered when incorporating auxiliary data into the prediction models

However, even if the auxiliary data can reweight energies enough to favor the native structure over the MFE structure, all other potential structures are also reweighted. This means that a third structure is sometimes predicted instead of the native and MFE structure. Which structure is favored depends heavily on the way that auxiliary data is incorporated into the NNTM. In this chapter, we will focus on the first, still prevalent model of incorporating data from [4], called the "Deigan model." The Deigan model converts data into pseudoenergies that are added to the energies of all potential structures. Like subsequent pseudoenergy models, the Deigan model has parameters which determine how auxiliary data reweights the energies assigned to structures. There are suggested default parameters for the Deigan model, but a variety of different parameter values are used in practice. Thus, we ask:

Question 2 How much do the parameters in the Deigan model affect predictions?

To answer this question, we analyze the influence of parameters on predictions for the same 100 sequences in the test set. We use empirically derived distributions for auxiliary data from [5] to generate simulated auxiliary data for each sequence. Then, we generate statistics on how the predictions change as the Deigan parameters vary. We find that for each RNA sequence, the parameter space is nearly partitioned into regions where specific structures dominate over all other structures. Moreover, for a substantial portion of the sequences, a small change in the default Deigan parameters leads to drastic changes in the predicted structures, suggesting that the precise choice of parameters plays a crucial role in the success of the methodology. We will show more details in Subheading 4. Before delving into the details, we first give some background information on auxiliary data and prediction models.

2 Materials

2.1 Sequences

We use a test set of 50 5S rRNA and 50 tRNA sequences originally from the Comparative RNA Web Site [6] and curated in [7]. We chose the well-studied 5S rRNA and tRNA families because they offer a broad range of prediction accuracies and sequence compositions. The sequences in the set were chosen so that their GC content and their accuracy (as given by `gtfold` [8]) were distributed nearly uniformly over the range of possible values, $[0, 1]$. The tRNA sequences range from 72 to 79 nucleotides in length, while the 5S rRNA sequences range from 117 to 133.

2.2 Secondary Structures

To measure the success of prediction algorithms, we will want to know the native *secondary structure* corresponding to an RNA sequence. A secondary structure is a two-dimensional approximation to a sequence's tertiary structure, where each nucleotide is either paired to exactly one other nucleotide, or it is unpaired. Typically, pairings form a Watson-Crick pair (G and C or A and U), or a wobble pair (G and U). As an added restriction, secondary structures are sometimes prohibited from having pseudoknots. This means that pairings cannot cross over each other. That is, whenever we consider paired nucleotides in positions (i, j) and another pair (i', j') with smaller indices written first and $i < i'$, then either $j < i'$, or $i < i' < j' < j$.

When considering secondary structures without pseudoknots, we will write structures in *dot-bracket notation*. Here, pairs of brackets correspond to paired nucleotides and dots correspond to unpaired nucleotides. For example, the word ((..).) corresponds to a structure where the first and last nucleotide are paired together, the second and fifth nucleotide are paired together, and the rest are unpaired. Dot-bracket notation was popularized by the ViennaRNA prediction software from [9].

There are online databases of known RNA secondary structures. Our data set of 50 tRNA and 50 5S rRNA sequences and their corresponding structures comes from the Comparative RNA Web Site [6]. This database was curated by using large collections of similar RNA sequences, aligning them, and deriving the structures of each sequence via their alignments. The comparative methods used here are described in [10] and were shown to be highly accurate in [11].

2.3 SHAPE Data

Auxiliary data is typically used to shift the energies of all structures by assigning additional energy terms to each structure, called *pseudoenergies*, described further in the next section. Experimentally, folded RNA sequences are exposed to chemical reagents in probing experiments, resulting in SHAPE data, named after the laboratory method, *selective 2′-hydroxyl acylation analyzed by primer extension*. There are several different reagents that are commonly used: N-methylisatoic anhydride (NMIA), 1-methyl-7-nitroisatoic anhydride (1M7), and 1-methyl-6-nitroisatoic anhydride (1M6). As discussed in [12], different chemicals give different structural signals.

Regardless of the chemical used, once this experiment is completed, each nucleotide in an RNA sequence is assigned a (typically) non-negative score reflecting its reactivity to the reagent. In general, the higher the reactivity, the less likely a nucleotide is to be paired. As described below, there are several ways of incorporating this information into the NNTM. The accuracy of single-conformation predictions is often raised substantially when SHAPE data is included in predictions.

There are online repositories of RNA sequences and corresponding SHAPE data, such as the RNA Mapping Database [13]. However, it will also be useful to have empirical distributions describing SHAPE data, both to analyze the data and to generate simulated data when experimental data is unavailable. In [5], the authors fit distributions for paired and unpaired nucleotides to SHAPE data from two *Escherichia coli* ribosomal sequences: a 16S rRNA sequence with 1542 nucleotides and a 23S rRNA sequence with 2904 nucleotides, both used in the Deigan analysis [4]. The authors formed both a binary model where they fitted the data separately for paired and unpaired nucleotides, and a ternary model where they broke paired nucleotides into *center-paired* and *edge-paired* nucleotides. A paired nucleotide at position i is center-paired if the nucleotides at positions $i-1, i,$ and $i+1$ are all paired and are all part of the same helix. All other paired nucleotides are called edge-paired.

The ternary model performed better in simulations and thus is what we will use to generate SHAPE data. However, the binary model will be simpler to use when analyzing the behavior of SHAPE data. In the binary model, we will let $g(x)$ be the probability distribution function for the paired distribution and $h(x)$ for the unpaired distribution. Then, $g(x)$ is a generalized extreme value distribution with parameters $\xi = 0.895341, \sigma = 0.0712473$, and $\mu = 0.054239$, while $h(x)$ is an exponential distribution with parameter $\lambda = \frac{1}{0.681211}$.

3 Methods

Predicting the secondary structure of an RNA structure involves several steps:

3.1 Assign Energies to All Possible Structures

Given an RNA sequence, the standard approach is to assign an energy to every potential structure into which the sequence can fold. Then, we search for the structure with the lowest energy and use it as our prediction. To assign energies, we use the *nearest neighbor thermodynamic model* (NNTM) [2]. The NNTM has hundreds of parameters to cover every possible structural feature and type of bond that can appear in an RNA secondary structure. These parameters are available in an online database [2]. The total energy of a structure is the sum of the energies of all of its components.

3.2 Use SHAPE Data to Adjust Energies

The lowest-energy structure according to the NNTM is often close to the native structure of an RNA sequence. This is impressive when considering that the number of potential structures for a sequence grows exponentially in the length of the sequence. However, there are families of sequences where the accuracy is improved by incorporating SHAPE data into predictions. One option is to convert SHAPE data into pseudoenergies that are added directly to the energies from the NNTM. Several methods of deriving pseudoenergies have been proposed, including [4, 14], and [15] (which are collectively reviewed in [16]). Another approach is to sample structures using the NNTM energies and to select a structure from the sample using the SHAPE data, as in [17]. More recently, Rsample [18] combines aspects of both approaches to search for structures in sequences that may have multiple native conformations.

In this chapter, we will focus on the Deigan method [4], which remains one of the most widely used methods for converting SHAPE data into pseudoenergies. Each structure is assigned pseudoenergies of the following form:

$$\Delta G_{\text{SHAPE}}(i) := m \ln(1 + a_i) + b. \tag{1}$$

Here, m is a positive constant, b is a negative constant, and a_i is the SHAPE reactivity value corresponding to nucleotide i. The penalty $\Delta G_{\text{SHAPE}}(i)$ is used only if the ith nucleotide is paired. This penalty is easily incorporated into the dynamic programming algorithm that computes the NNTM energy for a structure S. In [4], the authors showed that shifting the probabilities defined in Eq. 2 by incorporating these pseudoenergies dramatically improves structural predictions for some classes of structures.

When using the Deigan method to incorporate SHAPE data, we will need to decide on a parameterization of m and b. When the method was originally proposed in [4], default parameters of $m = 2.6$ and $b = -0.8$ were suggested. However, parameterizations of $m = 1.7$ or $m = 1.8$ and $b = -0.6$ were proposed in [19] and [20], and parameters as extreme as $m = 30$ and $b = -6$ have been used in [21]. As mentioned in the introduction, we will investigate how these parameters influence predictions in Subheading 4.

Due to the way the Deigan penalty is incorporated into dynamic programming algorithms, $\Delta G_{\text{SHAPE}}(i)$ is applied once for edge-paired nucleotides, twice for center-paired nucleotides, and never for unpaired nucleotides. We will *see* in Subheading 4 that on average, the SHAPE pseudoenergy applied to a structure S grows linearly in terms of the number of structural differences between S and the RNA sequence's actual structure. However, this technicality on how the SHAPE pseudoenergies are applied will break the strict linearity of the increase.

3.3 Find the Structure with Minimum Energy or Create a Sample of Structures

Despite the fact that the number of potential structures for a sequence grows exponentially in the length n of a sequence, there are dynamic programming algorithms that find the structure with minimum free energy in polynomial time in terms of n. Many such prediction programs are based on the Zuker algorithm from [22], and algorithms are continually being updated and improved. Some popular current implementations include UNAfold [23], RNAStructure [24], and the ViennaRNA package [25], all of which have webservers available online, as well as `gtfold` [8], which we will use in the Notes section of this chapter.

Sometimes, it is of interest to analyze the suboptimal structures predicted by the SHAPE-directed energy models. This could be because an RNA sequence folds into multiple stable structures, as described in [26]. Alternatively, as in this chapter, we could study suboptimal structures to evaluate what energetically favorable alternatives there are to the MFE. Given a fixed RNA sequence, the energies $\{\mathcal{E}(S)\}$ of the set of structures for a sequence, $\{S\}$, can be used to define a Boltzmann distribution of structures using the equation,

$$\mathbb{P}(S) := \frac{e^{-\mathcal{E}(S)/RT}}{Z}, \tag{2}$$

where Z is a partition function over all possible structures for the sequence, $R = 0.001987$ kcal/(mol K) is the gas constant, and T is the temperature, which we take to have default value 310K. Fortunately, in [27], a dynamic programming algorithm was introduced to compute the partition function Z (and base pair probabilities) in polynomial time. Additionally, efficient algorithms exist to sample structures from the distribution, as described in [28] and [29], which we will use later.

3.4 If Sampling, Analyze the Sample

Creating a sample of structures allows us to explore the suboptimal structures efficiently. This will give information about whether one structure clearly dominates over all other structures (in which case the sample will be comprised of almost entirely one structure) or whether there are competing structures close to the MFE structure.

A challenging aspect of analyzing a sample is that there are often several distinct structural choices in different regions of an RNA sequence. This means there are sometimes many possible structures for the same RNA sequence, leading any individual structure to appear in a sample with low probability. As proposed in [30], one option to amplify structural signals is to analyze the sample on a helix level. That is, we identify helices that appear in a high percentage of structures within the sample, and then categorize structures according to which helices they have present. If any one base pair from a helix is present in a structure, we consider that helix present in the structure, and therefore consolidate similar structures via helix classes. We will use this method of analyzing samples in Subheading 4. RNA profiling from [30] is an implementation of this strategy robust to the noise in stochastic sampling, and webservers are available online.

4 Notes

Quantifying energy changes due to SHAPE. SHAPE-directed predictions are useful in scenarios where the MFE structure for an RNA sequence has a lower energy than the native secondary structure. However, with the Deigan method, auxiliary data cannot always direct predictions towards the native structure. A necessary (but insufficient) condition for accurate predictions is that the auxiliary data needs to shift the energies of the MFE and native structures more than the difference in their energies.

We will check this condition for the 100 5S rRNA and tRNA sequences from the test set. We start by analyzing how much auxiliary data shifts the energies of the MFE and native structures with the Deigan method. Call the difference between the MFE and native structures' SHAPE pseudoenergies their pseudoenergy window. Then, we ask:

> **Question 1b** On average, what is the pseudoenergy window between an RNA sequence's MFE structure and its native structure?

Let M be the MFE structure for a sequence, and let N be the native structure for a sequence. We will show that the pseudoenergy window between the structures is approximately a linear function of the number of differences between M and N. With this in mind, let $\{s_i\}_{i=1}^{n}$ be a sequence of SHAPE values giving information about the native structure N.

We consider each nucleotide in the structures individually. For a nucleotide that is unpaired in both M and N, the SHAPE pseudoenergy is not applied in either structure. Similar intuition might lead one to believe that if a nucleotide is paired in both M and N, the same SHAPE pseudoenergy is applied to both structures and thus does not affect the relative energy. However, because the SHAPE pseudoenergy is applied once for edge-paired nucleotides and twice for center-paired nucleotides, the computation becomes more nuanced.

With this in mind, we denote the multiset of positions in the RNA sequence where N is paired and M is unpaired by $\mathrm{pu}(N-M)$ (with pu standing for paired and unpaired). If structure N is center-paired, the position appears twice in the set. Similarly, let $\mathrm{ce}(N-M)$ be the set of positions where N is center-paired and M is edge-paired. Define $\mathrm{pu}(M-N)$ and $\mathrm{ce}(M-N)$ similarly. Then, using Eq. 1, the pseudoenergy window between the structures is:

$$\sum_{i\in \mathrm{pu}(N-M)\cap \mathrm{ce}(N-M)} [m\ln(1+s_i)+b] - \sum_{i\in \mathrm{pu}(M-N)\cap \mathrm{ce}(M-N)} [m\ln(1+s_i)+b].$$

To compute the expected value of this expression, we will need the expected value of each SHAPE value, $\ln(1+s_i)$. We model this here with the binary distributions from [5] mentioned above, for simplicity. We assume that the distribution of each SHAPE value depends only on whether the nucleotide is paired or unpaired in the native structure N, and that all SHAPE values are otherwise generated independently of each other. We compute that $\mathbb{E}(1+s_i)=0.192$ if the nucleotide is paired, and $\mathbb{E}(1+s_i)=0.455$ if the nucleotide is unpaired.

As the parameters m and $-b$ increase, the expected pseudoenergy window increases. For the original default parameter values of $m=2.6$ and $b=-0.8$, we have the following:

$$\text{Expected pseudoenergy window} =$$
$$0.383|\mathrm{pu}(N-M)| + 0.301\big[|\mathrm{pu}(M-N)| + |\mathrm{ce}(M-N)| - |\mathrm{ce}(N-M)|\big] \quad (3)$$

This formula aligns with the intuition behind the Deigan pseudoenergy, as the pseudoenergy window generally increases as the number of structural differences between N and M increases. There is the very rare exception that pseudoenergy window does decrease in the scenario that there are many positions that are center-paired in M and edge-paired in N. This is because the SHAPE pseudoenergies are generally negative at paired positions, and in a position where M is center-paired while N is edge-paired, M will receive the favorable pseudoenergy twice. For example, consider when N and M are given by the following pairings in dot-bracket notation:

$$\mathrm{M} : ((..((((...))))..))$$
$$\mathrm{N} : ((..))((...))((..))$$

Then, the SHAPE pseudoenergies for M are applied in exactly the same positions as they are for N, but they are applied twice in positions $6, 7, 13$, and 14. Since a SHAPE sequence corresponding to N is likely to have low SHAPE values in these positions, structure M will likely have even lower SHAPE pseudoenergies than structure N.

Thus, overall, we have found a formula indicating that the pseudoenergy window increases as the number of structural differences between N and M increases. Now, we turn back to answering Question 1 in the introduction by using the test set of 100 5S rRNA and tRNA sequences. For each sequence, our goal is to compute whether a SHAPE sequence is likely to drive predictions away from the MFE structure towards a known, native structure for the sequence. To test this, we took each sequence and computed the corresponding MFE structure M by using `gtfold`. Then, we compared this structure to the native structure N obtained from the Comparative RNA Web Site [6]. As described in the previous section, we computed the structural differences by counting $\mathrm{pu}(N-M), \mathrm{pu}(M-N), \mathrm{ce}(N-M)$, and $\mathrm{ce}(M-N)$. Next, we used `gtfold` with added constraints to compute the energy of the native structures, N. With this information, we could compare the relative energies of the MFE and native structures to their expected pseudoenergy windows, by using Eq. 3.

As long as the expected pseudoenergy windows are larger in magnitude than the relative energies between the MFE and native structures, then SHAPE data is likely to direct the predictions. The results are given in Fig. 2 and also in Tables 1 and 2.

Only 3 out of 50 tRNA sequences were below the line $y = -x$, and these sequences were very close to the line. This implies that SHAPE data generally can shift predictions towards the native structure for tRNA sequences. For the rRNA sequences, on the other hand, 12 out of 50 sequences were below the line $y = -x$. In total, the default parameters are generally enough to direct the distributions away from the MFE structure, with an estimated 85% success rate in our data set.

How parameters direct predictions. At this point, we have found that for the majority of sequences in our test set, SHAPE sequences successfully direct predictions away from an MFE structure. But is it towards the native structure? For a given sequence, the SHAPE pseudoenergies reweight the energies for all possible structures corresponding to this sequence. As a result, the predicted structure is often neither the MFE structure nor the native structure. It turns out that the predicted structure is highly dependent on the parameters m and b in the Deigan pseudoenergy, Eq. 1. This leads us to investigate Question 2 from the introduction. We will see that the

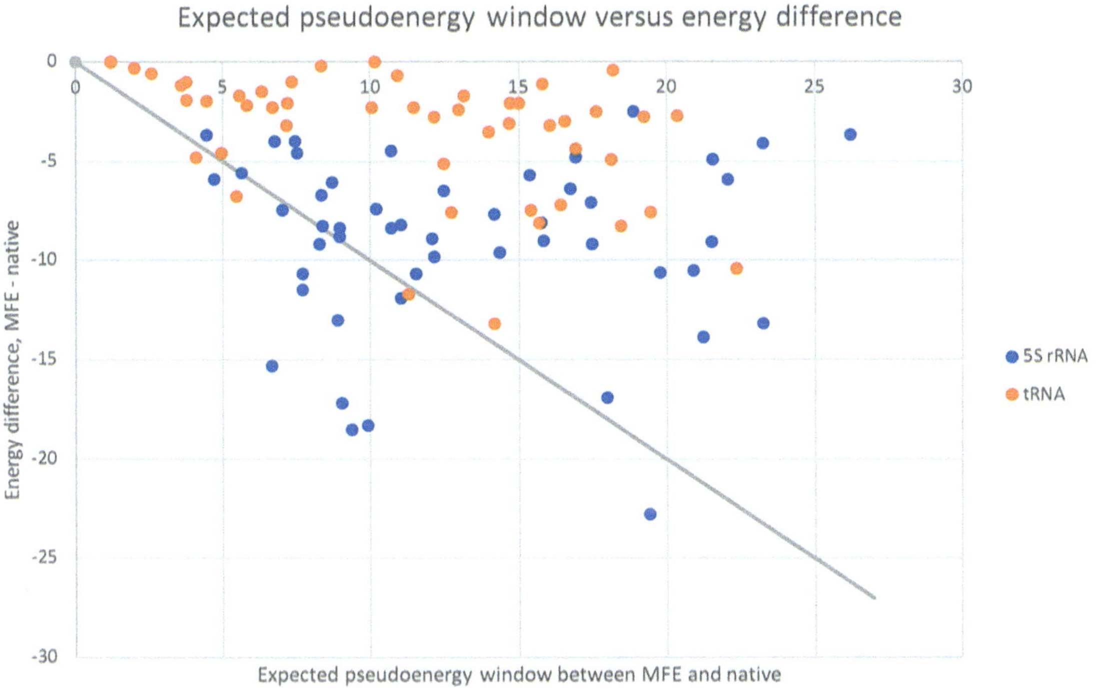

Fig. 2 Here, the horizontal axis represents the value of Eq. 3 for a particular structure, and the vertical axis represents the difference in energy between the MFE structure and the native structure. For structures above the line $y = -x$, it is expected that the SHAPE pseudoenergies are strong enough to drive predictions away from the MFE structure towards the native structure

parameter space is partitioned into regions where certain structures predominate. In particular, we find that for some fixed sequences and SHAPE data, a small perturbation of the parameters leads to drastically differing predictions.

Example We consider the 5S rRNA sequence from the organism, *Misgurnus fossilis* (accession number V00647), which was one of the 50 5S rRNA sequences from the test set used above. Using the native structure for the sequence from [6], we randomly generate a SHAPE sequence corresponding to the structure with the ternary model from [5] described above. Then, we consider each choice of parameters $0 \leq m \leq 5$ incremented by 0.5 and $-2 \leq b \leq 0$ incremented by 0.2. Overall, the parameter space is dominated by three conformations: the MFE conformation, the native conformation from [6], and a third alternative conformation. Diagrams of the three structures appear in Fig. 3.

In order to see how these three structures interact as the parameters m and b vary, we use sampling to measure how strong the signal is for each structure. In particular, for each set of parameters, we input the sequence and the SHAPE data into `gtfold` to produce a sample of 1000 structures. To account for other possible structures

Table 1
tRNA scnd GC content used for computations

	Accuracy	GC content	Length	Accession number	MFE—native energy	Expected relative SHAPE energy	Category
1	0	0.625	72	AY934184	−7.60	19.48	5
2	0.051	0.592	76	AE013169	−11.70	11.27	4
3	0.093	0.595	74	AY934018	−13.20	14.17	3
4	0.133	0.465	71	AJ010592	−2.80	19.23	2
5	0.133	0.583	72	U41549	−7.20	16.42	4
6	0.136	0.703	74	AE000520	−4.40	16.94	5
7	0.163	0.675	77	CP000143	−2.10	15.02	5
8	0.211	0.671	73	AY934254	−10.4	22.38	5
9	0.213	0.542	72	AE009952	−8.10	15.70	5
10	0.227	0.479	73	AE017159	−3.10	14.69	5
11	0.244	0.597	77	AY934351	−2.70	20.36	3
12	0.25	0.697	76	BA000023	−4.90	18.11	5
13	0.256	0.547	75	AY934387	−7.50	15.40	5
14	0.27	0.526	76	AY933864	−8.30	18.47	5
15	0.298	0.684	76	AJ248288	−3.50	14.01	3
16	0.304	0.64	75	CP000471	−0.40	18.19	5
17	0.304	0.644	73	BA000011	−3.20	16.06	3
18	0.311	0.605	76	BX321863	−3.00	16.58	2
19	0.326	0.554	74	CP000423	−2.50	17.62	3
20	0.341	0.648	71	CR936257	−7.60	12.75	5
21	0.35	0.528	72	AC004932	−2.10	14.72	3
22	0.465	0.584	77	AY933788	−0.20	8.37	3
23	0.5	0.405	74	X04465	−2.30	11.46	5
24	0.5	0.616	73	DQ093144	−2.40	13.00	5
25	0.522	0.73	74	AE014184	−1.50	6.32	3
26	0.533	0.618	76	CP000473	−1.70	13.16	2
27	0.545	0.539	76	K00349	−1.10	15.81	3
28	0.553	0.597	77	K00283	−5.10	12.48	5
29	0.585	0.581	74	CP000492	−2.80	12.15	4
30	0.591	0.548	73	CR382129	−2.30	6.68	3

(continued)

Table 1 (continued)

	Accuracy	GC content	Length	Accession number	MFE—native energy	Expected relative SHAPE energy	Category
31	0.638	0.558	77	X03154	−2.30	10.07	4
32	0.682	0.548	73	CP000254	−1.00	7.36	2
33	0.696	0.734	79	CP000493	−4.60	4.95	2
34	0.698	0.5	74	BA000021	0	10.18	5
35	0.698	0.507	73	CP000412	−0.70	10.94	2
36	0.711	0.558	77	AP006618	−1.20	3.58	3
37	0.711	0.562	73	CP000099	−3.20	7.17	3
38	0.727	0.675	77	CP000141	−4.80	4.10	5
39	0.762	0.616	73	AC006340	−0.60	2.57	3
40	0.773	0.63	73	BX569691	−1.70	5.55	5
41	0.8	0.514	72	AF137379	−2.20	5.83	3
42	0.837	0.68	75	K02972	−6.80	5.47	5
43	0.878	0.589	73	DQ396875	−2.10	7.20	1
44	0.913	0.688	77	CP000142	−2.00	4.43	1
45	0.95	0.581	74	BX640433	−1.90	3.78	4
46	0.976	0.425	73	AJ294725	−1.00	3.78	1
47	0.977	0.568	74	X04779	−0.30	1.97	1
48	1	0.487	76	AY934393	0	1.20	1
49	1	0.608	74	AE008623	0	1.20	1
50	1	0.726	73	AE000657	0	1.20	1

that may appear in the samples, we choose a set of helices that are likely to occur in competing structures. We do this by approximating the 50 helices compatible with the *Misgurnus fossilis* sequence that are the most energy-favorable. Our approximation assigns each GC pairing a score of 3, each AU pairing a score of 2, and each GU pairing a score of 1. We considered every possible string of consecutive pairings, summed the scores of these pairings, and picked the top 50 helices. This was a large enough number to usually include at least one helix from each arm of the tRNA sequences and 5S rRNA sequences in our test sets.

For each sample, we then check individually whether each of these helices was present in each structure. A helix was considered present in a structure if at least one base pair from the helix was in

Table 2
5S rRNA sequences of varying accuracy and GC content used for computations

	Accuracy	GC content	Length	Accession number	MFE—native energy	Expected relative SHAPE energy	Category
1	0.154	0.575	120	V00647	3.8	19.8	3
2	0.171	0.542	120	M24954	15.4	21.6	2
3	0.189	0.602	123	X14441	6.9	22.1	5
4	0.2	0.642	120	X72588	7.7	17.5	5
5	0.243	0.533	120	X00996	17.7	21.5	1
6	0.246	0.588	119	M24839	−15.7	19.4	5
7	0.267	0.642	120	V00336	7.6	16.8	5
8	0.293	0.6	120	L49397	16.2	15.4	4
9	0.306	0.6	120	M10432	19.5	16.9	3
10	0.333	0.65	120	X02627	14.4	26.2	4
11	0.361	0.504	119	Z75742	−17.3	23.3	5
12	0.384	0.558	120	Z93433	18.6	23.3	1
13	0.472	0.612	121	M10691	−3.9	9	5
14	0.493	0.575	120	U31855	6.8	18.9	3
15	0.553	0.551	127	X62864	0.4	18	2
16	0.583	0.546	119	M18170	−11.8	9.9	5
17	0.603	0.542	120	AB015591	−0.1	14.4	4
18	0.609	0.517	120	X00993	4.9	15.8	1
19	0.633	0.563	126	X62859	11.7	21.2	2
20	0.667	0.583	120	AB015590	7.7	12.5	4
21	0.676	0.533	120	Y14281	−3.2	12.1	1
22	0.694	0.542	120	X06094	−18.3	8.3	1
23	0.703	0.525	120	X13718	13.1	14.2	2
24	0.716	0.432	118	U39694	−15	8.9	2
25	0.722	0.567	120	M25016	−3.9	7.7	1
26	0.729	0.637	124	M26976	−7.5	6.6	1
27	0.74	0.513	119	X06102	6.5	10.7	1
28	0.74	0.575	120	X13035	0.4	11.5	1
29	0.746	0.624	117	AJ251080	3.3	20.9	5
30	0.75	0.567	120	Z33604	−6	11	2

(continued)

Table 2 (continued)

	Accuracy	GC content	Length	Accession number	MFE—native energy	Expected relative SHAPE energy	Category
31	0.757	0.683	120	U67518	−11.1	7	5
32	0.767	0.562	130	M36188	−3.6	15.8	3
33	0.773	0.533	120	X13037	6.8	9	1
34	0.789	0.583	120	Y00128	−0.1	10.7	1
35	0.794	0.632	117	M25591	3.1	17.5	4
36	0.806	0.533	120	X01004	2.9	8.7	1
37	0.816	0.558	120	D00076	6.5	8.4	5
38	0.833	0.55	120	AJ242949	7	7.4	1
39	0.838	0.6	120	M74438	4.9	7.5	1
40	0.849	0.517	120	K02343	−3.8	5.6	1
41	0.857	0.624	125	X72495	1.1	10.2	1
42	0.861	0.533	120	M34911	2.3	12.1	4
43	0.861	0.575	120	X06996	−5.4	6.8	1
44	0.867	0.717	127	X07692	−4.7	11	1
45	0.868	0.546	119	K00570	5.4	4.4	1
46	0.882	0.738	130	M21086	1.5	7.7	1
47	0.884	0.635	126	X05870	−11.1	9.1	1
48	0.889	0.513	119	X99087	−7.8	4.7	1
49	0.894	0.718	131	AE009942	−7.4	8.4	1
50	0.896	0.677	133	X07545	−12.6	9.4	1

the structure. If a helix was present in at least 80% of the structures, we marked it as present in the sample. In Fig. 4, different colors represent different collections of helices that were present as the Deigan parameters are changed.

Overall, the figure is dominated by regions corresponding to the MFE structure, the native structure, and the alternative structure. Colors close to the colors of the MFE, native, and alternative structure represent regions where some helices from the structures were dropped, and the green color represents transitional regions where only the helix 2 is present, which was present for all choices of parameters. The regions with fewer helices than the MFE, native, and alternative structures typically occur because there were several

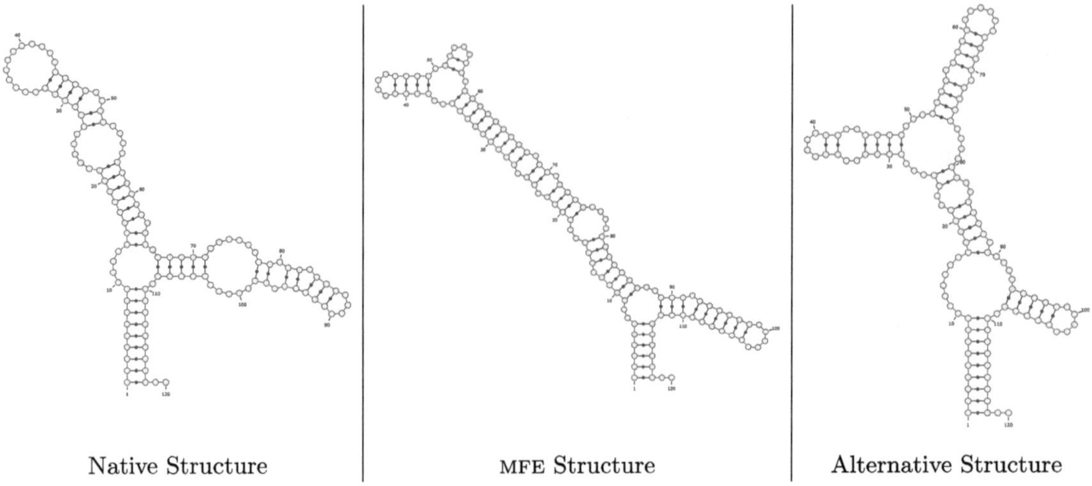

Fig. 3 Above are three structures for the 5S rRNA sequence from the organism, *Misgurnus fossilis*. When incorporating SHAPE data into predictions, there are Deigan parameters *m* and *b* which result in each of these three structures being the predicted structure

helices which appeared with frequency less than 80% in these regions. The exception to this is in the upper right corner of the figure, corresponding to an extreme choice of parameters with a high *m* value and a *b* value near zero. In this case, almost all SHAPE pseudoenergies are very positive even when the SHAPE values correspond to paired nucleotides. As a result, there were fewer pairings overall in these structures.

No energy-favorable helices appear besides the ones present in the MFE structure, the native structure, and the alternative structure. Overall, this indicates that the parameter space is partitioned into relatively few regions. Note that both sets of proposed default parameters, $[m, b] = [2.6, -0.8]$ and $[m, b] = [1.8, -0.6]$, lie within the yellow region corresponding to the native structure. This indicates that the SHAPE-directed predictions for *M. fossilis* would be correct on the helix level with either set of parameters.

However, besides in the upper right part of the figure, we see that there are often relatively quick transitions in the regions of the diagram, particularly between the MFE and native structures. In particular, the default parameters $[m, b] = [1.8, -0.6]$ are right along the border of the native and MFE structures. This leads to the following question:

Question 2b Are SHAPE-directed predictions using the default parameters $[m, b] = [1.8, -0.6]$ stable with respect to perturbations in the parameters?

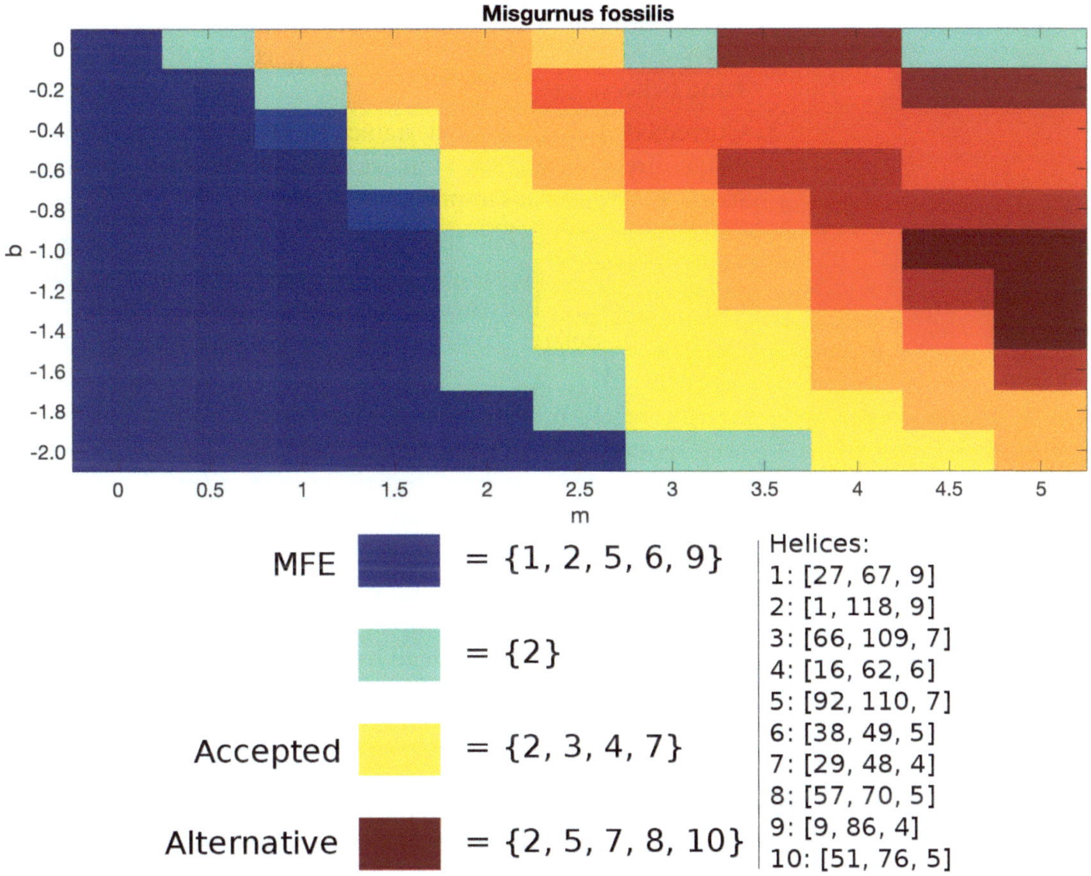

Fig. 4 For each set of parameters m and b, a sample of size 1000 was created using `gtfold`. Then, within each sample, we checked whether 50 of the most energy-favorable helices were present in over 80% of the sample. Each color represents a different collection of helices present in the corresponding sample. The helices are listed in $[i, j, k]$ notation, meaning that the nucleotides in positions $i, i+1, \ldots, i+k-1$ are paired to nucleotides $j, j-1, \ldots, j-k+1$. The MFE structure has the same helices as the blue, the native has the same helices as the yellow, and the alternative structure has the same helices as the deep red

We return to the test set of 50 tRNA and 50 5S rRNA sequences curated by [7] from [6]. First, for the tRNA sequences, we begin by computing the top 50 most energy-favorable helices for each sequence. Then, we repeat the procedure used in the *M. fossilis* example above to check whether or not the default SHAPE parameters are correct. We define a set of parameters $[m, b]$ to be *stable* if all other choices of parameters $[m', b']$ give the same collection of helices, where $|m' - m| \leq 0.5$ and $|b' - b| \leq 0.2$.

With respect to the default parameters $[m, b] = [1.8, -0.6]$, the tRNA sequences can be classified into five different groups:

1. Correct without SHAPE. 7/50 tRNA sequences. For these sequences, the MFE structure already had the same energy-favorable helices that the native structure had before SHAPE

was incorporated into predictions. In all of these cases, the SHAPE-directed sample with $[m, b] = [2.6, -0.8]$ also had the same helices.

2. Corrected with SHAPE and stable. 6/50 tRNA sequences. The SHAPE-directed sample had helices in agreement with the native structure, as did all samples created with nearby parameters.
3. Corrected with SHAPE and unstable. 14/50 tRNA sequences. The SHAPE-directed sample had helices in agreement with the native structure, but there exists a set of parameters $[m', b']$ close to the default parameters where the helices in the sample are different.
4. SHAPE-directed prediction incorrect, parameters unstable. 5/50 tRNA sequences. Here, the default parameters lead to a sample with the incorrect helices, but there exists a set of parameters $[m', b']$ close to the default that have helices in agreement with the native structure.
5. SHAPE-directed prediction incorrect, along with nearby parameters. 18/50 tRNA sequences. Here, the default parameters and all parameters nearby lead to samples with helices differing from the native structure. With few exceptions, there were no choices of parameters that yielded the native structure.

Details of the sequence categorizations are given in Table 1. For 19/50 of the tRNA sequences, the default parameters were on the edge of the region where the native structure was predicted, indicating that the parameter instability seen in *M. fossilis* above is not uncommon. This breakdown of categories was still the same numerically if the original default parameters $[m, b] = [2.6, -0.8]$ were used instead, although two sequences swapped between categories 2 and 3. Of course, sequences with high MFE accuracy tended to be in category 1. Otherwise, there was no strong connection between a sequence's classification into one of the remaining four categories and its MFE accuracy.

Next, for the 5S rRNA sequences, we change our criteria of correct slightly: the native structures often had non-canonical pairings that are not predicted by `gtfold`. As a result, we consider a helix as present in a structure as long as there is a pairing where both paired positions are within two nucleotides of a corresponding pairing in the helix. With this, we have the following breakdown:

1. Correct without SHAPE. 24/50 5S rRNA sequences
2. Corrected with SHAPE and stable. 6/50 5S rRNA sequences
3. Corrected with SHAPE and unstable. 4/50 5S rRNA sequences
4. SHAPE-directed prediction incorrect, parameters unstable. 6/50 5S rRNA sequences
5. SHAPE-directed prediction incorrect, along with nearby parameters. 10/50 5S rRNA sequences

Here, 10 out of 50 sequences were in categories 3 and 4, indicating that predictions for these sequences were still sensitive to parameter perturbations. Additionally, for several sequences in category 5, perturbations of the default parameters led to very different predictions that were still incorrect. This indicates that the precise choice of parameters influences the structural predictions for a large set of sequences.

References

1. Mathews DH, Sabina J, Zuker M, Turner DH (1999) Expanded sequence dependence of thermodynamic parameters improves prediction of RNA secondary structure. J Mol Biol 288(5):911–940
2. Turner DH, Mathews DH (2009) NNDB: the nearest neighbor parameter database for predicting stability of nucleic acid secondary structure. Nucleic Acids Res 38:D280–D282
3. Zuker M, Stiegler P (1981) Optimal computer folding of large RNA sequences using thermodynamics and auxiliary information. Nucleic Acids Res 9(1):133–148
4. Deigan KE, Li TW, Mathews DH, Weeks KM (2009) Accurate SHAPE-directed RNA structure determination. Proc Natl Acad Sci 106(1): 97–102
5. Sükösd Z, Swenson MS, Kjems J, Heitsch CE (2013) Evaluating the accuracy of SHAPE-directed RNA secondary structure predictions. Nucleic Acids Res 41(5):2807–2816
6. Cannone JJ, Subramanian S, Schnare MN, Collett JR, D'Souza LM, Du Y, Feng B, Lin N, Madabusi LV, Müller KM et al (2002) The comparative RNA web (CRW) site: an online database of comparative sequence and structure information for ribosomal, intron, and other RNAs. BMC Bioinform 3(1): 2, 2002.
7. Barrera-Cruz F, Heitsch C, Poznanović S (2018) On the structure of RNA branching polytopes. SIAM J Appl Algebra Geom 2(3): 444–461
8. Mathuriya A, Bader DA, Heitsch CE, Harvey SC (2009) GTfold: a scalable multicore code for RNA secondary structure prediction. In 24th Annual ACM Symposium on Applied Computing
9. Hofacker IL, Fontana W, Stadler PF, Bonhoeffer LS, Tacker M, Schuster P (1994) Fast folding and comparison of RNA secondary structures (the Vienna RNA package). Monatshefte für Chemie Chem Monthly 125(2): 167–188
10. Woese CR, Pace NR (1993) Probing RNA structure, function, and history by comparative analysis. In Gesteland RF, Atkins JF (eds) The RNA World, vol 24. Cold Spring Harbor Laboratory Press, pp 91–117
11. Gutell RR, Lee JC, Cannone JJ The accuracy of ribosomal RNA comparative structure models. Curr Opin Struct Biol 12(3):301–310
12. Rice GM, Leonard CW, Weeks KM (2014) RNA secondary structure modeling at consistent high accuracy using differential shape. RNA 20(6):846–854
13. Cordero P, Lucks JB, Das R (2012) An RNA mapping database for curating RNA structure mapping experiments. Bioinformatics 28(22): 3006–3008
14. Zarringhalam K, Meyer MM, Dotu I, Chuang JH, Clote P (2012) Integrating chemical footprinting data into RNA secondary structure prediction. PLOS ONE 7(10):1–13
15. Washietl S, Hofacker IL, Stadler PF, Kellis M (2012) RNA folding with soft constraints: reconciliation of probing data and thermodynamic secondary structure prediction. Nucleic Acids Res 40(10):4261–4272
16. Eddy SR (2014) Computational analysis of conserved RNA secondary structure in transcriptomes and genomes. Annu Rev Biophys 43(1):433–456. PMID: 24895857
17. Quarrier S, Martin JS, Davis-Neulander L, Beauregard A, Laederach A (2010) Evaluation of the information content of RNA structure mapping data for secondary structure prediction. RNA 16(6):1108–1117
18. Spasic A, Assmann SM, Bevilacqua PC, Mathews DH (2018) Modeling RNA secondary structure folding ensembles using SHAPE mapping data. Nucleic Acids Res 46(1): 314–323
19. Wilkinson KA, Gorelick RJ, Vasa SM, Guex N, Rein A, Mathews DH, Giddings MC, Weeks KM (2008) High-throughput SHAPE analysis reveals structures in HIV-1 genomic RNA strongly conserved across distinct biological states. PLoS Biol 6(4):e96

20. Hajdin CE, Bellaousov S, Huggins W, Leonard CW, Mathews DH, Weeks KM (2013) Accurate SHAPE-directed RNA secondary structure modeling, including pseudoknots. Proc Natl Acad Sci 110(14):5498–5503
21. Watts JM, Dang KK, Gorelick RJ, Leonard CW, Bess Jr JW, Swanstrom R, Burch CL, Weeks KM (2009) Architecture and secondary structure of an entire HIV-1 RNA genome. Nature 460(7256):711–716
22. Zuker M (1989) Computer prediction of RNA structure. In: RNA Processing Part A: General Methods, volume 180 of Methods in Enzymology. Academic Press, New Yrok, pp 262–288
23. Markham NR, Zuker M (2008) UNAFold. Methods Mol Biol 453:3–31
24. Xu ZZ, Mathews DH (2016) Secondary structure prediction of single sequences using rnastructure. Methods Mol Biol 1490:15–34
25. Lorenz R, Bernhart SH, Siederdissen CH, Tafer H, Flamm C, Stadler PF, Hofacker IL (2011) ViennaRNA package 2.0. Algorithms Mol Biol 6:26–26
26. Leonard CW, Hajdin CE, Karabiber F, Mathews DH, Favorov OV, Dokholyan NV, Weeks KM. Principles for understanding the accuracy of SHAPE-directed RNA structure modeling. Biochemistry 52(4):588–595. PMID: 23316814
27. McCaskill JS (1990) The equilibrium partition function and base pair binding probabilities for RNA secondary structure. Biopolymers 29(6–7):1105–1119
28. Ding Y, Lawrence CE (2003) A statistical sampling algorithm for RNA secondary structure prediction. Nucleic Acids Res 31(24): 7280–7301
29. Mathews DH (2006) Revolutions in RNA secondary structure prediction. J Mol Biol 359(3):526–532
30. Rogers E, Heitsch CE (2014) Profiling small RNA reveals multimodal substructural signals in a Boltzmann ensemble. Nucleic Acids Res 42(22):e171–e171

Chapter 6

Classified Dynamic Programming in RNA Structure Analysis

Björn Voß

Abstract

Analysis of the folding space of RNA generally suffers from its exponential size. With classified Dynamic Programming algorithms, it is possible to alleviate this burden and to analyse the folding space of RNA in great depth. Key to classified DP is that the search space is partitioned into classes based on an on-the-fly computed feature. A class-wise evaluation is then used to compute class-wide properties, such as the lowest free energy structure for each class, or aggregate properties, such as the class' probability. In this paper we describe the well-known shape and hishape abstraction of RNA structures, their power to help better understand RNA function and related methods that are based on these abstractions.

Key words Folding space, Abstraction, Classified Dynamic Programming, Abstract shapes, Hishapes

1 Introduction

The central difficulty in the analysis of RNA secondary structure is that the number of possible structures is exponential in sequence length. The explicit enumeration of the exponentially many solutions is commonly avoided by the use of dynamic programming. Thereby, the respective algorithms achieve polynomial complexity in space and time. This efficiency is of course desirable, but the downside is that most of the information present in the ensemble of all possible structures is lost. Partial solutions to the problem are the prediction of suboptimal structures and stochastic sampling of the structure space. The latter delivers robust results for large samples only, and the first suffers from the exponential growth of the structure space, even when restricted to near-optimal structures, e.g. within 10 kcal/mol of the optimum.

A holistic and still efficient analysis becomes possible, when we do not consider individual structures but rather classes or common structural features. For the latter, base pairs are a well-known example, and the McCaskill algorithm [1] can efficiently compute the probabilities of the base pairs. Essentially, these are the sum of the probabilities of all structures a base pair occurs in. Although

Ronny Lorenz (ed.), *RNA Folding: Methods and Protocols*, Methods in Molecular Biology, vol. 2726,
https://doi.org/10.1007/978-1-0716-3519-3_6,

base pairs are the central building blocks of RNA structure, their probabilities allow only limited insight into the full structure space. A reason for this is that the same base pair can be present in very different structures. But commonly, we are interested in grouping similar structures, which directly poses the problem of how to define similarity. Classic distance measures, such as the base pair distance [2], are well-suited, but expensive to compute w.r.t. the huge structure space. Another definition of similarity is that two structures "look" similar, i.e. that they have the same overall shape. In other words, if two structures are made up from the same helices and loops, and only differ in the length of the elements, we would consider them similar. A different loop composition would then imply dissimilarity. This is the central idea of shape abstraction [3], which is a powerful method to analyse the folding space of an RNA. The level of abstraction can be fine-tuned to the specific needs of the current analysis. Additionally, also the incorporation of new abstraction schemes is straightforward with the help of classified Dynamic Programming (cDP) implemented in Algebraic Dynamic Programming (ADP) [4, 5].

2 Background

The concept of classified Dynamic Programming is historically heavily intertwined with the development of RNAshapes [3]. Its implementation was carried out using Algebraic Dynamic Programming. Originally in the Haskell ADP framework and later ported to Bellmans GAP [6] and integrated into the RNA shapes studio [7]. In the following I will very briefly review the concept of ADP and how this eases, especially due to the use of algebra products, the implementation of cDP algorithms.

2.1 Algebraic Dynamic Programming

Algebraic Dynamic Programming (ADP) is a discipline of Dynamic Programming (DP) [8–10], which is one of the major computational paradigms used in bioinformatics. Typically, DP algorithms are formulated with recurrences and tables for storing intermediate results. ADP approaches DP problems by separating concerns. In ADP, a grammar $\mathcal{G}$ describes the search space, e.g. the set of valid RNA secondary structures as exemplified by $\mathcal{G}_{\text{RNAfold}}$ in Fig. 1a.

Candidates derived from the grammar are evaluated by an evaluation algebra $\mathcal{A}^h$ with its associated choice function h, which selects valid candidates w.r.t. the optimisation objective. As an illustrative example Fig. 1b shows an algebra for base pair maximisation. In the area of RNA folding, the most common objective is free energy (ΔG) minimisation, which is achieved with the algebra $\mathcal{A}_{\Delta G}^{min}$ (details not shown) and its choice function $h = min$. The algorithm for free energy minimisation of RNA S can then be

```
struct = comps |
         region |
         empty
block  = region closed |
comps  = block comps |
         block |
         block region
closed = base          min3          base |
         base         closed         base |
         base    region  closed      base |
         base    closed  region      base |
         base region closed region   base |
         base    block    comps      base
min3   = base base region
region = base |
         base region
base   = 'A' | 'C' | 'G' | 'U'
```

(a)

```
sc(x)     = x
sc(r)     = 0
sc(e)     = 0
bk(r,x) = x
bc(x,y) = x + y
bb(x)     = x
bn(x,r) = x
hl(b1,r,b2) = 1
sp(b1,x,b1) = x + 1
bl(b1,r,x,b2) = x + 1
br(b1,x,r,b2) = x + 1
il(b1,r1,x,r2,b2 = x + 1
ml(b1,x,y,b1) = x + y +1
m(b1,b2,r) = 0
rb(b) = 0
rr(b,r) = 0
ba(b) = 0
h = sum
```

(b)

Fig. 1 (**a**) Grammar for RNA folding in EBNF notation. (**b**) Corresponding algebra functions for the base pair maximisation algebra A_{bp}^{sum} with sum as the choice function

written as $\mathcal{G}_{\text{RNAfold}}(\mathcal{A}_{\Delta G}^{min}, S)$. In order to get an actual representation of the optimal structure, we have to define an additional algebra $\mathcal{A}_{\text{dotBracket}}^{id}$ that produces the dot-bracket notation for RNA secondary structure. It uses identity (*id*) as the choice function *h*. In order to get a reimplementation of the RNAfold algorithm, the algebras for energy minimisation and dot-bracket notation have to be combined, which in ADP is easily achieved with the help of algebra products [11]. The full algorithm is then $\mathcal{G}_{\text{RNAfold}}(\mathcal{A}_{\Delta G}^{min} * \mathcal{A}_{\text{dotBracket}}^{id}, S)$. For details about algebra products and their variants, we refer the interested reader to the respective publications [5, 6].

2.2 Classified Dynamic Programming

The central idea in cDP is that we can classify candidates according to a feature that we can compute from the candidate itself. For example, we could assign a candidate structure with a single-hairpin loop the feature value "[]" and one with two hairpin loops the "[] []". Not to affect efficiency, the feature should be computable recursively.

2.2.1 Classification

In ADP, candidates are evaluated by algebras and thus, also the classification feature can be computed by an algebra, the so-called classifying algebra C^h. Classifying algebras also have an associated choice function *h*. Let's consider the simple example that we want to compute the best structure with and without a multiloop for a given RNA. This constitutes a binary classification problem and we have to design the corresponding algebra C^{h}_{ml}, such that its result is 1 when a multiloop is present and 0 otherwise. Skipping implementation details, we further have to consider which choice function h to use. In the context of classification, the concept of a choice function is irrelevant, as the purpose of a classification algebra is to assign each candidate to a specific class, not to select the optimal one based on a particular criterion. Therefore, we set $h = id$ (i.e. no choice). But how does this help to achieve our goal. Until now, an algorithm $\mathcal{G}_{\text{RNAfold}}(\mathcal{A}^{min}_{\Delta G} * \mathcal{A}^{id}_{\text{dotBracket}} * C^{h}_{ml}, S)$ would compute the mfe structure of RNA S in dot-bracket notation and with the additional information whether it contains a multiloop or not. How can we turn this into an algorithm that computes for each class the mfe structure? The trick is to apply the choice function of the scoring algebra to subsets with identical class, i.e. to compute class-wise.

In more technical terms the class-wise computation is performed in the way that candidate lists are evaluated with the scoring and the classifying algebra, forming lists of result tuples $[(a_1, c_1), \ldots, (a_k, c_k)]$, where a_i is the score and c_i the class of a candidate. The optimal solution for each class is then found with $[(r, s) | s \leftarrow set([c_i, \ldots, c_k]),\ r \leftarrow h_{\mathcal{A}}[t | (t, u) \leftarrow [(a_1, c_1), \ldots, (a_k, c_k)], u = s]]$, where $h_{\mathcal{A}}$ is the choice function of the scoring algebra and $set()$ turns a list into a proper set. In ADP the combination of algebras in this way is automised via the *lexicographic* product, details of which are described in [6].

3 Materials

The tools used to produce the results shown in Subheading 4 are available at the following sites: RNAshapes, RNAlishapes and RNAcast are available as part of the RNA shapes studio at https://bibiserv.cebitec.uni-bielefeld.de/rnashapesstudio; RNAHelices together with HiTed, HiPath and HiKinetics is available for download at https://github.com/bjoernvoss/RNAHeliCes. The RNA sequences used as examples can be found in the corresponding software packages.

4 Methods

4.1 Shape Abstraction

The basic idea of shape abstraction for RNA stems from the observation that for a single RNA sequence, many of the possible secondary structures differ by only a few base pairs, while their overall shape is the same. In other words, opening or closing a base pair creates a different structure, but not a different shape. Thus, the shape of an RNA folding is defined by the composition and

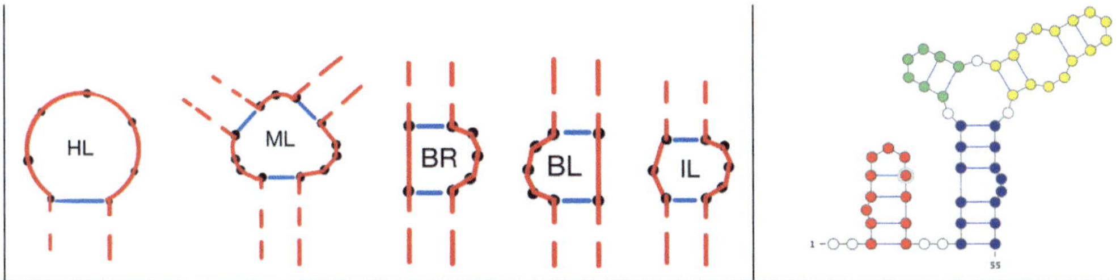

lvl						
5	[]	[◇ ◇]	◇	◇	◇	[] [[] []]
4	[]	[◇ ◇]	◇	◇	[◇]	[] [[] [[]]]
3	[]	[◇ ◇]	[◇]	[◇]	[◇]	[[]] [[[] [[]]]]
2	[]	[_◇_◇_]	[◇]	[◇]	[◇]	[[]] [[_[]_[[]]_]]
1	[_]	[_◇_◇_]	[◇_]	[_◇]	[_◇_]	_[_[]]_[[_[]_[_[]_]_]_]

Fig. 2 Schematic representation of shape levels for individual loops and for an example structure. The ◇ symbol represents nested substructures of any kind (lvl = shape level)

arrangement of its loops, namely, hairpin, bulge, internal and multiple loop. The level of abstraction can be adjusted and the five levels available with RNAshapes are shown in Fig. 2.

The first application of shape abstraction is the deep exploration of the structure space of an RNA. The number of possible RNA secondary structures is enormous and even close to the mfe, e.g. within 5 kcal/mol, this number is huge. Additionally, these numbers grow exponentially with sequence length. As mentioned before many of these structures differ by only a few base pairs, while they share the same overall shape. With shape abstraction it is possible to explicitly compute those structures that differ in shape, which provides a holistic overview over the structure space of an RNA. The latest implementation of shape abstraction is available with RNA shapes studio [7]. It offers the five abstraction levels shown in Fig. 2 and different grammars for RNA folding (*see* [12] for details about the grammars).

In the following we will present some use cases and lines of action. First, think of the scenario that you identified a tRNA gene in some newly sequenced genome and want to do further studies. Typically, you would expect that the secondary structure of the tRNA forms the well-known cloverleaf. But, in the case of this new tRNA, it actually does not as can be seen in Fig. 3. The mfe structure of this tRNA has a single-hairpin shape. Interestingly with shape analysis the second-best shape corresponds to the expected cloverleaf. This example illustrates the power of shape abstraction for structure space exploration. It was based on the most abstract shape level 5, but this is not always the best choice. The spliced leader RNA of*Leptomonas collosoma* has two conformations that are important for its function. Obviously, predicting the mfe structure cannot give insight into both conformations, and, thus, we directly start with predicting abstract shapes. The result in Fig. 4a shows two distinct shapes, but the second differs by an additional hairpin

```
  dG    GGGCCCAUAGCUCAGUGGUAGAGUGCCUCCUUUGCAAGGAGGAUGCCCUGGGUUCGAAUCCCAGUGGGUCCA  shape
-33.00  (((((((((((((((.((((.....(((((((...))))))).))))))))))).........)))))))).  []
```

(a) mfe

```
  dG    GGGCCCAUAGCUCAGUGGUAGAGUGCCUCCUUUGCAAGGAGGAUGCCCUGGGUUCGAAUCCCAGUGGGUCCA  shape
-33.00  (((((((((((((((.((((.....(((((((...))))))).))))))))))).........)))))))).  []
-31.30  ((((((...((((.......)))).(((((((...))))))).....(((((.......))))).)))))).  [[][][]]
```

(b) Shapes within 3kcal/mol above the mfe. dG = free energy in kcal/mol

Fig. 3 Shape analysis of a *Natronomonas pharaonis* tRNA. (**a**) mfe. (**b**) Shapes within 3 kcal/mol above the mfe. dG = free energy in kcal/mol

```
  dG   AACUAAAACAAUUUUUGAAGAACAGUUUCUGUACUUCAUUGGUAUGUAGAGACUUC  shape
-9.30  .......................((((((((((((.....)))))..)))))))..  []
-7.90  ........(((...)))......((((((((((((.....)))))..)))))))..  [][]
```

(a) Shape level 5

```
  dG   AACUAAAACAAUUUUUGAAGAACAGUUUCUGUACUUCAUUGGUAUGUAGAGACUUC  shape
-9.30  .......................((((((((((((.....)))))..)))))))..  []
-9.20  .......((((((..(((((.((((...)))).)))))..))).))).........  [[[]]]
-9.00  ..((...((((((..(((((.((((...)))).)))))..))).)))..)).....  [[[[]]]]
```

(b) Shape level 4

Fig. 4 Shape analysis of the *L. collosoma* spliced leader RNA with different abstraction levels. (**a**) Shape level 5. (**b**) Shape level 4

only. Furthermore, this additional hairpin has a positive free energy contribution of 2 kcal/mol, rendering it unstable. The proper alternative conformation shows up when switching to a less abstract shape level. With level 4, like shown in Fig. 4b, the second lowest energy shape represents the alternative conformation.

4.1.1 Size of the Shape Space

RNA secondary structure prediction is a combinatorial optimisation problem, where the number of possible structures is exponential in the sequence length N [13]. In [3] we derived 1.4^N as a rough estimate for the number of possible structures, which means that a sequence with $N = 100$ already has more than 4×10^{14} possible structures. To what extent does shape abstraction reduce or circumvent this exponential explosion? We want to clearly state that the number of shapes suffers from the same exponential growth, but to much lesser extent. Estimates from empirical measurements result in 1.1^N possible shapes (level 5) [3]. Furthermore, theoretical derivations of shape space sizes were presented in [14, 15]. It is important to note that in practical applications the number of all possible shapes is often irrelevant, because the user is interested in near-optimal shapes, e.g. within 5 kcal/mol above the mfe. Although also the number of shapes in this range grows exponentially with N, the actual numbers are reasonably small, e.g. ~ 100 shapes for $N = 300$.

4.1.2 Shape Probabilities

The power of classification is not only to filter out unwanted candidates, but more that the search space is partitioned into disjoint classes. And for these classes we can compute features from their members. For example, we could count the number of candidates in each class, which would give us class sizes. In the case of RNA secondary structures, it is more meaningful to also take the free energy into account. This has been pioneered by McCaskill for computing base pair probabilities [1] and is based on the partition function $Q = \sum_{i \in S} e^{\frac{-E_i}{RT}}$, where E_i is the free energy of structure i, S the set of all possible structures, R the universal gas constant and T the temperature. This is the sum of the Boltzmann factors over all structures; thus, the partition function can be interpreted as Boltzmann weighted structure counting. The fraction of an individual structure j and the partition function $\frac{e^{\frac{-E_j}{RT}}}{Q}$ gives us the probability $P(j)$ of j in the ensemble of all structures. Structure probabilities are small and are essentially only transformed free energy values. Therefore, they do not provide further information. This changes when shape abstraction comes into play. Each shape represents a class of similar structures, and the more structures fall into a certain class and the lower their free energies, the more likely it is that the RNA takes the corresponding shape. In order to compute shape probabilities, all we have to do is to sum structure probabilities, or better Boltzmann factors, per shape (class). This is straightforward with cDP: First, we have to construct an Algebra $\mathcal{A}_{pf}^{sum}$ to compute Boltzmann factors and sum them up to the full partition function. Second, we combine this algebra with a classifying algebra for shape abstraction C_{shape}^{id} via the algebra product $\mathcal{A}_{pf}^{sum} * C_{shape}^{id}$. Finally, the algorithm to compute shape probabilities is $\mathcal{G}_{RNAfold'}(\mathcal{A}_{pf}^{sum} * C_{shape}^{id})$. The grammar $\mathcal{G}_{RNAfold'}$ is a variant of $\mathcal{G}_{RNAfold}$ introduced in [16] that removes ambiguities, which would result in incorrect partition function calculation. The topic has been nicely studied in [12] and we refer the interested user to this publication for further reading.

Complete Analysis In the above described way we can compute probabilities for all possible shapes of an RNA. Figure 5 shows the results for the *L. collosoma* spliced leader RNA. The first seven shapes make up 98.4% of all shapes. Further manual inspection based on the representative structure suggests that shapes 1,3,5 and 2,4,6,7 are similar and could be further grouped. Doing so results in probabilities $P_{1,3,5} = 0.5538878$ and $P_{2,4,6,7} = 0.4313884$, which nicely fit the expectation for a bistable RNA.

```
  dG    AACUAAAACAAUUUUUGAAGAACAGUUUCUGUACUUCAUUGGUAUGUAGAGACUUC      P        shape
-9.20   .......((((((..(((((.((((...)))).)))))..))).))).........  0.2735281  [[[]]]
-9.30   .......................((((((((((((.....)))))..)))))))..  0.2482983  []
-9.00   ..((...((((((..(((((.((((...)))).)))))..))).)))..)).....  0.1955581  [[[[]]]]
-7.90   ........(((...)))......((((((((((((.....)))))..)))))))..  0.1353308  [][]
-8.00   .(((((.........(((((.((((...)))).))))))))))).............  0.0847991  [[]]
-6.60   ........(((...)))......(((((((.((((.....))))...)))))))..  0.0304695  [][[]]
-6.90   ..((..((......))..))...((((((((((((.....)))))..)))))))..  0.0172882  [[]][]
-6.80   ..((...(((.((..(((((.((((...)))).)))))..))..)))..)).....  0.0067029  [[[[[]]]]]
-6.80   .......((((((..(((((.((((...)))).)))))..))).))).((....))  0.0045262  [[[]]][]
-5.60   ..((..((......))..))...(((((((.((((.....))))...)))))))..  0.0027100  [[]][[]]
-4.20   ........(((...)))..((...((((((.((((.....))))...)))))).))  0.0004628  [][[[]]]
-4.60   .......(((.((..(((((.((((...)))).)))))..))..))).((....))  0.0001807  [[[[]]]][]
-3.10   ....(((....))).(((((.((((...)))).)))))..(((........)))..  0.0000723  [][[]][]
-3.20   ................((((.((((...)))).((.(((....))).))...))))  0.0000249  [[][[]]]
-1.80   ....(((....)))((....)).((((((((((((.....)))))..)))))))..  0.0000121  [][][]
-1.80   ..((..((......))..))...(((.(((.((((.....))))...))).)))..  0.0000103  [[]][[[]]]
-2.20   ................((((.((((...)))).((((((....)))..))).))))  0.0000067  [[][]]
-1.30   ........(((...)))....((((...)))).((.(((....))).)).......  0.0000045  [][][[]]
-2.30   ..((...(((.((..((.((.((((...)))).)).))..))..)))..)).....  0.0000043  [[[[[[]]]]]]
-2.40   .((..((((..((((...))))..))))..)).((.(((....))).)).......  0.0000040  [[[]]][[]]
-1.10   ..((...........(((((.((((...)))).))))).((.....)).)).....  0.0000017  [[[]][]]
-0.40   ........(((...)))..((...((.(((.((((.....))))...))).)).))  0.0000011  [][[[[]]]]
```

Fig. 5 Shape probabilities (P, level 4) for the *L. collosoma* spliced leader RNA

```
  dG    AUCCAGGAGGCUAGCGCGUGAGAAGAGAAACGGAAAACAGCGCCUGAAAGCCUCCCAGUGGAGGCUUUUUUUG      P        shape
-23.20  ...((((..(((....(((..........)))......))).)))).((((((((....))))))))......  0.8318860  [[[]]][]
-21.50  ..((....))...((((((.................)).))))..((((((((((....))))))))))....  0.0511103  [][[]][]
-20.90  .............((((((.................)).))))..((((((((((....))))))))))....  0.0294958  [[]][]
-21.30  ..((....))...((((......................))))..((((((((((....))))))))))....  0.0283276  [][][]
-20.70  .(((((((((((((.((((....................))))))...)))))))...))))...........  0.0267715  [[[[]]]]
...
```

Fig. 6 First five shapes of the pheS-pheT attenuator sequence

A powerful property of shape probabilities is illustrated by the example in Fig. 6. Here, the fourth best shape in terms of energy (3rd row) achieves a higher probability than the third best (4th row). Such "overtaking" indicates that a certain shape might be more relevant for the function as reflected by the free energy.

Computation of shape probabilities in reasonable time is possible for sequences of moderate length (< 100 nt). Beyond that length the exponentially growing number of possible shapes dominates the computation. In the case of simple shape abstraction based on free energy minimisation (as introduced in Subheading 4.1), it is no problem to study longer sequences by restricting the depth (in terms of energy range above the mfe) of the analysis. This strategy is not possible when computing shape probabilities, because here we sum Boltzmann factors, and only if we take into account all contributions, the results will be correct.

ΔG	P	P_f	P_{1k}	P_{10k}	P_{100k}	shape
-24.30	0.8285376	0.8285419	0.8260000	0.8297000	0.8281300	[][][]
-22.80	0.1437373	0.1437380	0.1440000	0.1419000	0.1441700	[][]
-21.60	0.0171483	0.0171482	0.0140000	0.0173000	0.0166900	[[][]][]
-20.70	0.0036501	0.0036501	0.0060000	0.0042000	0.0041300	[][][][]
-20.20	0.0029953	0.0029953	0.0080000	0.0034000	0.0029900	[[][]]
-20.40	0.0023932	0.0023930	n.s.	0.0018000	0.0023200	[[][][]]
-19.90	0.0006908	0.0006907	0.0010000	0.0009000	0.0006900	[[][]][][]
-19.40	0.0003342	0.0003342	0.0010000	0.0005000	0.0003300	[]
-19.90	0.0002461	0.0002461	n.s.	0.0002000	0.0002000	[[][][]][]
			Performance measures			
CPU [s]	482	1.5	1.1	6	58	
Mem [MB]	1,492	39.5	20.0	67.2	547	

Fig. 7 Probabilities of the shapes within 5 kcal/mol above the mfe of the 113 nt long adenine riboswitch from *Vibrio vulnificus*. Exact computation (P), low probability ($< 10^{-6}$) filtering (P_f) and estimation via sampling of 1,000 (P_{1k}), 10,000 (P_{10k}) and 100,000 (P_{100k}) structures. The last two rows give the runtime in CPU seconds and the maximum memory in MB. n.s. means that the shape was not sampled

Nevertheless, for longer sequences we want to at least estimate shape probabilities or restrict the computation to the "important" shapes, whatever important means. In the following we describe three strategies for this purpose.

Low Probability Filtering In the recursive computation, whenever we have a list of solutions for a certain subword, we sum up the Boltzmann factors per shape. The result list will likely hold many entries with very low values and thus very low probabilities that are unlikely to contribute significantly to the overall result. Thus, pruning these entries, e.g. with probabilities below 10^{-6}, reduces runtime and memory consumption while only slightly affecting the accuracy, as can be seen in Fig. 7. The probabilities with filtering (P_f) differ only marginally (at most $4.2 * 10^{-6}$) from the exact values, while the speed-up is more than 200-fold.

Sampling Ding and Lawrence [17] presented an algorithm for sampling structures according to their probability. In ADP this can be implemented via the choice function, which in the case of structure sampling, draws a sample according to the Boltzmann distribution of the actual candidate list. Let *sample*() be such a sampling function, then the algorithm for probabilistic structure sampling is $\mathcal{G}_{RNAfold'}(\mathcal{A}_{pf}^{sample} * C_{shape}^{id})$. In detail, this algorithm draws a structure from the Boltzmann distribution for the given RNA sequence together with its shape. Repeated sampling, e.g. 1,000 times, can then be used to estimate shape probabilities via their occurrence frequency in the sample. The example in Fig. 7 shows how accuracy and runtime are affected by the sample size.

As expected, the larger the sample, the more accurate the estimation, but also the longer the runtime. Interestingly, the sampling is always less accurate than the low probability filtering described in the previous paragraph and also less efficient in terms of runtime and memory, which is mainly caused by postprocessing for shape counting.

The efficiency lag of the sampling turns into a gain with longer sequences. For example, for a 281 nt long sequence, the filtering approach takes 621 s and needs 1,520 MB of RAM, while sampling 100,000 structures and deriving shape probabilities take 231 s and 715 MB. This shows that the sampling approach does not suffer from the exponential explosion in the way the exact and filtering methods do.

Single Shape Analysis The sampling strategy has the drawback that for longer sequences and/or increased accuracy larger samples are needed, which increases computational demands. Furthermore, the results contain a lot of shapes that may actually be of no interest to the user. Ideally, we should compute exact shape probabilities for only those shapes we are interested in, efficiently. The first step is then to define which are the interesting shapes. In general, there will be no a priori shape(s) of interest, but we rather have to derive them in a first step. A reasonable choice are all shapes within a certain energy range above the mfe, which can be computed efficiently.

But how can we compute the probability for a specific shape of interest? There is a theoretical solution that has never been implemented and an actual solution known as RapidShapes [18]. The theoretical approach is to use cDP in terms of $\mathcal{G}_{RNAfold'}(\mathcal{A}_{pf}^{sum} \times C_{shape}^{\subseteq})$, where the classifying algebra $C_{shape}^{\subseteq}$ keeps only those (partial) solutions whose shape is a subset of the shape of interest. The solution of RapidShapes in contrast is to generate a distinct grammar for the shape of interest $\mathcal{G}_{shoi}$ and to use the standard partition function algebra, i.e. $\mathcal{G}_{shoi}(\mathcal{A}_{pf}^{sum})$. In order to compute the probability, computation of the full partition function is needed, which is easily available with $\mathcal{G}_{RNAfold'}(\mathcal{A}_{pf}^{sum})$. RapidShapes has to compile an executable for each shape of interest, which is also reflected by the runtime.

RapidShapes starts with computing, e.g. the *k* best shapes using the standard approach, then for each of these it generates a specific grammar, compiles it together with the partition function algebra into an executable and uses the latter to compute the restricted partition function for the actual RNA sequence. Finally, RapidShapes computes the full partition function for the RNA to calculate the shape probabilities. Although this sounds quiet time consuming, computing exact probabilities for the 9 shapes shown in Fig. 7 takes 74.47 s, from which 73.39 s are used for compiling and 1.08 s for the actual computations, and requires 300 MB of memory. Furthermore, the approach scales nicely with sequence length, such that computing probabilities for the 10 best shapes of a 281 nt long sequence are done in 3 minutes, requiring 350 MB RAM.

Note that the number of shapes to be analysed is an arbitrary choice and the initial guess might not be perfect. In order to assess, if the chosen number makes sense, it is best to check whether the accumulated probability of the chosen shapes is close to 1. If this is not the case, one or more shapes with a significant share have been missed and the number should be increased.

Which Tool to Use When? At the end of this section, we want to give some advice on when to use which of the mentioned tools. At first hand, the choice depends on the length of the sequence. For sequences shorter than 100 nt, the exact computation is favourable and can be done on a standard PC or laptop. If the sequence length is in the range of 100–250 nt, the low probability filtering approach is the best choice, followed by sampling for sequences longer than 250 nt. In the case that one wants to assure perfect accuracy, the RapidShapes approach is to be used for sequences > 100 nt.

4.1.3 Consensus Approaches

RNA structure prediction for a single sequence is not very reliable, especially for long sequences. Thus, taking into account related sequences is a reasonable approach to improve the results and this can be done in various ways: first, by aligning the sequences and computing a common structure based on the alignment; second, by predicting individual structures and aligning them afterwards; and third, by doing both alignment and folding simultaneously. The latter is the well-known Sankoff algorithm [19], which due to its algorithmic complexity was never implemented entirely, but only in restricted versions [20–22].

The first approach was pioneered by Hofacker et al. [23] with the program RNAalifold. The central idea of this approach is that regions of structural importance exhibit covariation in the sequence alignment. Thus, if two aligned positions of an alignment show covariation and are able to base pair, the free energy score for the base pair is augmented with an arbitrary bonus energy for the covariation. For details we refer the reader to the original publication.

RNAlishapes Although the incorporation of evolutionary information is thought to improve the prediction, the arbitrary covariation scoring introduces a new source of uncertainty, which makes it difficult to trust the single "optimal" prediction. Furthermore, some types of RNAs, e.g. riboswitches and ribozymes, have two or more conformations that are important for their function. Analogous to the single sequence case, the analysis of suboptimal solutions or even the complete folding space is thus required. And again, like it is for single sequences, the size of the folding space is exponential in the alignment length. Thus, it is straightforward to also use shape abstraction as implemented in the tool RNAlishapes [24].

RNAcast The authors of RNAcast [25] termed the approach an alternative to the Sankoff algorithm, although alignment and folding are not done simultaneously. Nevertheless, what justifies this statement is the fact that for each sequence multiple possible foldings are considered and the best conserved among all sequences is selected. The approach used is based on shape abstraction: For each sequence in the dataset, the *k* best shapes for a user-defined abstraction level are computed. Next, the lists for the individual sequences are scanned for shapes common to all of them. If several common shapes are found, the one that minimises the sum of ranks is the consensus shape.

4.2 Positional Abstraction

Shape abstraction is powerful and enables in-depth analyses of the RNA folding space. Nevertheless, the abstraction method has one significant shortcoming, which is that the shape is independent of the actual position of the respective structural elements. This fact is already nicely reflected in Fig. 4. In shape level 5, we see the same stem-loop at the 3' end in both shapes. When we switch to shape level 4, a second stem-loop with comparable free energy shows up. In this case, the different pattern of internal loops between the two hairpins leads to their segregation in different shapes. This is unproblematic in the case of free energy minimisation, as long as the hairpins at the different positions differ in their composition of bulge and/or internal loops. But, it turns into a problem when we do probabilistic shape analysis, where Boltzmann factors are summed up. Here, hairpins from different locations within the sequence may contribute to the same shapes, such that the shape probabilities are a mixture of different hairpins.

The straightforward solution is to use a positional abstraction like it was introduced with RNAHeliCes [26]. The name is the short form of RNA Helix Centers, and this already describes how the positional abstraction is implemented. We take the view that each loop (hairpin, internal, bulge or multiple) is enclosed by a helix, which is at least the closing base pair of the loop. Unfortunately, the position of a helix has no natural representation. One could choose the first or the last base pair and use their positions, but then a helix shortened by one base pair will result in a different location, which is counterintuitive. A measure that is robust under helix extension and reduction is the central position of the base

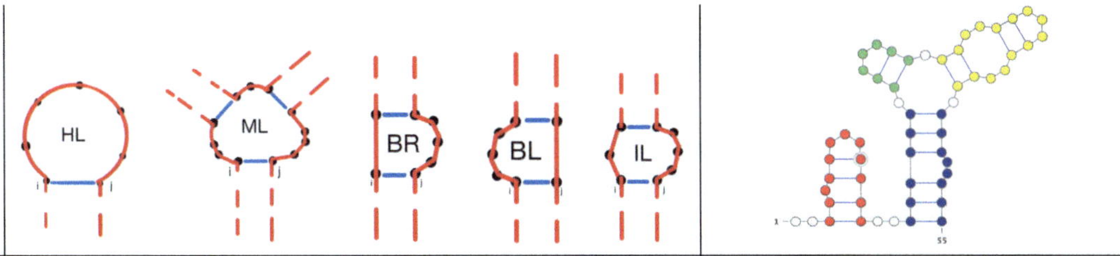

lvl	HL	ML	BR	BL	IL	
h	$\frac{i+j}{2}$	◇ ◇	◇	◇	◇	9,27,39
h+	$\frac{i+j}{2}$	(◇ ◇)	◇	◇	◇	9,(27,39)
m	$\frac{i+j}{2}$	$\frac{i+j}{2}$m,(◇ ◇)	◇	◇	◇	9,35m,(27,39)
a	$\frac{i+j}{2}$	$\frac{i+j}{2}$m,(◇ ◇)	$\frac{i+j}{2}$b,◇	$\frac{i+j}{2}$b,◇	$\frac{i+j}{2}$i,◇	8.5b,9,36b,35m,(27,39i,39)

Fig. 8 The four different abstraction levels, so-called hishapes, of RNAHeliCes

pairs of the helix. Together with an indicator for the loop type (*i*, internal; *b*, bulge; *m*, multiple; none, hairpin), they form the so-called helix index, which is a suitable position-dependent structure abstraction and was termed hishapes (helix index shapes). Like in classic shape abstraction, the resolution can be fine-tuned by handling helices from different loop types differently as shown in Fig. 8.

The different abstraction levels are implemented as distinct algebras $\mathcal{A}_h^{id}, \mathcal{A}_{h+}^{id}, \mathcal{A}_m^{id}, \mathcal{A}_a^{id}$. The cDP algorithm denoted by $\mathcal{G}(\mathcal{A}_{mfe}^{min} * \mathcal{A}_{\{h,h+,m,a\}}^{id})$ computes all possible hishapes together with their minimum free energy structure. And again, by exchanging the mfe algebra with the partition function algebra, we can compute probabilities of hishapes.

Analysing the spliced leader sequence from Fig. 4 using the abstraction level "h" results in the prediction shown in Fig. 9. The positional abstraction nicely separates the two single-hairpin shapes into the hishapes `[38]` and `[27]` and interestingly the latter is energetically less stable but more probable. Another feature of the positional abstraction is that it allows to accumulate probabilities of individual helices from different hishapes. For helices `[38]` and `[27]` this results in probability estimates of 0.501635 and 0.478218, respectively.

4.2.1 Size of the Hishape Space

The increased resolution of hishape abstraction is not for free and comes with an increased size of the search space and, thus, increased demand for computational resources, compared to classic shape abstraction. For a sequence of length *N*, the number of possible structures can be roughly estimated to 1.43^N and the number of level 5 shapes to 1.13^N. For hishapes the numbers are in the range between those two ($1.23^N - 1.39^N$) [26].

```
AACUAAAACAAUUUUUGAAGAACAGUUUCUGUACUUCAUUGGUAUGUAGAGACUUC
.......................((((((((((((.....)))))..)))))))..   -9.30          [38]   0.345271
.......((((((..(((((.((((...)))).)))))..))).))).........   -9.20          [27]   0.470394
........(((...)))......((((((((((((.....)))))..)))))))..   -7.90       [13,38]   0.061698
...(((((....))))).......((((((((((((.....)))))..)))))))..   -7.80     [10.5,38]   0.066247
..((..((......))..))...((((((((((((.....)))))..)))))))..   -6.90     [11.5,38]   0.009375
.......((((((..(((((.((((...)))).)))))..))).))).((....))   -6.80   [27,52.5]   0.004868
..............(((((.((((...)))).)))))..(((.........)))..   -6.70   [27,47.5]   0.002956
..((.(((.....)))..))...((((((((((((.....)))))..)))))))..   -6.60       [11,38]   0.006642
....((((...)))).........((((((((((((.....)))))..)))))))..   -6.40       [10,38]   0.008893
...........(((....)))..((((((((((((.....)))))..)))))))..   -6.30     [16.5,38]   0.003509
```

Fig. 9 The ten energetically most favourable hishapes of the spliced leader RNA together with their probabilities

4.2.2 Hishape-Based Algorithms

Distance Measure The positional information provided by hishapes is ideally suited to compute distances of structures. In [26] we introduced the tool HiTED to compute the tree-edit distance of RNA secondary structures. Hishapes have a natural tree representation, just as secondary structures have, and this makes it straightforward to implement a distance measure based on the tree-edit model. The distance has two contributions: σ_h scores the deletion, insertion and relabeling of helices and is taken from [27]; σ_p is the absolute positional distance of the two helices, e.g. $\sigma_p(34b, 23m) = |34 - 23| = 11$. For two hishape trees T_1 and T_2 and a series of edit operations $S = s_1, \ldots s_k$, the distance is defined as:

$$HiTED(T_1, T_2) = \sum_{\forall s_i \in S} D_{TED}(s_i) \tag{1}$$

$$D_{TED}(H_1, H_2) = \sigma_h(H_1, H_2) + \lambda \times \sigma_p(H_1, H_2) \tag{2}$$

The weighting factor λ is used to scale the positional contribution and $\lambda = 5$ performed best in our benchmark [26]. Please note that in Eq. (2) for deletion and insertion λ is always zero.

HiTED can be used to compare different hishapes from the same sequence as well as from different sequences.

Energy Barrier Estimation RNA structures are dynamic molecules where base pairs open and close to finally reach thermodynamic equilibrium. It is therefore straightforward to model the distance of two conformations as the energy needed to convert one into the other. For this conversion, base pairs have to be opened and closed, for which energy is needed. The amount of energy needed for this conversion constitutes the energy barrier between the two conformations and it can be estimated from the highest energy structure of the conversion path. The problem is that there are exponentially many possible conversion paths, but we are interested in the one with the minimum energy barrier. An exhaustive approach is available with `barriers` [28] but due to the exponential complexity only applicable to sequences up to 100 nt. For this reason the authors also provide a heuristic method which is based on a breadth-first search (BFS) [29]. We extended this idea by using hishapes as guides for the BFS-based computation. In order to do this efficiently, we introduced the notion of related hishapes that are defined in [26] as follows: Given two hishapes H_1 and H_2 and a function ϕ that extracts hairpin loop helix indices, related hishapes H_r are those that satisfy $\phi(H_r) \subseteq \phi(H_1) \cap \phi(H_2)$. In words, related hishapes are all hishapes that contain at least one of the hairpins present in H_1 or H_2. The computation of related hishapes can be implemented using cDP. The key is to use a modified classifying algebra $\mathcal{A}^{h}_{\{h,h+,m,a\}}$ where the choice function h removes intermediate solutions without hairpin helices of interest. It is important to note that this does not overcome the exponential complexity of hishape computation, but results in a dramatic speed-up in practice. The full algorithm, called HiPath, takes two structures (from the same sequence), converts them into hishapes, computes their related hishapes and for the latter computes all pairwise folding pathways based on BFS. The corresponding energy barriers are stored in a matrix. Then Dijkstra's algorithm [30] is used to find the shortest (least free energy consuming) path between the start and the target structure.

Benchmarks against barriers and BFS showed that in terms of computation time, HiPath is faster but less accurate than barriers and slower but more accurate than BFS. In summary, HiPath represents a good trade-off between speed and accuracy and makes it possible to compute energy barriers for structures from 500 nt long sequences within a few minutes.

Folding Kinetics In order to study the folding kinetics of an RNA, we need to know all possible states (structures) and their transition probabilities, which can be derived from energy barriers via the Arrhenius equation. With RNAsubopt [31] and barriers a combination of tools exist that compute these, but again suffer from the exponential growth. By combining hishape analysis with the fast energy barrier estimation of HiPath, we reduce the computational burden, such that the resulting tool HiKinetics [32] can be used to study the folding kinetics of moderately sized RNAs.

We exemplify the application of HiKinetics with an artificial bistable sequence [33] and the results are shown in Fig. 10. Starting from the open chain (light blue), the second (blue) and third (green) best hishapes are populated first. Especially the blue one dominates the folding space before the mfe hishape (red) finally takes over. At the end of the simulation, thermodynamic equilibrium is reached and the population densities more or less perfectly

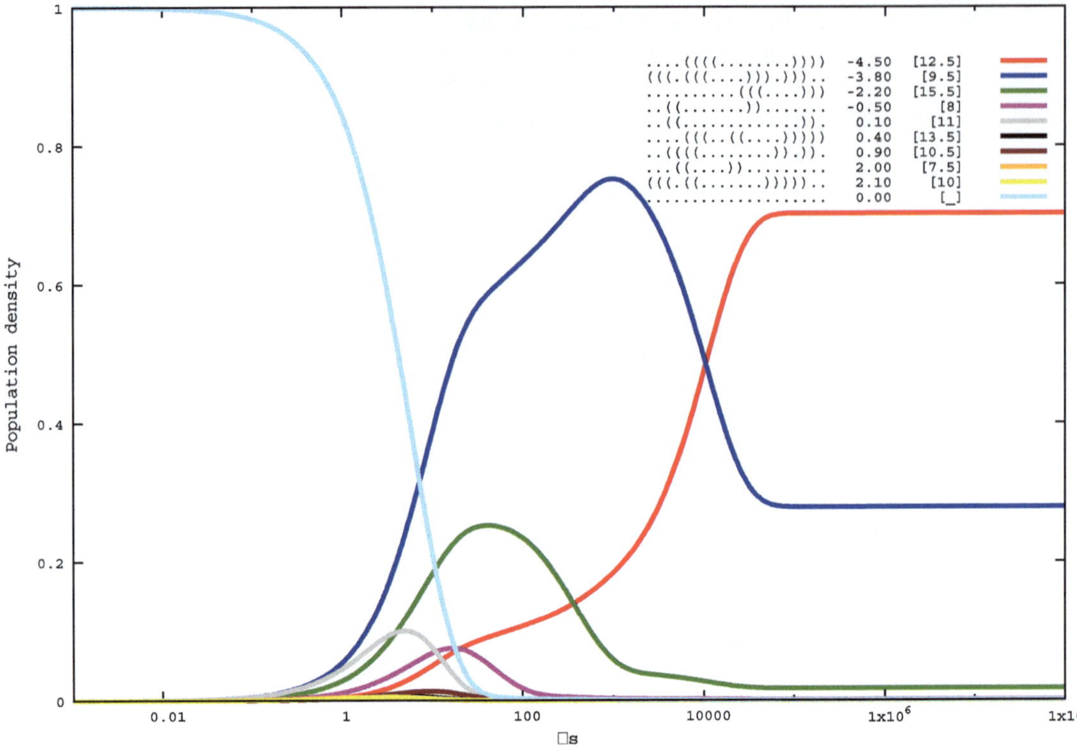

Fig. 10 Folding kinetics of the artificial "xbix" sequence CUGCGGCUUUGGCUCUAGCC simulated with HiKinetics for the 10 best level m hishapes

agree with the hishape probabilities, which are 0.703630 for hishape [12.5], 0.276945 for hishape [9.5] and 0.017032 for hishape [15.5]. Simulated folding kinetics for real-world example are shown and described in the original publication [32] to which we refer the interested reader.

References

1. McCaskill JS (1990) The equilibrium partition function and base pair binding probabilities for RNA secondary structure. Biopolymers 29: 1105–1119
2. Hofacker IL et al. (1994) Fast folding and comparison of RNA secondary structures. Monatshefte Für Chem Chem Mon 125:167–188
3. Giegerich R, Voß B, Rehmsmeier M (2004) Abstract shapes of RNA. Nucl Acids Res 32: 4843–4851
4. Giegerich R (2000) A systematic approach to dynamic programming in bioinformatics. Bioinformatics 16:665–677
5. Giegerich R, Meyer C, Steffen P (2004) A discipline of dynamic programming over sequence data. Sci Comput Programm 51: 215–263
6. Sauthoff G, Janssen S, Giegerich R (2011) Bellman's GAP: a declarative language for dynamic programming. In: Proceedings of the 13th International ACM SIGPLAN Symposium on Principles and Practices of Declarative Programming. ACM, New York (2011), pp 29–40
7. Janssen S, Giegerich R (2015) The RNA shapes studio. Bioinformatics 31:423–425

8. Bellman R (1952) On the theory of dynamic programming. Proc Natl Acad Sci U S A 38: 716–719
9. Bellman R (1954) The theory of dynamic programming. Bull Amer Math Soc 60:503–515
10. Bellman R (1957) Dynamic Programming, 1st edn. Princeton University Press, Princeton
11. Steffen P, Giegerich R (2005) Versatile and declarative dynamic programming using pair algebras. BMC Bioinformat 6:224
12. Janssen S, Schudoma C, Steger G, Giegerich R, Lost in folding space? comparing four variants of the thermodynamic model for RNA secondary structure prediction. BMC Bioinformat 12: 429
13. Waterman MS (1995) Introduction to Computational Biology: Maps, Sequences and Genomes. CRC Press, Boca Raton
14. Lorenz W, Ponty Y, Clote P (2008) Asymptotics of RNA shapes. J Comput Biol 15:31–63
15. Nebel ME, Scheid A (2009) On quantitative effects of RNA shape abstraction. Theory Biosci 128:211
16. Voß B, Giegerich R, Rehmsmeier M (2006) Complete probabilistic analysis of RNA shapes. BMC Biol 4:5
17. Ding Y, Lawrence CE (2003) A statistical sampling algorithm for RNA secondary structure prediction. Nucl Acids Res 31:7280–7301
18. Janssen S, Giegerich R (2010) Faster computation of exact RNA shape probabilities. Bioinformatics 26:632–639
19. Sankoff D (1985) Simultaneous solution of the RNA folding, alignment and protosequence problems. SIAM J Appl Math 45:810–825
20. Mathews DH, Turner DH (2002) Dynalign: an algorithm for finding the secondary structure common to two RNA sequences. J Mol Biol 317:191–203
21. Havgaard JH, Lyngso RB, Gorodkin J (2005) The FOLDALIGN web server for pair-wise structural RNA alignment and mutual motif search. Nucl Acids Res 33(Suppl_2): W650–653
22. Will S, Otto C, Miladi M, Möhl M, Backofen R (2015) SPARSE: quadratic time simultaneous alignment and folding of RNAs without sequence-based heuristics. Bioinformatics btv185:2489–2496
23. Hofacker IL, Fekete M, Stadler PF (2002) Secondary structure prediction for aligned RNA sequences. J Mol Biol 319:1059–1066
24. Voß B (2006) Structural analysis of aligned RNAs. Nucl Acids Res 34:5471–5481
25. Reeder J, Giegerich R (2005) Consensus shapes: an alternative to the Sankoff algorithm for RNA consensus structure prediction. Bioinformatics 21:3516–3523
26. Huang J, Backofen R, Voß B (2012) Abstract folding space analysis based on helices. RNA 18:2135–2147
27. Shapiro BA, Zhang KZ (1990) Comparing multiple RNA secondary structures using tree comparisons. Comput Appl Biosci 6:309–318
28. Flamm C, Fontana W, Hofacker IL, Schuster P (2000) RNA folding at elementary step resolution. RNA. 6:325–338
29. Flamm C, Hofacker IL, Maurer-Stroh S, Stadler PF, Zehl M (2001) Design of multistable RNA molecules. RNA 7:254–265
30. Dijkstra EW (1959) A note on two problems in connexion with graphs. Numer Math 1:269–271
31. Wuchty S, Fontana W, Hofacker IL, Schuster P (1999) Complete suboptimal folding of RNA and the stability of secondary structures. Biopolymers 49:145–165
32. Huang J, Voß B (2014) Analysing RNA-kinetics based on folding space abstraction. BMC Bioinformat 15:60
33. Wolfinger MT, Svrcek-Seiler WA, Flamm C, Hofacker IL, Stadler PF (2004) Efficient computation of RNA folding dynamics. J Phys A Math Gen 37:4731

Chapter 7

Classification and Identification of Non-canonical Base Pairs and Structural Motifs

Roman Sarrazin-Gendron, Jérôme Waldispühl, and Vladimir Reinharz

Abstract

The 3D structures of many ribonucleic acid (RNA) loops are characterized by highly organized networks of non-canonical interactions. Multiple computational methods have been developed to annotate structures with those interactions or automatically identify recurrent interaction networks. By contrast, the reverse problem that aims to retrieve the geometry of a look from its sequence or ensemble of interactions remains much less explored. In this chapter, we will describe how to retrieve and build families of conserved structural motifs using their underlying network of non-canonical interactions. Then, we will show how to assign sequence alignments to those families and use the software `BayesPairing` to build statistical models of structural motifs with their associated sequence alignments. From this model, we will apply `BayesPairing` to identify in new sequences regions where those loop geometries can occur.

Key words RNA, Structure, Non-canonical base pairs, Motif, Module, Graph, Bayesian net

1 Introduction

RNA achieve their functions through complex geometries, which are obtained through hierarchical folding [2]. Initially, the bulk of the secondary structure is assembled through the stacking of energetically favorable canonical and Wobble Watson-Crick/Watson-Crick base pairs. Unconstrained regions of the secondary structure form the loops and often contain organized networks of non-canonical base pairs [11], which create a scaffold for complex loop geometries [12]. Non-canonical base pairs can involve any pair of nucleotides in 12 different configurations. The Leontis-Westhof representation, shown in Fig. 1, is a convenient way to classify those different interactions, independently from their composition.

To understand the rules governing those networks of interactions, high-quality 3D structures are needed. Although the resolution and number grew exponentially in the last 10 years (Fig. 2 shows the number of structures available in the Protein Data Bank

Ronny Lorenz (ed.), *RNA Folding: Methods and Protocols*, Methods in Molecular Biology, vol. 2726,
https://doi.org/10.1007/978-1-0716-3519-3_7,

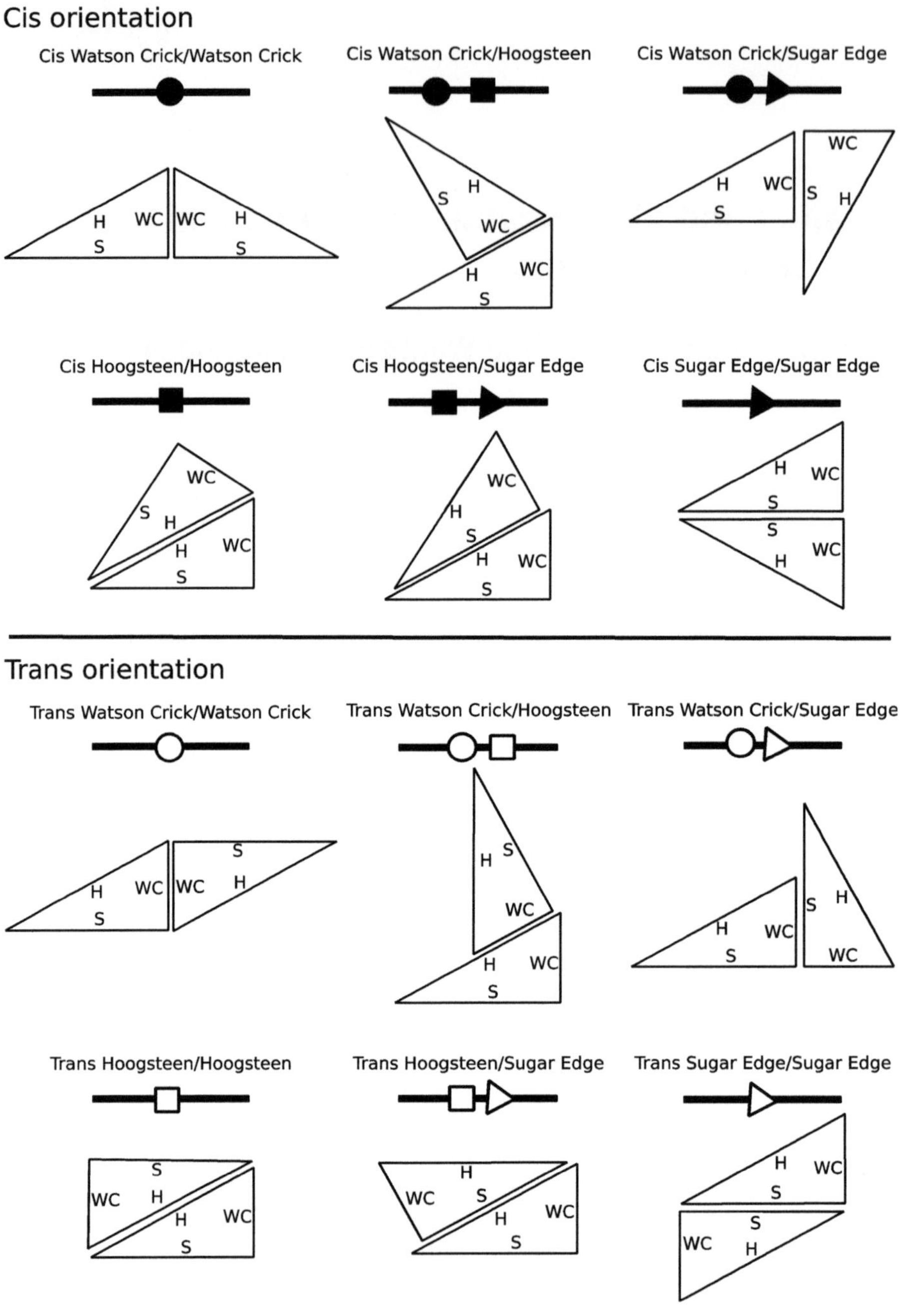

Fig. 1 Non-canonical base pairs. The 12 different geometries and their Leontis-Westhof annotation. Each nucleotide is represented as a right-angled triangle, each side uniquely determined. An interaction can occur between any two sides of the triangles

(wwPDB) [1]), they are still outpaced by the amount of sequence data generated by NGS methods. For instance, as of 2019, the European Nucleotide Archive contains over 2.2 billion sequences with a total of more than 5.5 trillion nucleotides.

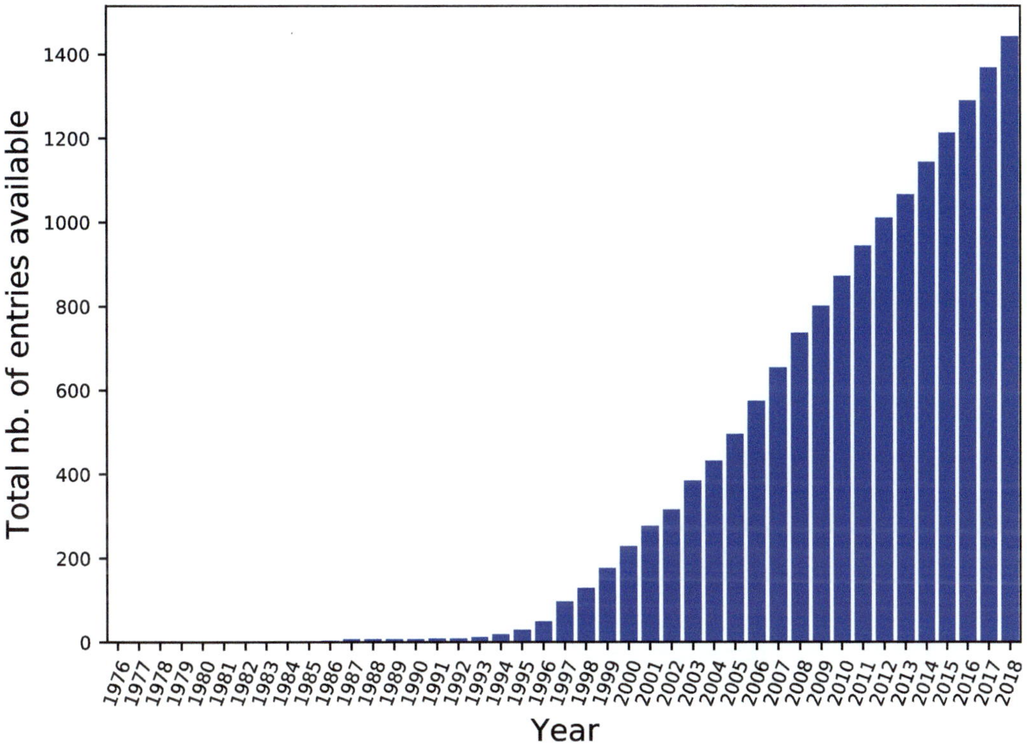

Fig. 2 Increase in the amount of RNA structures available in the PDB database through the years

To translate full-atom structures to their Leontis-Westhof representations, we need methods that automatically compute the angles and contact points between nucleotides. Many tools such as `FR3D` [19], `MC-Annotate` [7], `RNAview` [25], or `ClaRNA` [24] have been created to that effect. Conveniently, the RNA Structure Atlas[1] [16] is an up-to-date database annotating all interactions in all structures in the `wwPDB`. This particular resource uses the annotations produced by `FR3D`.

With the availability of sets of annotated structures arise several questions. Do these structures share similar patterns? How should we classify them? Have structural patterns been conserved through evolution? Are these patterns associated with important functions? For instance, we know that specific tertiary structure conformations are necessary for riboswitches to bind to their ligands [21]. Conversely, important structural motifs such as the kink-turn motif cannot be obtained through classical canonical base pairs and instead require non-canonical interactions [10].

The automatic classification of locally conserved substructures is now conducted through different computational methods like `RNA 3D Motif Atlas` [16], `RNA Bricks` [3], or `Rna3Dmotif`

[1] http://rna.bgsu.edu/rna3dhub/pdb

[6]. However, these methods only consider structural modules that are present inside one element of the secondary structure (i.e., hairpins, interior loops, or multi-loops). Yet, many networks of non-canonical interactions are used to shape the global conformation of the structure by bridging distant secondary structure elements. Several of these long-range motifs have already been identified by the community. A classic example is the A-minor motif [15]. `caRNAval` is the first automatic method for the discovery and annotation of such *long-range* non-canonical networks of interactions [17], and helped to reveal new patterns.

Although the classification and analysis of these structural sub-motifs is far from being completed, the ever-increasing amount of 3D structures combined with the development of new algorithms allows us to progressively unravel the complex ensemble of conformations that permit RNA to achieve its many functions. The prediction of the geometries of those sequences is one of the important challenges in computational molecular biology. Hence, the identification of known substructures in new sequences is a milestone for which the first algorithm designed for this purpose is `RMDetect` [5]. Here, we provide a practical guide to `BayesPairing` [18], a method that expands `RMDetect` through many algorithmic improvements. We will show how we can use pre-built database of motifs, generate new ones, and predict in novel sequences the location or lack thereof of those motifs.

2 Materials

Full RNA structures and associated sequences.

The Protein Data Bank (`wwPDB`) [1] is a repository of experimentally determined three-dimensional structures of proteins, nucleic acids, and complex assemblies. The Nucleic Acid Database [4] is a specialized, RNA-focused version of that archive.[2] RNA 3D structure files are generally found in two formats. For decades, the only commonly used format was the Protein Data Bank (`.pdb`) format, which contains the spatial coordinates of every atom in the molecule, as well as complementary information about the molecule and the experimental process, in text format. One of its main limitations was the maximum number of 99,999 atoms that can be present in one file. Since 2014, the standard archive model is now the macromolecular Crystallographic Information File (`.mmCIF`) that supersedes the PDB format and removes its previous limits. This was needed to accommodate the growing amount of large complexes for which we have 3D structures. Both formats, as well

[2] http://ndbserver.rutgers.edu

as a FASTA file containing plain sequences, can be obtained from either database.

In this chapter, we illustrate our methods on the sequence and experimentally determined structures of the thiamine pyrophosphate (TPP)-specific riboswitch. The *Arabidopsis thaliana* structure is found under the PDB code `3D2V` [23]. The *Escherichia coli* structure is found under PDB code `2GDI` [21].

Structural motif prediction.

The structural motif prediction tools presented in this chapter leverage non-canonical base pair annotations of three-dimensional structures. `FR3D` [19][3] is commonly used to produce such annotations. All RNA 3D structures in the PDB database are annotated with FR3D and updated on a weekly basis by RNA3DHub.[4]

`Rna3Dmotif` [6][5] identifies local structural 3D motifs in PDB files. All motifs found in the PDB database as of February 2018 can be downloaded directly from the `BayesPairing` website.[6]

`caRNAval` [17][7] is a collection of recurrent interaction networks containing long-range interactions within RNA 3D structures.

RNA family multiple sequence alignments.

`Rfam` 14.1 [9][8] is a collection of functional RNA families. Each family is represented by a multiple sequence alignment (in Stockholm or FASTA format), a consensus secondary structure, and a covariance model.

Some experimentally determined three-dimensional structures belong to known RNA families. At positions for which we know a structural motif exists, we can generate an ensemble of suitable sequences for that motif looking at the family. Basic mappings between a family alignment and 3D structures are made available by `Rfam`.

Sometimes, it is convenient to align sequences to multiple sequence alignments. If an RFAM family is known, `Infernal` is often the best option [14]. A new alignment can also be done with `Clustal Omega` [22]. This tool is freely available as a downloadable software[9] or as a web server.[10]

[3] https://www.bgsu.edu/research/rna/help/rna-3d-hub-help/annotating-pdb-files-with-fr3d.html
[4] http://rna.bgsu.edu/rna3dhub/pdb
[5] https://rna3dmotif.lri.fr/
[6] http://bayespairing.cs.mcgill.ca/resources
[7] http://carnaval.lri.fr/
[8] https://rfam.xfam.org/
[9] http://www.clustal.org/omega/
[10] https://www.ebi.ac.uk/Tools/msa/clustalo/

RNA secondary structure prediction tools.

The identification of 3D motifs in sequence requires a known or predicted secondary structure prediction. These predictions are performed with the RNAfold program, available as part of the ViennaRNA package [8, 13]. The ViennaRNA package[11] is frequently updated. The experiments presented in this chapter were executed with version 2.4.12. A web server is also available.

RNA module training prediction from sequence.

We perform the identification of RNA 3D motifs in sequence with the `BayesPairing` software [18], available as a downloadable software and a web server.[12]

To install `BayesPairing`, a computer with Python 3 is required as a text editor. The source code must be downloaded as a zip file. Unzip this folder, navigate to the root of the folder, and execute the `setup.py` python script for installing python dependencies. The software is then ready to run. More ample details are available on the `README` file.

Note: a different version of the software, `BayesPairing2`, is also available for specifically identifying local modules. It works similarly to the version presented in this chapter; see more in the `README` file.

3 Methods

In this section, we describe in detail a typical pipeline for the identification of structural motifs in sequences. We will describe how to use various software tools to find structural motifs in a RNA structure, map structure positions to multiple sequence alignments, and then train a `BayesPairing` motif model from this data.

We will illustrate the use of the tools presented in this chapter through four code examples. Those code examples are in the python language and should be executed from the **test** directory of the `BayesPairing` python package.

3.1 Non-canonical Base Pairs

The Leontis-Westhof base pairing nomenclature, shown in Fig. 1, represents nucleotides as right-angled triangles. The side names are as follows: Hoogsteen edge (H) for the hypotenuse, Watson-Crick edge (W) for the adjacent side, and Sugar edge (S) for the opposite side.

Additionally pairs can interact in cis (c) or trans (t) orientation of the glycosidic bond, characterizing whether or not they are oriented in the same direction.

[11] https://www.tbi.univie.ac.at/RNA/

[12] http://bayespairing.cs.mcgill.ca/

An interaction can therefore be described as a triplet in $\{c|w\} \times \{W|H|S\} \times \{W|H|S\}$. The order matters; a *cWH* is different from a *cHW*.

The canonical and Wobble base pairs that compose the backbone of an RNA structure are specific. A canonical base pair is an `A–U` or `C–G` in the cWW orientation, while a Wobble base pair is a `U–G` in cWW orientation.

The Leontis-Westhof classification retains sufficient information to reconstruct full 3D structures and is coarse enough to simplify comparisons between similar base pairs. Consequently we look at those annotation to annotate recurrent base pairing patterns between RNA structures.

3.1.1 Annotating Non-canonical Base Pairs in the TPP Riboswitch Structure with `FR3D`

The annotation of non-canonical base pairs in a PDB file can be done with several software tools like `FR3D`, `RNAview`, `MC-Annotate`, and `ClaRNA`. However, for the large majority of common tasks, no software is necessary. Precomputed `FR3D` annotations are available for all publicly available RNA 3D structures on the RNA Structure Atlas[13] which is updated weekly. On that website from a PDB code, we obtain two main pieces of information.

First, a customizable, interactive graph of interactions between nucleotides allows the user to look for specific types of non-canonical interactions and to observe the interaction patterns in the context of the secondary structure. In this graph, displayed in Fig. 3, it can be observed that a lot of the non-canonical base pairs revolve around a multibranched loop in the center of the structure.

Also provided is the complete list of interactions in text form, including base pairs, but also stackings, and base-ribose and base-phosphate interactions. A sample of the interaction list is shown in Fig. 4.

The left and right identifiers represent each partner of the interaction. Most of the time, they have five entries characterizing a base, separated by pipes. In order, these entries correspond to the PDB code, the PDB model number, the chain identifier, the nucleotide, and the sequence position of the base. The center identifier contains the interaction type, in a code of three to five letters (*see* **Note 1**). The Leontis-Westhof interactions are encoded with three letters, the first one representing the orientation (c for cis, t for trans) and the latter two the respective edges of the left and right base. For instance, the residues 8 and 72 share a cis Sugar-Sugar (cSS) interaction. When two nucleotides are very close to the cutoff for the annotation of an interaction, the letter n is added at the start. Hence, two nucleotides that barely qualify to be considered to form a trans Sugar-Hoogsteen base pair would be annotated as

[13] http://rna.bgsu.edu/rna3dhub/pdb

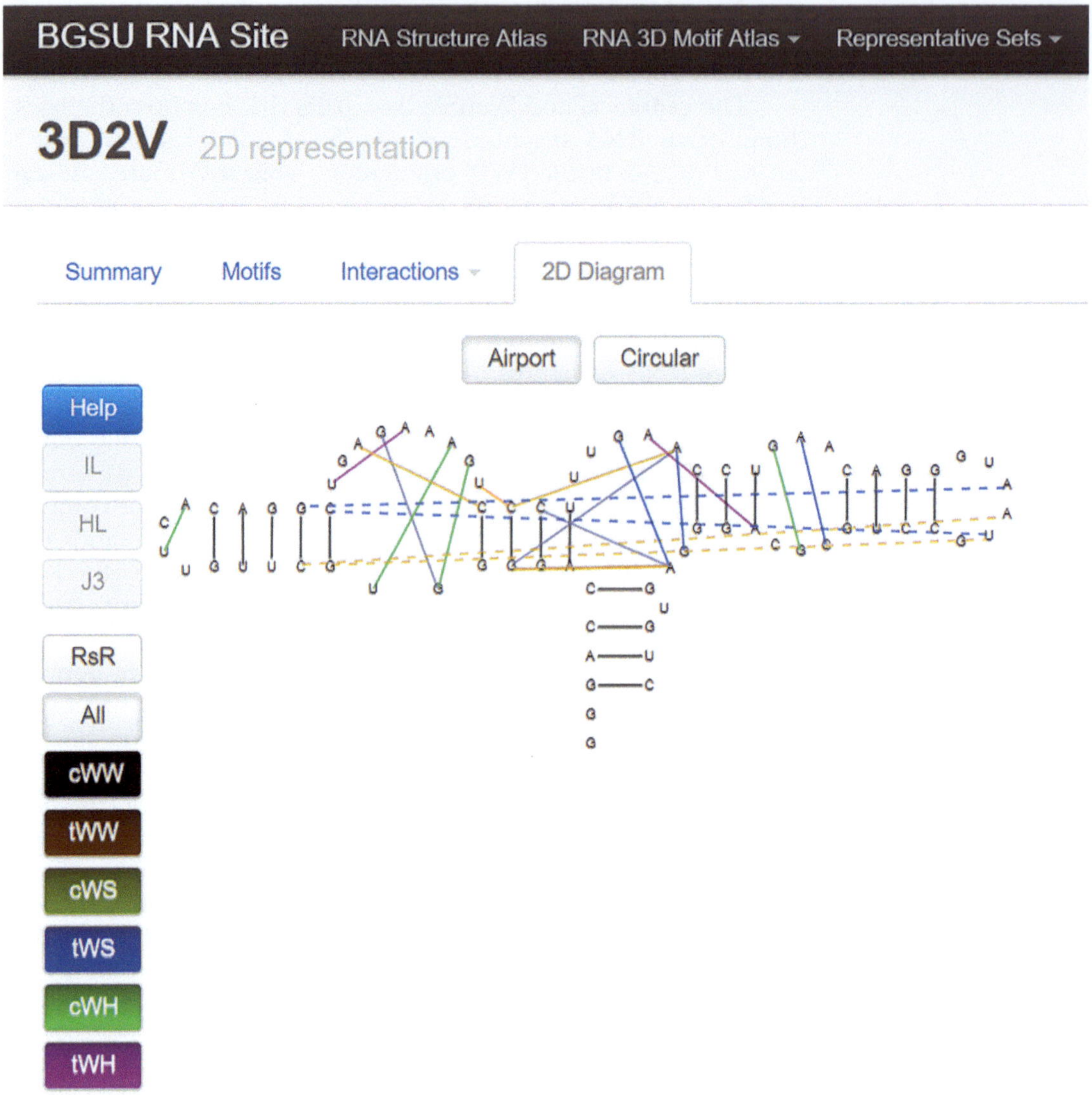

Fig. 3 The FR3D base pair graph view of PDB 3D2V, a TPP riboswitch of *Arabidopsis thaliana*. We can notice how the center of the structure is a multi-loop with many non-canonical interactions

3D2V|1|A|A|7	cWW	3D2V|1|A|U|39
3D2V|1|A|G|8	s53	3D2V|1|A|A|7
3D2V|1|A|G|8	cWW	3D2V|1|A|C|38
3D2V|1|A|G|8	s55	3D2V|1|A|U|39
3D2V|1|A|G|8	cSS	3D2V|1|A|A|72

Fig. 4 Subset of FR3D annotations of the 3D2V PDB structure of the TPP riboswitch

ntSH. It should be noted that, at the tertiary structure level, nucleotides can have many interactions. Indeed, in this short sample of interactions, the nucleotide at position 8 is involved in 4 interactions, including 2 base pairs.

Other notable interactions include base stackings, noted as s followed by two numbers, corresponding to the respective sides of the bases that are interacting (5' or 3'). We can observe that nucleotide 8 is part of two base stackings of different types, with nucleotides 7 and 39.

The patterns of base pair and base stacking interactions constitute the structural motifs described in this chapter.

3.2 Structural Motifs in RNA

As an RNA folds into its secondary structure, loops of unpaired nucleotides form. Those loops are then stabilized by non-canonical interactions within the loop, as well as with other loops. Interactions between bases of the same loop are the most common as they shape the local three-dimensional structure.

While non-canonical interaction patterns within a loop impact the final structure of the molecule, they are not always directly relevant to the function.

There are two main ways to characterize the functional relevance of some network of interactions. It can be assessed experimentally, namely, through NMR, that the loop of interest forms a key interaction with another loop or molecule. Experimentally establishing that some part of a biomolecule's structure is key to its function would warrant searching other molecules for similar substructures.

Unfortunately, such experiments can be very expensive in time and resources, and this type of data is not available for a majority of known RNA structures. In this case, functional relevance needs to be inferred. We rely on the assumption that a specific network of non-canonical base pairs is very unlikely to occur by chance in distinct RNA structures. Thus, networks of interactions that are recurrent through multiple RNA structures are hypothesized to be functionally meaningful.

Several definitions are used to classify those recurrent substructures. RNA modules were defined in seminal work `RMDetect` as an *ensemble of non-canonical base pairs embedded between canonical base pairs*. Those RNA modules, which include sequence and base pair information, are separated in two main classes:

- Local modules, in which all the interactions are located within a single loop
- Interaction modules, which connect nucleotides from two or more loops

Additionally, `caRNAval` defines a recurrent interaction network (RIN) as a network of non-canonical interactions, with complete abstraction of the sequence. Moreover, in the context of `caRNAval`, all RINs are required to include at least one interaction between two loops, called long-range interaction. Thus, a RIN can represent the structure of several modules, each having the same interactions but with different sequences.

In this chapter, we will show (1) how to identify local modules, (2) how to identify interaction modules, (3) map the modules to a family alignment, (4) build a model of the modules with `BayesPairing`, and (5) search for this module in new sequences.

3.3 Identifying Local Structural Motifs in the TPP Riboswitch Structure with `Rna3Dmotif`

The local structural motifs in some RNA structures can be thought of as the behavior of each of their secondary structure loops at the tertiary structure level. In this section we show how to identify local structural motifs with `Rna3Dmotif` and save them in a format that can be used by `BayesPairing` to build models and identify them in sequences (*see* Subheadings 3.6 and 3.7).

1. First, we will select a software for local motif identification. In this chapter, we will use `Rna3Dmotif`, but other options are available (*see* **Note 2**).
2. We download the full `Rna3Dmotif` annotations, as well as motif graph collection assembled on the `BayesPairing` website, at http://bayespairing.cs.mcgill.ca/resources (*see* **Note 3**).
3. If applicable, we choose a number of strands for our motif of interest. In this case, we will be looking for a three-way junction (*see* **Note 4**).
4. `BayesPairing` provides a python function to find a three-way junction motif in some PDB structures. We can use it to search for motifs with three strands on graphs associated with PDB code `2GDI` as indicated in Code example 7.1. The snippet of code will print out the name of the `Rna3Dmotif` DESC files containing a motif matching the requirements.
5. In this case, the TPP riboswitch only contains one junction, so a single file name is returned: `2GDI.X.1`. This motif is also returned as a graph stored in the `graphs` variable, in which the nucleotides are the nodes, and the interactions, the edges.
6. From the full annotations folder (obtained from **step 2**), we can fetch the file `2GDI.X.1.DESC`. Its contents are shown in Fig. 5 (*see* **Note 5**).
7. This file, which contains the necessary structural information to build a model for this 3D motif, will be used as an input for `BayesPairing` in Subheading 3.6 (Fig. 6).

```
id: 1
Bases: 14_U  15_C  51_G  52_U  53_A  54_U  55_C  56_A
57_C  82_G  83_G  84_A  85_A
(   82_G )--- C/C - --- (   83_G )
(   57_C )--- 5/5 s --- (   83_G )
(   57_C )--- +/+ c --- (   82_G )
(   56_A )--- H/S t --- (   83_G )
(   56_A )--- C/C - --- (   57_C )
(   55_C )--- C/C - --- (   56_A )
(   56_A )--- 5/5 s --- (   84_A )
(   83_G )--- C/C - --- (   84_A )
(   54_U )--- C/C - --- (   55_C )
(   53_A )--- H/h t --- (   84_A )
(   53_A )--- C/C - --- (   54_U )
(   52_U )--- C/C - --- (   53_A )
(   51_G )--- C/C - --- (   52_U )
(   84_A )--- C/C - --- (   85_A )
(   14_U )--- -/- c --- (   85_A )
(   15_C )--- +/+ c --- (   51_G )
(   14_U )--- C/C - --- (   15_C )
```

Fig. 5 Rna3Dmotif annotation of TPP riboswitch three-way junction

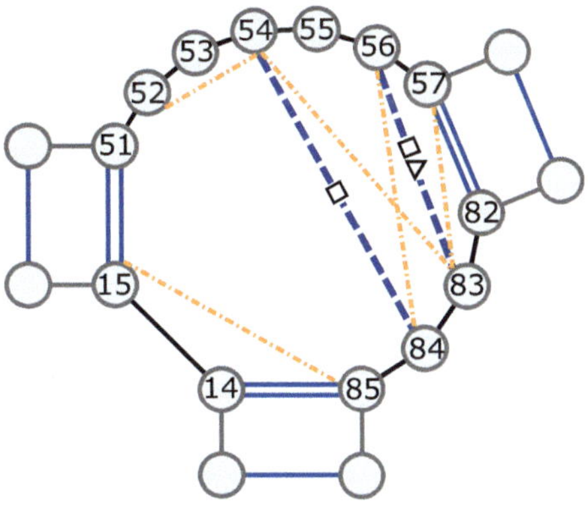

Fig. 6 The graph representation of the three-way junction found in TPP riboswitch structure 3D2V, generated from Rna3Dmotif annotations. The nodes represent the sequence positions. Thin black edges are backbone interactions, and blue full single or double lines are canonical base pairs. Blue dashed lines represent non-canonical base pairs and yellow dashed lines stackings. The first base pair extending into each helix is represented for context, but is not part of the motif

Code example 7.1 Parsing the Rna3Dmotif dataset for local motifs

```
from bayespairing . src . find_rna3dmotif_modules\
    import parse_dataset_for_modules as find
import pickle
from pathlib import Path
```

```
path_models = Path( ' . . / models ' )
path_rna3dmotif_graphs = path_models / ' all_rna3dmotif_graphs . cPickle '

dataset = pickle . load ( open ( path_rna3dmotif_graphs , ' rb ' ))
PDBs = [ "2GDI" ]
graphs , ids = find ( dataset , PDBs, n_strands=3)
print ( ids )
```

3.4 Identifying Interaction Motifs in the TPP Riboswitch with caRNAval

In the previous section we identified a local motif from Rna3Dmotif.

We will now show how to obtain a recurrent interaction network model identified by caRNAval in the TPP riboswitch. Like the local motif, it will be used to build BayesPairing models in Subheadings 3.6 and 3.7.

1. We search for TPP riboswitch on the RCSB PDB database to find PDB IDs matching this structure (*see* **Note 6**). Two bound structures are found with PDB identifiers 2GDI and 3D2V. The former is bacterial and the latter eukaryotic.
2. We will now identify recurrent interaction networks in the TPP riboswitch. For it to be relevant to its function, we would ideally find it in both known structures (*see* **Note 7**). We can take advantage of the Filter by PDB option on the caRNAval website to search for 2GDI,3D2V.
3. The search will redirect us to a new version of the front page. Clicking *All RINs* will show all RINs that appear in both searched PDB codes. Clicking on the RINs will show their profile page (*see* **Note 8**) (Fig. 7).
4. Select *All occurrences detailed information* to show the information needed to construct the motif (*see* Fig. 8, **Note 9**). The output of this page should be saved in a text file; we will call it carnaval_output.FASTA.
5. In order to be used later to train a BayesPairing model, a python networkx DirectedGraph object is required. This can be done with the file previously created carnaval_output.FASTA. Code example 7.2 shows how to build the graph using the get_carnaval_graph function of BayesPairing (*see* **Note 10**). This graph is saved in a file called carnaval_tpp.pickle.
6. We now have a graph object representing a recurrent interaction network that appears in both known occurrences of the bound TPP riboswitch. It will be used as input to BayesPairing in Subheading 3.6.

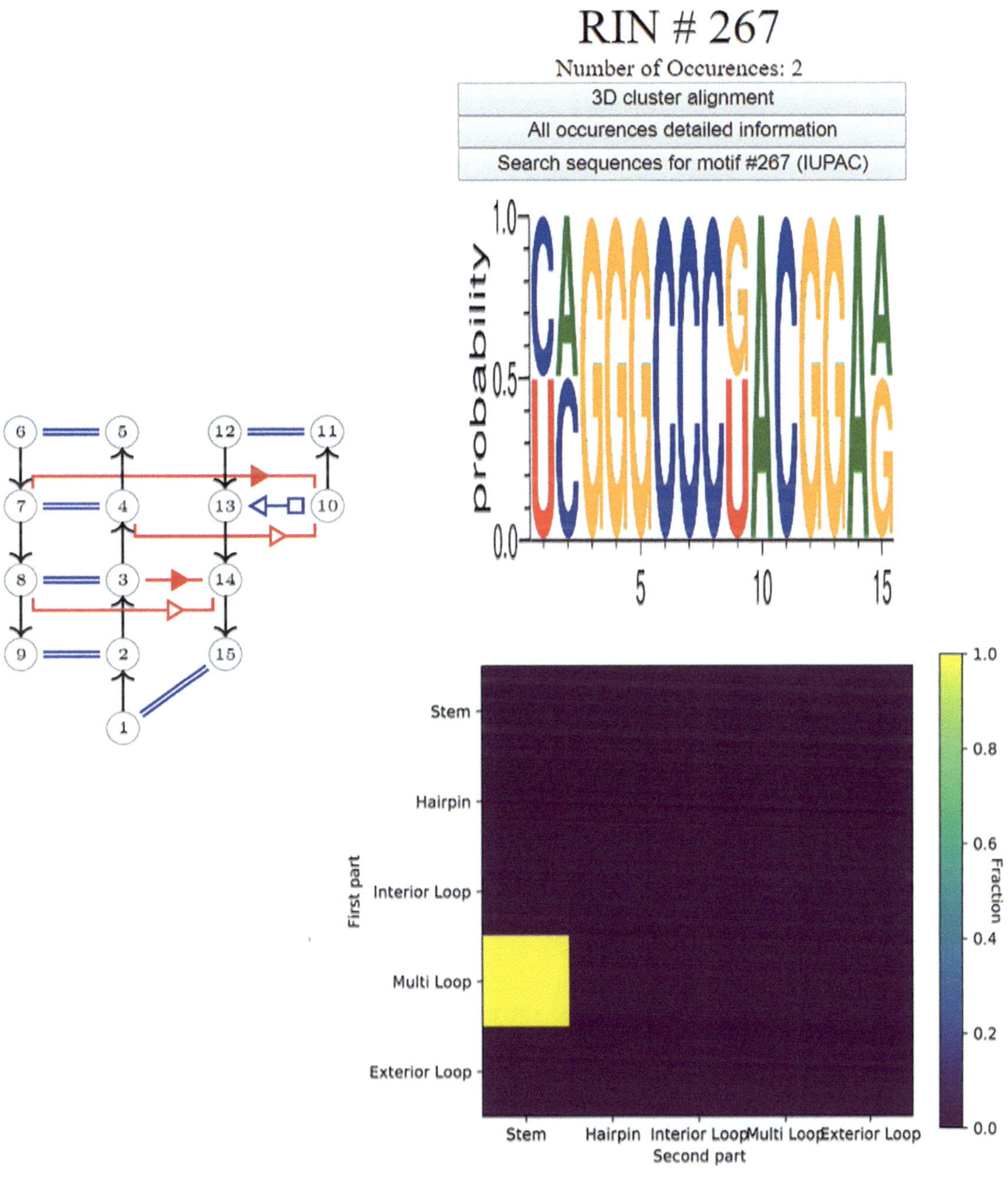

Fig. 7 The `caRNAval` information page of RIN #257 which belongs to structure 3D2V

Code example 7.2 Generating a networkx graph from a caRNAval description

```
import  pickle
from  bayespairing . src . find_rna3dmotif_modules import get_carnaval_graph

g  =  get_carnaval_graph ( "carnaval_output . txt ")
pickle .dump(g , open ( "carnaval_tpp . pickle " , "wb" ))
```

```
RNA: 3D2V, Chain: A
Structure of the eukaryotic tpp-specific riboswitch bound to /
        the antibacterial compound pyrithiamine pyrophosphate
+++Long Range Interactions+++
8 G CSS 72 A
9 G TSS 44 A
37 C CSS 44 A
38 C TSS 72 A
44 A TSS 9 G
44 A CSS 37 C
72 A CSS 8 G
72 A TSS 38 C
+++Local Motif: Stem+++
8 G B53 9 G
8 G CWW 38 C
9 G B53 10 G
9 G CWW 37 C
10 G CWW 36 C
36 C B53 37 C
37 C B53 38 C
+++Local Motif: Multi Loop+++
6 C B53 7 A
6 C CWW 73 G
7 A CWW 39 U
44 A B53 45 C
44 A THS 71 G
45 C CWW 70 G
70 G B53 71 G
71 G B53 72 A
72 A B53 73 G
```

Fig. 8 caRNAval annotation of a recurrent interaction network in the TPP riboswitch, the one shown in Fig. 7

3.5 Mapping Structural Motifs to Rfam Family Alignments

A large number of sequences are necessary to create a statistical model of a structural motif. Since motifs come from 3D structures, and there are a few of them, it is common for the number of sequences to not be sufficient.

In that case, we can often leverage the alignments of RNA families provided by Rfam. Those families are RNAs that fulfill the same function, with a similar 3D structure, and share a common ancestor.

We assume that a functionally significant structural motif present in one member of a family will be found in the others. Therefore, looking at the positions in the alignment where our structural motif of interest is found will give us a diversity of sequences with the same structure. In this section, we will show how to locate the two motifs we selected in the previous sections in the alignments of their respective families in the Rfam database (*see* **Note 11**).

1. To match part of a PDB structure to a family alignment, a PDB sequence is required. We can download it from the *Sequence* tab on the RCSB website. It is important to double-check that it starts at 1. If it does not, we must adjust the `first_n` variable in Code example 7.3 (*see* **Note 12**).

Code example 7.3 Mapping a `Rna3Dmotif` motif to a `Rfam` alignment

```
import bayespairing . src . map_PDB_to_Rfam as map
from bayespairing . src . find_rna3dmotif_modules \
           import parse_dataset_for_modules as find
from bayespairing . src . Rfam_to_module import rfam_to_module
import pickle

path_rna3dmotif_graphs = path_models / ' all_rna3dmotif_graphs . cPickle '
dataset = pickle . load ( open ( path_rna3dmotif_graphs , ' rb ' ))

graphs , ids = find ( dataset ,PDBs, n_strands=3)
junction_graph = graphs [ 0 ] #there is only one

f i r s t _ n = 10
PDB_pos = [ pos- f i r s t _ n +1 for pos in junction_graph . nodes ( ) ]

extra_seqs = map. PDB_to_Rfam ( [ "2GDI.Y " ] , [ PDB_pos ] )

f i l e = "rna3dmotif_tpp_seqs . fasta "
rfam_to_module ( output_name =f i l e , a s _ l i s t= True , seqs=extra_seqs )
```

2. We execute the code described in Code example 7.3. The function `PDB to Rfam` takes as input a PDB code, including a chain and a list of positions, and outputs a list of sequences. The `rfam_to_module` method then generates a FASTA file of motif sequences (*see* **Note 13**).
3. With the graph object and the FASTA files, all the required inputs to train a `BayesPairing` model have been assembled (*see* **Note 14**). Code example 7.4 contains the python code described in **steps 3–8** of this section. We start by loading the graph object in memory from its pickle file.
4. We will now apply a similar process to the `caRNAval` RIN, with one difference. Instead of only learning sequences from mapping one of the structures to the alignment, we will be mapping both structures where the RIN is observed in `caRNAval` (*see* **Note 15**). Starting with `3D2V`, we adjust the `first_n` variable to the residue number of the first PDB position in `3D2V`. The python script then uses this, along with the node information from the graph we loaded at the previous

step to build a list of PDB positions. Along with this list of positions, we send `3D2V.B` (the DESC file prefix) as input to the `PDB_to_Rfam` method, which returns a list of module sequences from `Rfam` (*see* Code example 7.4). The next steps will show how to process the other occurrence of the RIN, `2GDI`.

5. We repeat **step 5** of the previous section, by saving the output from `caRNAval` to a file, but this time for the other PDB structure, with code `2GDI`. The carnaval annotations for `2GDI` must be saved to a different text file (here `carnaval_output2.txt`).
6. We then repeat **step 4** of this section with the `2GDI` structure (*see* Code example 7.4, **Note 16**).
7. We combine the two lists to obtain extra module sequences produced from both occurrences of the RIN (*see* Code example 7.4).
8. Finally, `rfam_to_module` will produce the final FASTA file, like it did at **step 3** (*see* Code example 7.4). We now have extra sequences from alignment for both the local motif and the recurrent interaction network, and `BayesPairing` is ready to run.

Code example 7.4 Mapping a `caRNAval` RIN to a `Rfam` alignment

```
import  bayespairing . src . map_PDB_to_Rfam as map
from  bayespairing . src . find_rna3dmotif_modules \
                      import parse_dataset_for_modules as  find
from  bayespairing . src . find_rna3dmotif_modules import get_carnaval_graph
from  bayespairing . src . Rfam_to_module import rfam_to_module
import  pickle

#carnaval  3 D2V,   loaded  from  earlier
carnaval_g  =   pickle . load ( open ( "carnaval_tpp . pickle " , "rb" ))

# Step 4
f i r s t _ n   =  1
PDB_pos = [ pos- f i r s t _ n +1   for   pos   in   sorted ( carnaval_g . nodes ( ) ) ]
extra_seqs_3D2V  =  map. PDB_to_Rfam ( [ "3 D2V.B " ] , [ PDB_pos ] )

# Step 5 , with  other   carnaval   output
g  =  get_carnaval_graph ( "carnaval_output2 . txt ")

# Step 6
f i r s t _ n   =  10
PDB_pos  =   [ pos- f i r s t _ n +1   for   pos   in   sorted ( g . nodes ( ) ) ]
extra_seqs_2GDI  =  map. PDB_to_Rfam ( [ "2GDI.Y " ] , [ PDB_pos ] )
```

```
output_name = "carnaval_RIN_seqs . fasta "

# Step 7
extra_seqs = extra_seqs_2GDI + extra_seqs_3D2V

# Step 8
rfam_to_module ( output_name = output_name , a s _ l i s t= True , seqs=extra_seqs )
```

3.6 Building a Structural Motif Model with `BayesPairing`

`BayesPairing` is a software that identifies structural motifs in RNA sequences. More than 150 structural motifs are provided in a basic library that comes with `BayesPairing`. We will now show how to create a model for a new module that can be added into the `BayesPairing` library.

To create a statistical model for a 3D motif, `BayesPairing` requires two inputs: a python3 networkx `DirectedGraph` object and a FASTA file of sequences (*see* **Note 17**). This `DirectedGraph` object can be generated directly from a `Rna3Dmotif` DESC file (as shown in this section) or from `caRNAval` base pair output (*see* Subheading 3.4). The FASTA file, which is optional, contains the motif sequences obtained from alignments. We have generated these files in Subheading 3.5.

We have obtained or generated those requirements in the two previous sections and are now ready to create modules.

1. In the command line environment, we navigate to the `bayespairing/src` directory (*see* **Note 18**).
2. At this point, the four files listed in the previous two code examples should be in the `test` directory. `BayesPairing` can learn models from either a DESC file or a precomputed graph object in pickle format. We will illustrate this with the files generated in Subheadings 3.3–3.5:

   2GDI.X.1.desc, the DESC file containing the base pair annotations for the three-way junction in a TPP riboswitch 3D structure as identified by `Rna3Dmotif` (*see* Subheading 3.3).

   rna3dmotif_tpp_seqs.fasta, a FASTA file containing more sequences matching the three-way junction obtained from mapping the 3D structure to the TPP riboswitch `Rfam` family in Subheading 3.5.

   carnaval_tpp.pickle, the pickle file containing the RIN motif graph, as obtained from `caRNAval` in Subheading 3.4. This graph can also be stored as a DESC file when such file is available, as it was for the three-way junction (*see* **Note 19**).

   carnaval_RIN_seqs.fasta, the FASTA file containing the sequences obtained from mapping the RIN to `Rfam` Subheading 3.5.

3. From the **test** directory, we can execute the following two commands to create a new `BayesPairing` dataset, named TPP (*see* **Notes 20** and **21**):

```
python ../src/module_from_desc.py -g 2GDI.X.1.desc \
-seq rna3dmotif_tpp_seqs.fasta -n TPP

python ../src/module_from_desc.py -g carnaval_tpp.pickle \
-seq carnaval_RIN_seqs.fasta -n TPP
```

   This dataset will be stored in memory as a set of four files in the `models` directory. All four files will be identified by the prefix `TPP_` (*see* **Note 22**).

4. `BayesPairing` can draw the graphs of all modules within a dataset, with the following command, from the **src** directory:

```
python display_modules.py -n TPP
```

   It can be used to verify that exactly two graphs and two sequence logos have been created in the Graphs folder.

5. There should be exactly two modules in the dataset, and the graph of module 1 should have more nodes than the graph of module 0. If that is the case, we are ready to identify the modules in new sequences (*see* **Note 23**).

3.7 Identifying Structural Motifs in the TPP Riboswitch Sequences

In this section, we use `BayesPairing` with the dataset built in previously. The two main criteria in `BayesPairing` to identify the location of a module are as follows: how well a subsequence matches the module and whether the secondary structure at this position can accommodate the loop required to insert the module.

`BayesPairing` has five key parameters. It has other options which can affect results, but their default values will be appropriate for the vast majority of users. The advanced parameters are beyond the scope of this chapter, but we suggest that interested readers look at the `BayesPairing` manual where all options are described.

The five key parameters of the main `BayesPairing` python script, `parse_sequences.py`, are the following:

- -**d**: the name of the dataset of RNA modules to search for. By default, it will use a standard dataset of common modules aggregated with `Rna3Dmotif`. Here, we will set it to TPP, the name of the model we trained in Subheading 3.6.
- -**mod**: the list of modules to search for, as a sequence separated by spaces (*see* **Note 24**). Module indexes start at 0. If this argument is not included, all modules of the dataset are searched for.
- -**seq**: the sequence to parse, or the name of a FASTA file containing sequences (*see* **Note 25**). This argument is required.

- **-sw**: the weight of the secondary structure compatibility in the score evaluation. The standard value is 1. Increasing the value for this argument would increase the penalty for a slightly imperfect structure and overall decrease the score.
- **-t**: the score threshold for a potential motif location to be printed out to the user. This threshold should depend on the task and the value given to `-sw`. In practice, a score above 20 with `sw=1` is on average associated to a false discovery rate of roughly 0.05.

For this test, we can aggregate the two sequences from PDB models `2GDI` and `3D2V` in a single FASTA file named `TPPseqs.fasta`. It should be saved in the `test` directory.

We can first search for pure sequence matches on the respective sequences of `2GDI` and `3D2V`, by setting the structure weight to 0. This is done by entering the following command, from the **src** directory:

```
python parse_sequences.py -d TPP -mod 0 1 \
-sw 0 -t 25 -seq ../test/TPPseqs.fasta
```

We can observe the output of the run of `BayesPairing` on a sequence in Fig. 9. The output has one row per identified module location and four columns: the module ID, the probabilistic score associated with its identification (or how likely it is to be found), the insertion position, and the input sequence at those positions.

Notice that for both motifs in both sequence, the correct location (as identified in previous sections) receives the highest score. They are successfully identified. We also observe that the

```
RESULTS FOR ID 2GDI
=====================================================================
| MODULE     SCORE       MODULE POSITIONS            SEQUENCE         |
|  0         28.81      41-47,72-75,78-79          GUAUCAC*GGAA*UC    |
|  0         28.81      4-5,41-47,72-75            UC*GUAUCAC*GGAA    |
|  1         36.04      4-8,38-41,46-47,72-75      UCGGG*CCCG*AC*GGAA |
|  1         34.91      4-8,37-40,46-47,72-75      UCGGG*ACCC*AC*GGAA |
|  1         33.00      5-9,38-41,46-47,73-76      CGGGG*CCCG*AC*GAAG |
=====================================================================

RESULTS FOR ID 3D2V
=====================================================================
| MODULE     SCORE        MODULE POSITIONS           SEQUENCE         |
|  0         34.34      5-6,38-44,69-72            CA*UUUGAAC*GGAG    |
|  0         34.34      19-20,38-44,69-72          CA*UUUGAAC*GGAG    |
|  0         34.34      21-22,38-44,69-72          CA*UUUGAAC*GGAG    |
|  1         35.62      5-9,35-38,43-44,69-72      CAGGG*CCCU*AC*GGAG |
=====================================================================

TOTAL TIME USED: 2.335
```

Fig. 9 `BayesPairing` output on TPP sequences with structure weight equal to 0. Module 0 is the `Rna3Dmotif` junction and module 1 is the `caRNAval` RIN

```
RESULTS FOR ID 2GDI
==========================================================================
| MODULE     SCORE        MODULE POSITIONS            SEQUENCE            |
| 0          26.94      4-5,41-47,72-75           UC*GUAUCAC*GGAA         |
| 0          26.05      41-47,72-75,78-79         GUAUCAC*GGAA*UC         |
| 1          32.01      4-8,38-41,46-47,72-75     UCGGG*CCCG*AC*GGAA      |
| 1          20.00      4-8,37-40,46-47,72-75     UCGGG*ACCC*AC*GGAA      |
| 1          19.84      3-7,37-40,46-47,72-75     CUCGG*ACCC*AC*GGAA      |
==========================================================================

RESULTS FOR ID 3D2V
==========================================================================
| MODULE     SCORE         MODULE POSITIONS          SEQUENCE             |
| 0          12.15      5-6,38-44,69-72           CA*UUUGAAC*GGAG         |
| 0           5.94      21-22,38-44,69-72         CA*UUUGAAC*GGAG         |
| 0           4.84      19-20,38-44,69-72         CA*UUUGAAC*GGAG         |
| 1          14.16      5-9,35-38,43-44,69-72     CAGGG*CCCU*AC*GGAG      |
==========================================================================

TOTAL TIME USED: 2.335
```

Fig. 10 `BayesPairing` output on TPP sequences with structure weight equal to 2. Module 0 is the `Rna3Dmotif` junction and module 1 is the `caRNAval` RIN

alternate suggestions, with slightly lower or equal score, have either the exact same sequence or a close sequence. All suggested insertion sites are, in any case, in the correct vicinity as the majority of their sequence positions were correctly identified. Indeed, the erroneous strand positions are at only off by one nucleotide. We also notice strands can appear in different order in a junction. For instance, the first output for 2GDI has its shortest strand at the 3' end rather than the 5' end, but the circular order of the strands is conserved.

Nevertheless, this output is not very practical as it contains a lot of false positives which, in practice, could not be inserted correctly in the secondary structure. For instance, moving the start of the junction from the 5–6 region to the 19–20 region as suggested by the `3D2V` output is energetically highly unfavorable, even though the sequences are the same.

Repeating this experiment with a structure weight value of 2 can be done as follows, from the **src** directory:

```
python parse_sequences.py -d TPP -mod 0 1 \
-sw 2 -t 0 -seq ../test/TPPseqs.fasta
```

(also *see* **Note 26**), scores higher the secondary structure compatibility. The results are shown in Fig. 10.

We can now observe that the strong secondary structure constraints drastically reduce the overall score, especially in the `3D2V` output for module 0. This is because module 0 was trained on `2GDI` and is overfitted to that structure. However, it can still be correctly identified in its analog. Most notably, the three-way tie as the top insertion site for module 0 in the 3D2V sequence has been broken. The actual location of the junction is easily distinguished from the other sites with identical sequences. The TPP riboswitch will be used as an example.

3.8 Discussion

The methods presented in this chapter are designed to allow for the prediction of *any RNA module*. Allowing this degree of flexibility in the predicted structures is a difficult task and as such, while BayesPairing represents the state of the art for flexible module prediction, it is not perfect. There exist specialized methods that are more accurate for the prediction of specific types of modules. One of such methods is implemented in the software BayesPairing2, which identifies *secondary structure loops* in sequences. If all the RNA modules of interest happen to constitute full secondary structure loops, this software, whose installation, execution, input, and output are almost identical to those presented in this chapter, should be preferred (see Materials).

4 Notes

1. It should be noted that while this format is found in a majority of FR3D annotations, the number of entries for each nucleotide can go up to eight. Further format details and edge cases are listed on the software website.[14]
2. There are several readily available databases of known local structural motifs ready to search in new sequences, namely, the RNA 3D Motif Atlas and RNA Bricks. Those databases can be indexed by PDB ID and potential motifs of interest in some known structure can be identified that way. Rna3Dmotif provides annotation on the non-canonical interactions within each loop of a PDB structure file. It has the upside of being a downloadable software, which allows it to be used to custom or modified PDB files if needed. Additionally, in the context of searching for local motifs, it is a bit more specialized than the RNA 3D Motif Atlas, which only includes hairpins and internal loops, and RNA Bricks, which includes interaction motifs. This is why BayesPairing, which currently focuses on local motifs, uses the Rna3Dmotif format.
3. For convenience, Rna3Dmotif was executed on the full PDB database, annotating over 45,000 loops. This dataset is accessible as a repository of individual files or a collection of python graph objects, using the networkx library.

[14] https://www.bgsu.edu/research/rna/help/rna-3d-hub-help/unit-ids.html

4. In riboswitches, three-way junctions are often an interesting type of structural motif to look for, as this class of motifs is often closely linked to function [20].
5. The dedicated annotation format of `Rna3Dmotif`, the .DESC format, has a well-defined title. The title includes the PDB code and chain of the structure the motif was identified in, as well as a unique ID, separated by dots. Thus, the name of this file is `2GDI.X.1`. The first line of the file reiterates its unique ID. The second line contains the list of the bases that are part of the motif. Each base is defined as a position in the PDB structure (indexed at 1), and the nucleotide at that position. The rest of the file is comparable to `FR3D`. All interactions between bases in the motif are listed. All interactions are encoded by three characters. The + and − characters represent canonical base pairs (+ for G − − C). W, S, and H are used to describe the two edges of each non-canonical base pair following the Leontis-Westhof nomenclature. The third character indicates the orientation of the base pair (c for cis or t for trans). Backbone interactions are encoded as C/C, and base stackings are shown as they are by FR3D, with the side of each partner indicated by 3 or 5.
6. This is a quick example of a starting point for such search. Advanced search on the Nucleic Acid Database is recommended for a more specific search.
7. Base pairing and motif annotations are not perfect. It is thus possible for a RIN or module found in only some or one of the known structures of an RNA to be functionally significant. If no match is found for all occurrences of the structure, searching structures one by one is acceptable.
8. The RIN profile includes a detailed graph of the interactions. In this graph, double blue lines are base pairs, black lines with arrows are directed backbone interactions, and other lines encode non-canonical interactions in the Leontis-Westhof notation. Red lines are long-range interactions. The node numbers represent the order of the nucleotides in sequence. A sequence probability logo and a heat map indicating the type of secondary structure contexts the motif is found in are also available.

   In this case, we are looking at a part of a multi-loop interacting with a stem.
9. The information provided by `caRNAval` is quite similar to `Rna3Dmotif`, but is only limited to backbone and base pairs. The identifiers on both sides represent the respective bases of the pair, and the label in the center indicates the type of interaction, either directed backbone (`B53`) or Leontis-Westhof notation.

10. The `get_carnaval_graph` function takes as input a text file with `caRNAval` information as shown in Fig. 8 and output a `DirectedGraph` object. We can then use the `pickle` library to save it. This library is used to save python objects without converting them to text.
11. The matching works as follows: the PDB sequence is obtained from the RSCB database. This sequence is aligned to the `Rfam` family with `Clustal Omega`. The columns corresponding to the nucleotides that are part of the motif are then extracted and considered new examples of the motif.
12. This is required because we are trying to match the PDB sequence to the family alignment. However, the nucleotide numbering in `Rna3Dmotif` uses PDB ID, and PDB IDs do not always start at 1. It is important for the mapping to work that we use the exact position of the nucleotide in the PDB sequence, which is why we are subtracting the position number of the first node and then correcting for the indexing of PDBs, which starts at 1, whereas string indexing in Python starts at 0. In this case, we notice the `3D2V` sequence on the RCSB website starts at 1, so we set the first position value in consequence.
13. A FASTA file is necessary because the creation of a `BayesPairing` RNA module takes as inputs a graph and a FASTA file of sequences.
14. If no `Rfam` match is found, we can repeat those steps, but with other chains of the PDB file. When a PDB file contains identical chains, the matching method might work for only one of them. If this fails, then repeat those steps with another PDB that contains the module, if applicable. In any case, it is possible to train a model with only a few sequences, so a `Rfam` alignment is not essential. However, the predictions will not be as robust.
15. Because we are working with a RIN and not an interaction motif, there is no sequence information attached to it. A RIN can thus be decomposed into multiple interaction motifs, and there might be significant differences between the sequences of the two occurrences of the RIN. This will affect how they align to the family. In order to avoid overfitting to either occurrence, we include the sequences generated by the mapping of both. If the mapping was perfect, those sequences would be exactly the same (as the two motifs would align) and would not affect the results. In practice, they sometimes are too different to align perfectly, and this process avoids overfitting to imperfections in the mapping process.
16. Unlike the `3D2V` sequence, the `2GDI` sequence starts at index 10, so the value of first_n must be adjusted to 10.

17. Those *sequences* are not full-length natural sequences, but rather the ensemble of nucleotides that are part of each observed or predicted occurrence of the module, in sequence order. Each sequence has a length equal to the size of the module. If a FASTA file is not provided, then only the graph nucleotides are used as sequence data.
18. The only difference is how we acquire the graph. Here, the graph was not obtained from the `Rna3Dmotif` graph database, it was stored in a pickle file in Subheading 3.2, from `caRNA-val`. We can simply open it from the same file.
19. We can also call the functions described in this section from another directory, by adding the relative path of the src folder before every python file.
20. Note that from here to the rest of the chapter, the code should be executed directly from command line (rather than within a python environment), as we are now running the main scripts of the software package.
21. The `module_from_desc` method takes as input a graph (-g) as a DESC file or pickle, as well as a FASTA file (-seq). The name of the dataset is defined with -n. After running this snippet of code, module 0 will represent the `2GDI` `Rna3Dmo-tif` junction, and module 1, the `caRNAval` TPP RIN.
22. The four pickle files are the following. All four files contain a python list, in which each index matches a module ID. Only the first two files are essential for basic `BayesPairing` function.
    - `TPP_one_of_each_graph.pickle`, which contains a list of lists of all known occurrences of the RNA motif, each stored as a directed graph, indexed by module ID.
    - `TPP_aligned_modulegraphs.pickle`, which contains a list of dictionaries, indexed by module ID. In those dictionaries, the keys are the nodes of one module graph (the first of the list, chosen arbitrarily), and the values are the respective node of every other graph that matches this node. For sequence information to be extracted from the graphs, the alignment needs to be known, and this is how this information is stored.
    - `TPP_PDB_names.pickle`, which contains a list of lists of PDB IDs of 3D structures containing the module. If none are known, the lists are empty.
    - `TPP_PDB_positions.pickle`, which contains a list of lists of tuples. The tuples include the PDB IDs and the nucleotide positions of the motif in 3D structures containing it. If none are known, the lists are empty.

23. If there are more or fewer than two graphs with the name TPP, it means something went wrong. It can be addressed by going into the models folder, deleting all files starting with the dataset name (here TPP), and starting over from the start of Subheading 3.5.
24. As an example, if we were to want to search for modules number 3, 4, and 6, then the option would read as `-mod 3 4 6`.
25. The FASTA file can also contain both a sequence and a secondary structure. In such a file, for each ID line, there would be two lines. The first one would have the sequence, and the second one the secondary structure.
26. The value of sw=2 is at the other extreme of 0. This means that instead of being inflated, the scores will be considerably deflated. In this context, a score above 10 should be considered significant. We set the threshold to 0 to show suboptimal suggestions.

References

1. Berman H, Henrick K, Nakamura H, Markley JL (2006) The worldwide Protein Data Bank (wwPDB): ensuring a single, uniform archive of PDB data. Nucleic Acids Res 35(suppl_1): D301–D303
2. Brion P, Westhof E (1997) Hierarchy and dynamics of RNA folding. Ann Rev Biophys Biomol Struct 26(1):113–137
3. Chojnowski G, Waleń T, Bujnicki JM (2014) RNA bricks—a database of RNA 3D motifs and their interactions. Nucl Acids Res 42 (D1):D123–D131
4. Coimbatore Narayanan B, Westbrook J, Ghosh S, Petrov AI, Sweeney B, Zirbel CL, Leontis NB, Berman HM (2013) The nucleic acid database: new features and capabilities. Nucl Acids Res 42(D1):D114–D122
5. Cruz JA, Westhof E (2011) Sequence-based identification of 3D structural modules in RNA with RMDetect. Nat Methods 8(6):513
6. Djelloul M, Denise A (2008) Automated motif extraction and classification in RNA tertiary structures. RNA 14(12):2489–2497
7. Gendron P, Lemieux S, Major F (2001) Quantitative analysis of nucleic acid three-dimensional structures. J Mol Biol 308(5): 919–936
8. Hofacker IL, Fontana W, Stadler PF, Bonhoeffer LS, Tacker M, Schuster P (1994) Fast folding and comparison of rna secondary structures. Monatshefte für Chem/Chem Monthly 125(2):167–188
9. Kalvari I, Argasinska J, Quinones-Olvera N, Nawrocki EP, Rivas E, Eddy SR, Bateman A, Finn RD, Petrov AI (2017) Rfam 13.0: shifting to a genome-centric resource for non-coding rna families. Nucl Acids Res 46(D1):D335–D342
10. Klein D, Schmeing T, Moore P, Steitz T (2001) The kink-turn: a new RNA secondary structure motif. EMBO J 20(15):4214–4221
11. Leontis NB, Westhof E (2001) Geometric nomenclature and classification of RNA base pairs. RNA 7(4):499–512
12. Leontis NB, Lescoute A, Westhof E (2006) The building blocks and motifs of RNA architecture. Current Opin Struct Biol 16(3): 279–287
13. Lorenz R, Bernhart SH, Zu Siederdissen CH, Tafer H, Flamm C, Stadler PF, Hofacker IL (2011) Viennarna package 2.0. Algorithms Mol Biol 6(1):26
14. Nawrocki EP, Eddy SR (2013) Infernal 1.1: 100-fold faster RNA homology searches. Bioinformatics 29(22):2933–2935
15. Nissen P, Ippolito JA, Ban N, Moore PB, Steitz TA (2001) RNA tertiary interactions in the large ribosomal subunit: the A-minor motif. Proc Natl Acad Sci 98(9):4899–4903
16. Petrov AI, Zirbel CL, Leontis NB (2013) Automated classification of RNA 3D motifs and the RNA 3D Motif Atlas. RNA 19(10): 1327–1340

17. Reinharz V, Soulé A, Westhof E, Waldispühl J, Denise A (2018) Mining for recurrent long-range interactions in RNA structures reveals embedded hierarchies in network families. Nucl Acids Res 46(8):3833–3840. 10.1093/nar/gky19.

18. Sarrazin-Gendron R, Reinharz V, Oliver CG, Moitessier N, Waldispühl J (2019) Automated, customizable and efficient identification of 3D base pair modules with BayesPairing. Nucl Acids Res 47:3321–3332. 10.1093/nar/gkz102.

19. Sarver M, Zirbel CL, Stombaugh J, Mokdad A, Leontis NB (2008) FR3D: finding local and composite recurrent structural motifs in RNA 3D structures. J Math Biol 56(1-2):215–252

20. Serganov A, Nudler E (2013) A decade of riboswitches. Cell 152(1–2):17–24

21. Serganov A, Polonskaia A, Phan AT, Breaker RR, Patel DJ (2006) Structural basis for gene regulation by a thiamine pyrophosphate-sensing riboswitch. Nature 441(7097):1167

22. Sievers F, Higgins DG (2014) Clustal omega. Current Protocols Bioinformat 48(1):3–13

23. Thore S, Frick C, Ban N (2008) Structural basis of thiamine pyrophosphate analogues binding to the eukaryotic riboswitch. J Amer Chem Soc 130(26):8116–8117

24. Waleń T, Chojnowski G, Gierski P, Bujnicki JM (2014) Clarna: a classifier of contacts in rna 3d structures based on a comparative analysis of various classification schemes. Nucl Acids Res 42(19):e151–e151

25. Yang H, Jossinet F, Leontis N, Chen L, Westbrook J, Berman H, Westhof E (2003) Tools for the automatic identification and classification of RNA base pairs. Nucl Acids Res 31(13):3450–3460

Chapter 8

Modified Nucleotides and RNA Structure Prediction

Yuliia Varenyk and Ronny Lorenz

Abstract

Nucleotide modifications are occurrent in all types of RNA and play an important role in RNA structure formation and stability. Modified bases not only possess the ability to shift the RNA structure ensemble towards desired functional confirmations. By changes in the base pairing partner preference, they may even enlarge or reduce the conformational space, i.e., the number and types of structures the RNA molecule can adopt. However, most methods to predict RNA secondary structure do not provide the means to include the effect of modifications on the result. With the help of a heavily modified transfer RNA (tRNA) molecule, this chapter demonstrates how to include the effect of different base modifications into secondary structure prediction using the `ViennaRNA Package`. The constructive approach demonstrated here allows for the calculation of minimum free energy structure and suboptimal structures at different levels of modified base support. In particular we, show how to incorporate the isomerization of uridine to pseudouridine (Ψ) and the reduction of uridine to dihydrouridine (D).

Key words Modified nucleotides, tRNA, pseudouridine, dihydrouridine, Energy directed folding, Minimum free energy structure, Suboptimal structures

1 Introduction

Post-transcriptional RNA modifications are found in all regions of all types of RNA, most notably mRNA (messenger RNA), tRNA (transfer RNA), snRNA (small nuclear RNA), miRNA (micro-RNA), and lncRNA (long non-coding RNA) in all domains of life [1, 2]. They can be as simple as an addition of small functional groups (e.g., methylation) or of more complex side chains [3]. Moreover, nucleotide modifications may include *substitutions* (e.g., converting uridine to 4-thiouridine), *isomerization* (e.g., converting uridine to pseudouridine), and *reductions* (e.g., converting uridine to dihydrouridine) [4].

A growing amount of research demonstrates that modified nucleotides of RNA are functionally indispensable for RNA processing and translation [2]. The presence of modified nucleotides impacts folding, stability, structure, and consequently the fitness

Ronny Lorenz (ed.), *RNA Folding: Methods and Protocols*, Methods in Molecular Biology, vol. 2726,
https://doi.org/10.1007/978-1-0716-3519-3_8,

of RNA sequence to perform a certain function [1, 5]. For example, in eukaryotic U2 snRNA, the extra hydrogen bond of pseudouridine significantly alters the branch-site architecture, with important consequences for splicing activity [6]. Nucleotide modifications found in mRNA sequences have been shown to have various effects on structure and play roles in the splicing, maturation, stability, expression, and degradation of mRNA [7]. The majority of known modifications, however, was found in tRNA, in which the fraction of modified nucleotides in a sequence can range from 1.7 to 23.7% [8].

Having realized the importance of modified nucleotides in RNA structure formation, it becomes inevitable to take their effect into consideration while studying RNA molecules. Currently, though, there is a significantly limited amount of technologies sensitive enough to explore modifications and their functions in RNAs both experimentally and computationally [2]. In the current chapter, however, we explore the possibility of including the modified bases information into computational secondary structure prediction methods. The majority of existing secondary structure prediction programs works exclusively with the standard "AUGC" alphabet and are therefore able to take into account only Watson-Crick and Wobble (G-C, A-U, G-U) base pairs, which constitutes a major obstacle for predicting RNA structures for sequences with modified nucleotides [1, 9]. At the time of writing, only two software packages for RNA secondary structure prediction allow the possibility for the inclusion of modified nucleotides in their algorithms: `RNAstructure` [10] and `ViennaRNA Package` [11].

In this chapter, we concentrate on demonstrating the method of structure prediction for RNA sequences with modified nucleotides using the `ViennaRNA Package`. The prediction will be performed on human tRNA Phe (tdbR00000103). tRNA molecules pose a particular interest in structure prediction with modifications since it is known to be the most modified type of RNA. Mature tRNAs range in size from around 70 to 90 nucleotides with, on average, 13 modifications per molecule [12]. In the majority of cases, tRNAs adopt a cloverleaf structure, which consists of five major structural elements: the acceptor stem, D-arm, anticodon arm, variable loop, and T-arm (TΨC-arm). Each arm is composed of a double-stranded stem and a single-stranded loop [13]. Modifications are located in different regions of tRNA sequence (*see* Fig. 1), with majority in two regions: the anticodon loop and the core region, in which the D-loop and the T-loop are interacting, contributing to the structural stability and functionality of tRNA [5]. The number of modifications in tRNA sequences and their significance in the folding process make tRNAs a good candidate to investigate structure prediction performance (Table 1).

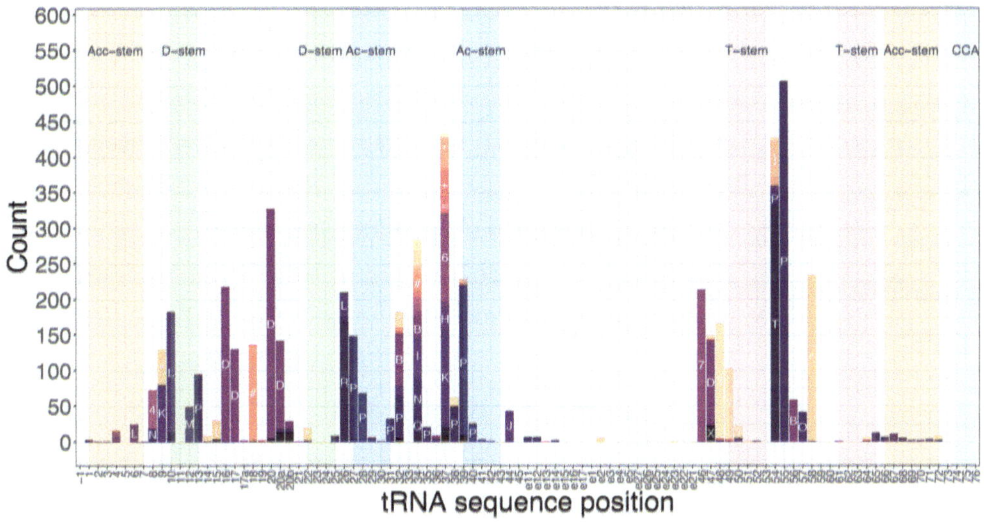

Fig. 1 Frequency of modified nucleotides in tRNA sequences. The X-axis denotes the location of modified bases mapped to the standard tRNA nucleotide consensus positions indicated by the `tRNAdb` database [14]. Specific modifications are denoted in the bar plot by different colors and one-letter code according to `tRNAdb`, *see* also Table 1

Table 1
One-letter codes for the ten most common modified nucleotides according to the `tRNAdb` naming system. The full table can be found and downloaded from the tRNAdb website http:/trnadb.bioinf.uni-leipzig.de/DataOutput/images/mod.pdf

Symbol	Modification	Symbol	Modification
P	pseudouridine	L	N2-methylguanosine
D	dihydrouridine	7	7-methylguanosine
T	ribosylthymine	K	1-methylguanosine
?	5-methylcytidine	R	N2,N2-dimethylguanosine
”	1-methyladenosine	#	2’-O-methylguanosine

2 Materials

2.1 Hardware

- A Linux- or MacOS X-based computer

2.2 Software

The following software is required to follow the examples within this chapter:

1. `ViennaRNA Package` (http://www.tbi.univie.ac.at/RNA)
2. A working `Python` 3 environment

3. Text Editor (**not** a word processor)
4. A `Postscript` Viewer (e.g., `gv`, `preview`, `gsview`)

2.3 Data Resources

Several available resources with information on modified RNA bases provide a listing of post-transcriptionally modified nucleotides in RNA:

1. Transfer RNA database (`tRNAdb`) [14]
2. `MODOMICS` [15, 16]
3. RNA Modification Database (`RNAMDB`) [17]
4. Small Subunit rRNA Modification Database (`SSUmods`) [18]

Within this chapter, we focus on 1.

3 Methods

3.1 General Remarks

In the following, we will make extensive use of the `Python` interface of the `ViennaRNA Package` to access its implemented prediction algorithms. This requires a working installation of the `ViennaRNA Package` version 2.4.17 or more recent. All below `Python` code can (i) be run interactively, (ii) stored in text files to be executed as `Python` scripts, or even (iii) be used from within an `IPython` notebook run by `JupyterLab` (https://jupyter.org/) or similar.

`Python` code within this chapter will be presented in the form of boxes with line numbers on the left side:

```
print("Hellow World")
```

Instructions to execute programs within a command line terminal are always preceeded by a dollar sign (`$`). Note here that the dollar sign itself is *not* part of the actual command line call:

```
$ echo "Hello World"
```

3.2 Installation and Usage

The most convenient way to install the `ViennaRNA Package` is to use the binary installer images provided at the official website https://www.tbi.univie.ac.at/RNA/#binary_packages. Alternatively, one may use `conda` and the respective `bioconda` [19] repository. Installation instructions for `conda` are available at https://bioconda.github.io/ and https://anaconda.org/bioconda/viennarna.

3.2.1 Installing the ViennaRNA Package from Source

If none of the above installation instructions are suitable, the recommended way to install the `ViennaRNA Package` is to download, unpack, compile, and install the latest source code tar-ball available at https://www.tbi.univie.ac.at/RNA/#source_code_packages using the following commands in your `Linux` or `Mac OSX` terminal:

```
$ tar xzvf ViennaRNA-2.4.17.tar.gz
$ cd ViennaRNA-2.4.17
$ ./configure
$ make
$ sudo make install
```

If users do not have *sudo* privileges and, therefore, cannot install the `ViennaRNA Package` system-wide, the installation can be directed to a non-default path (*see* **Note** 1). However, this may require additional set-up of the `PYTHONPATH` environment variable to follow the examples below.

3.2.2 Sequence and Structure Format

If not noted otherwise, RNA sequences should be stored in `FASTA` format, where each nucleotide sequence is preceded by a *fasta header* line starting with the " > " character. The sequence itself may span multiple lines. The `ViennaRNA Package` and some other software tools and data bases extend this format to include secondary structure annotation in dot-bracket format that follows immediately after the sequence. This format uses matching parenthesis (or brackets) to annotate base pairs while unpaired nucleotides are marked by the dot "." character. An example of such an *extended FASTA format* is shown in Fig. 2.

3.3 First Steps with the RNA Python Interface

The algorithms for RNA secondary structure prediction implemented within the `ViennaRNA Package` are not only available through executable programs. They are also exposed in the form of *RNAlib*, a library written in `C` that can be bound to any third-party program. Moreover, this library is available through a `Python` scripting

```
>tdbR00000103|Homo_sapiens|9606|Phe|#AA
GCCGAAAUALCUC"GDDGGGAGAGCRPPAGABU#AAWAPCUAAAG7DC?CUGGTPCG"UCCCGGGUUUCGGCACCA
(((((((..((((........)))).(((((.......))))).....(((((.......))))))))))))....
```

Fig. 2 Extended FASTA format with dot-bracket structure. Shown here is the data set of human tRNA Phe (`tdbR00000103`) in `FASTA` format as obtained from `tRNAdb` 2009. The first and second rows denote the header line with sequence identifier and the actual RNA sequence with modified bases, respectively. In contrast to *regular* `FASTA` format, this file contains extra information in the form of a dot-bracket secondary structure annotation in the third row following the sequence. Here, matching parentheses denote pairing bases while dots represent nucleotides that remain unpaired in the structure language interface and thus easily accessible for rapid prototyping and Python-based scripting pipelines.

Usually, this Python interface is part of the standard installation of the ViennaRNA Package and can be loaded in an interactive Python shell or any script with a simple

```
import RNA
```

statement. After this, all utilities, algorithms, and other implemented features of the ViennaRNA Package are accessible within the RNA module. For instance, to predict the minimum free energy (MFE) and corresponding structure for an input sequence, use the fold() function:

```
sequence = "GAGUAGUGGAACCAGGCUAU"
ss, mfe  = RNA.fold(sequence)

print("{}\n{} ({:6.2f})".format(sequence, ss, mfe))
```

In the above code, we store the secondary structure in dot-bracket format in the variable ss and the MFE in units of $kcal \cdot mol^{-1}$ in the variable mfe. Finally, the print() function produces the following output:

```
GAGUAGUGGAACCAGGCUAU
..(((((.......))))) ( -2.70)
```

As an alternative to the simplified *"flat"* interface as shown above, we can use the *object-oriented* interface instead. This becomes most useful when multiple computations are to be done for the same RNA sequence or if computations deviate from default settings. For this purpose, we create an object termed fold_compound that *"glues together"* all relevant parts of the computations. This is, amongst other things, the dynamic programming matrices, all model settings and parameters, as well as the input sequence. In Python, these objects come with a variety of methods attached, e.g., the mfe() method. The above MFE prediction then becomes:

```
sequence = "GAGUAGUGGAACCAGGCUAU"
fc       = RNA.fold_compound(sequence)
ss, mfe  = fc.mfe()

print("{}\n{} ({:6.2f})".format(sequence, ss, mfe))
```

A detailed overview of the ViennaRNA Package API and all available implementations can be found in the corresponding reference manual at https://viennarna.readthedocs.io/.

3.4 RNAs with Modified Bases

Annotations for modified bases in RNA sequences are found in several data bases. However, most data bases provide their own particular annotation codes. The `tRNAdb`, for instance, uses a one-letter code for about 60 distinguished modifications.

Most structure prediction programs, on the other hand, do not accept input sequences with a sequence alphabet other than the standard four letters *A, C, G*, and *U*. Consequently, if such additional one-letter codes are present in their input, the sequences might be treated incorrectly during the entire structure prediction process. The programs of the `ViennaRNA Package`, for instance, do treat such special characters as if they were unknown nucleotides, i.e., these bases are not allowed to be involved in base pairing.

This behavior may be correct for some modifications such as particular methylations, and the bases wyosine or wybutosine, since their effect on RNA structure is indeed known to prevent base pairing. However, some modifications may have the opposite effect and may act as stabilizing factor or simply change their pairing partner preference. Generally speaking, it is best to not simply feed prediction programs with input sequences that contain annotated modified bases. Instead, a more elaborate pre-processing step is required. Such processing can then adapt the sequence (and possibly the prediction program) according to all information known about the modifications present in the sequence.

In the following, we will use the specific example of a heavily modified human tRNA Phe sequence to investigate how structural constraints applied to existing prediction algorithms can improve their prediction performance. This sequence of 76 nucleotides consists of 17 modified bases in its native, functional state and seems a perfect candidate since regular MFE structure prediction with all modified bases converted to their unmodified counterparts does not yield the well known reference cloverleaf shape, cf. Fig. 3.

3.4.1 Retrieval and Preparation of the Input Sequence

The tRNA sequence required for all computations below can be obtained from `tRNAdb` 2009 [14] hosted at http://trnadb.bioinf.uni-leipzig.de/DataOutput/. There, use the *Search Database* functionality and search for the *tRNAdb ID* “tdbR00000103”. At the results page, select “Download FASTA structure” and click the “Submit” button to download a `FASTA` file of the annotated sequence and its reference structure. The content of the resulting file should look as follows:

```
>tdbR00000103|Homo_sapiens|9606|Phe|#AA
GCCGAAAUALCUC"GDDGGGAGAGCRPPAGABU#AAWAPCUAAAG7DC?CUGGTPCG"UCCCGGGUUUCGGCACCA
(((((((..((((........)))).(((((.......))))).....(((((.......))))))))))))....
```

Store this file at the relative path “*files/tRNA_phe.fasta*” and load the sequence block into `Python` using the `RNA.file_fasta_read()` function:

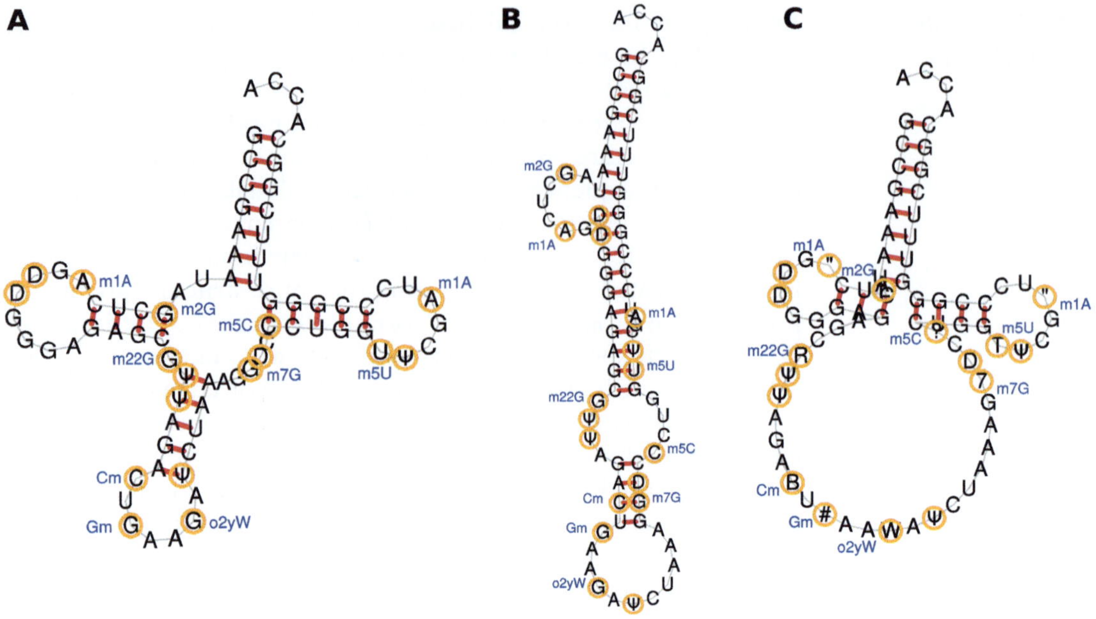

Fig. 3 Human tRNA Phe (tdbR00000103) with annotated base modification. Dihydrouridines are labeled by sequence character *D* and pseudouridines by character *Ψ*. Other modified bases are denoted by the sequence letter of their unmodified counterpart together with an additional label denoting the actual modification. (**a**) is the tRNA in its reference structure as obtained from `tRNAdb`, (**b**) shows the predicted ground state (MFE) structure for the unmodified sequence, and (**c**) the predicted MFE structure for the untreated sequence with special one-letter code for each modified base, i.e., each modification is treated as "*must not pair.*"

```python
"""
Read the first FASTA block from the input file
and store header, sequence and reference structure
"""
f = open("files/tRNA_phe.fasta")
retval, fasta_header, sequence, rest = RNA.file_fasta_read(f)
f.close()

# obtain reference structure
structure_ref = rest[0]
```

Here, every row in the input file that follows the actual RNA sequence is stored as an entry in the `rest` tuple. In the above code, we extract the reference structure from the first line that follows the input sequence. Alternatively, the `SeqIO.parse()` of the `biopython` package may be used to read `FASTA` formatted sequence files, *see* **Note** 2.

The sequence string read from the file consists of additional non-standard characters, which follow the special one-letter code for modified bases as used in `tRNAdb`. A complete list of the codes can be obtained from http://trnadb.bioinf.uni-leipzig.de/DataOutput/images/mod.pdf.

To use the one-letter annotation codes within the predictions below, we store them in lists, separately for each of the unmodified counterparts. Additionally, we write a convenience function to convert any RNA sequence that contains such annotations into its unmodified equivalent:

```
mod_A = ['H', '"', '/', '+', '*', '=', '6', 'E', '[', ':', 'I', 'O', '^', \
         '`']
mod_C = ['<', '%', 'B', 'M', '?', '\'', '}', '>', '\u00b0']
mod_G = [';', 'K', 'L', '#', 'R', '|', '7', '(', 'Q', '8', '9', 'Y', 'W']
mod_U = ['N', '{', '2', 'J', '4', '&', '1', 'S', '3', 'V', '5', '!', '$',
         'X', ',', ')', '~', 'D', 'P', ']', 'Z', 'T', 'F', '\\']

# Returns the unmodified version of the input sequence
def unmodify(sequence):
    # split sequence into individual bases
    unmod_sequence = list(sequence)
    # replace nucleotide if necessary
    for i in range(0, len(unmod_sequence)):
        base = unmod_sequence[i]
        if base in mod_A:
            unmod_sequence[i] = 'A'
        elif base in mod_C:
            unmod_sequence[i] = 'C'
        elif base in mod_G:
            unmod_sequence[i] = 'G'
        elif base in mod_U:
            unmod_sequence[i] = 'U'
    # return all bases glued together again
    return "".join(unmod_sequence)
```

3.5 Predicting MFE Structures

As mentioned above and shown in Fig. 3, the MFE algorithm implemented in the `ViennaRNA Package` does not predict the reference structure for this particular tRNA. In fact, structure-wise, the predicted ground state is rather distant to the reference. To quantify the structural distance, we can resort to the base pair distance

$$d_{\mathrm{BP}}(s,t) = |s \cup t| - |s \cap t| \tag{1}$$

between any two structures s and t, i.e., the number of base pairs that is not shared between s and t. The `Python` code below shows how to compute MFE structures for both input sequences: one with annotated modifications and another that resembles its unmodified counterpart. The resulting structures are then compared to the reference structure using the `bp_distance()` function, and their base pair distances to the reference conformation are finally printed.

```
# compute MFE and corresponding structure
ss, mfe = RNA.fold(sequence)

# compute MFE and corresponding structure for unmodified sequence
ss_unmod, mfe_unmod = RNA.fold(unmodify(sequence))

# compute base pair distances between reference structure
# and the predicted MFE structures
dist        = RNA.bp_distance(structure_ref, ss)
dist_unmod  = RNA.bp_distance(structure_ref, ss_unmod)

print("A: " + structure_ref + " (reference)")
print('B: {} ({:6.2f}) dist={:d}'.format(ss_unmod, mfe_unmod, dist_unmod))
print('C: {} ({:6.2f}) dist={:d}'.format(ss, mfe, dist))
```

The corresponding output is

```
A: (((((((..((((........)))).(((((.......))))).....(((((.......)))))))))))).... (reference)
B: ((((((((.......(((((((((((....((((..........))))...))))..)))))))))))))))).... (-23.60) dist=29
C: ((((((((..(((......)))...........................((((.......)))))))))))).... (-14.80) dist=14
```

and the individual structures can be conveniently visualized as in Fig. 3a–c. *See* **Note 3** for further details on annotation of secondary structure plots. At this point, however, the prediction results seem rather unconvincing. After all, the only part that has been predicted correctly for both, the sequence without modifications (B) and the raw input sequence (C), is the acceptor stem, though with an extra base pair between U8 and G65.

3.6 Suboptimal Structures

A major problem in RNA secondary structure prediction is that a single structure representative, e.g., the MFE structure, may not properly represent the vast ensemble of alternative conformations. At the same time, except for toy examples with very short RNA sequences, it is impossible to consider all possible structures. A potential solution to this dilemma is to closely investigate sub-optimal structures, i.e., alternative conformations that are close to the MFE structure in terms of free energy [20, 21]. For instance, the set

$$S^{\delta} = \{s \in \Omega \mid E(s) \leq \text{MFE} + \delta\} \tag{2}$$

of all structures in an energy band δ around the MFE might reveal structurally different alternatives.

Let us restrict ourselves to the 20 energetically best structures within an energy band of $\delta = 3$ kcal · mol^{-1}. The `Python` code to predict and enumerate the corresponding structures by energy in ascending order is as follows:

```python
# Set global switch for unique ML decomposition
RNA.cvar.uniq_ML = 1
# Sort subopt() output by free energy
RNA.cvar.subopt_sorted = 1

print("20 Suboptimal Structures for unmodifed input sequence")
for i, s in enumerate(RNA.subopt(unmodify(sequence), 300), start=1):
    print('{:2d} {} ({:6.2f}) dist={:d}'.format(\
        i, \
        s.structure, \
        s.energy, \
        RNA.bp_distance(structure_ref, s.structure)))
    if i == 20:
        break

print("20 Suboptimal Structures for raw input sequence")
for i, s in enumerate(RNA.subopt(sequence, 300), start=1):
    print('{:2d} {} ({:6.2f}) dist={:d}'.format(\
        i, \
        s.structure, \
        s.energy, \
        RNA.bp_distance(structure_ref, s.structure)))
    if i == 20:
        break
```

Note here that the implementation of the `RNA.subopt()` function requires us to pass the energy band in units of deca-calories per mol (dacal · mol^{-1}), i.e., $\delta = 3$ kcal · mol^{-1} = 300 dacal · mol^{-1}. In each line, we print the rank according to free energy, the structure in dot-bracket notation, the corresponding free energy, and the base pair distance to the reference structure. Then, the output for the tRNA sequence *without base modifications* is:

```
20 Suboptimal Structures for unmodified input sequence
 1 (((((((.......(((((((((....((((..........))))....))))..))))))))))))))).... (-23.60) dist=29
 2 ((((((...((((....((((((((....((((..........))))....))))..))))))))))))))).... (-23.30) dist=31
 3 (((((...(((((....((((((((....((((..........))))....))))..))))))))))))))).... (-23.00) dist=33
 4 (((((((..((((........))))....((((..........))))(((((.......)))))))))))).... (-22.80) dist=9
 5 (((((((..((((........))))...........((((....))))(((((.......)))))))))))).... (-22.50) dist=9
 6 ((((((((.((((........))))....((((..........))))((.((......)).)))))))))).... (-22.40) dist=19
 7 ((((((...((((........))))....((((..........))))(((((.......))))).)))))).... (-22.30) dist=10
 8 ((((((((.......(((((.((((....((((..........))))....))))...))))))))))))).... (-22.20) dist=28
 9 (((((((...(((....((((((((....((((..........))))....))))..))))))))))))))).... (-22.20) dist=29
10 ((((((...(((....(((((((((....((((..........))))....))))..))))))))))))))).... (-22.20) dist=31
11 ((((((((.((((........))))...........((((....))))((.((......)).)))))))))).... (-22.10) dist=19
12 (((((((..((((........)))).(((((.......)))))....(((.((......)).)))))))))).... (-22.10) dist=10
13 (((((((....((((((.............)))))).((((....))))(((((.......)))))))))))).... (-22.00) dist=19
14 ((((((...((((........))))...........((((....))))(((((.......))))).)))))).... (-22.00) dist=10
15 ((((((.......(.((((((((((....((((..........))))....))))..)))))).))))))).... (-22.00) dist=30
16 ((((((((.((((........))))....((((..........)))).((((.......)))))))))))).... (-21.90) dist=11
17 ((((((((........(((((((((....((((..........))))....))))..))))).)))))))).... (-21.90) dist=28
18 (((((((..((((........)))).(((((.......))))).....(((((.......)))))))))))).... (-21.90) dist=0
19 (((((((....((((((..((.....)).)))))).((((....))))(((((.......)))))))))))).... (-21.90) dist=21
20 ((((((...((((....(((.((((....((((..........))))....))))...))))))))))))).... (-21.90) dist=30
```

A more convenient way to analyze the results might be to visualize them as structure drawings as shown in Figs. 4 and 5. For the sequence without modified bases, the reference structure can be found at position 18 of the energy sorted list of solutions, while we do not find the reference structure in the list for the raw input sequence. This highlights the drawback of just using the raw input sequence: since the prediction method does not recognize the special annotation characters of the modified bases, all of them are treated as unknown nucleotides, effectively prohibiting them to form base pairs. Consequently, the exact tRNA Phe reference structure is not even among the set of all possible solutions, since some of the modified bases are actually required to base pair.

3.7 Thermodynamical Ensemble Properties

A more thorough check for our previous predictions can be made by comparing the MFE and reference structures against the *entire* ensemble of structures that are compatible with the input sequence. In thermal equilibrium, the probability $p(F)$ of any structural feature F is proportional to the partition function Z^F of the subset of structures that exhibit F, i.e.,

$$p(F) \propto Z^F \qquad \text{with} \qquad Z^F = \sum_{s|F \in s} exp(-\beta E(s)) \tag{3}$$

where $E(s)$ is the free energy of structure s and $\beta = RT$, with gas constant R and temperature T. The actual probability $p(F)$ can be computed once the partition function Z for the entire ensemble

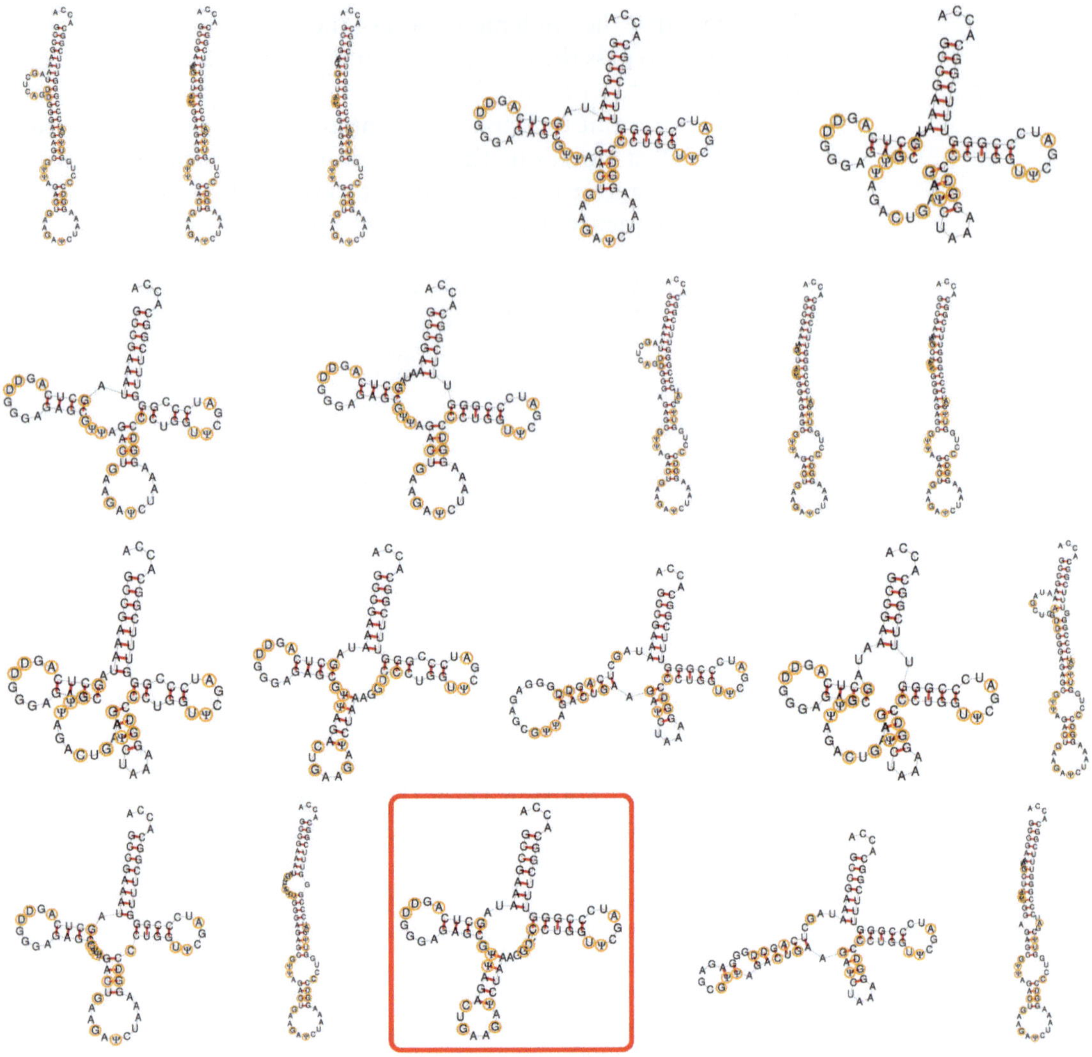

Fig. 4 Predicted suboptimal structures for the unmodified tRNA sequence. Shown are the 20 energetically best secondary structures, sorted by free energy (left to right, top to bottom). Modified bases in the source sequence are marked by yellow circles. The reference structure, i.e., native state of the tRNA, resides at position 18 and is highlighted by a red rectangle

$$p(F) = \frac{Z^F}{Z} \qquad \text{with} \qquad Z = \sum_s exp(-\beta E(s)) \tag{4}$$

is known. Equilibrium probabilities that turn out to be very useful for further analysis are, for instance,

$$p(s) = \frac{1}{Z} exp(-\beta E(s)), \qquad \text{and} \qquad p_{ij} = \frac{1}{Z} \sum_{s|(i,j) \in s} p(s) \tag{5}$$

where $p(s)$ is the probability of a single secondary structure s and p_{ij} is the pairing probability of a base pair (i, j). In particular, the latter,

Fig. 5 Predicted suboptimal structures for the tRNA sequence where all modified bases are prohibited to base pair. Shown are the 20 energetically best secondary structures, sorted by free energy (left to right, top to bottom). Modified bases in the source sequence are marked by yellow circles. The reference structure is not among the solution set, as some of the modified bases are still required to base pair in the native state

the base pair probabilities p_{ij}, which can be computed efficiently for all pairs (i, j) [22], enable us to compute various useful structural ensemble measures.

One such measure, the *positional entropy* $S(i)$, allows us to judge the uncertainty for any nucleotide i. For this, let us have a look at structures from an information theory perspective. Then, the structural entropy

$$S(i) = -\sum_j p_{ij}\log(p_{ij}) - q_i\log(q_i) \tag{6}$$

of a particular base at position i is determined from the states where i is paired, and those where it remains unpaired, i.e., $q_i = 1 - \sum_j p_{ij}$. The higher $S(i)$, the more structural alternatives are present in the ensemble. Consequently, any structure we select to represent the entire ensemble, e.g., the MFE structure, must be considered less reliable with respect to positions i where $S(i)$ is high. In contrast, positions i with low values $S(i)$ indicate reliable pairing states, as there are not many alternatives among the Boltzmann-weighted ensemble.

In our previous prediction, we computed base pair distances d_{BP} between MFE- and reference structures. This measure, however, restricts itself to only two conformations, thereby neglecting all structural alternatives in the predicted ensemble. In thermodynamic equilibrium, one would rather like to know the expected distance

$$\langle d_{\mathrm{BP}}(s)\rangle \quad = \sum_t p(t) d_{\mathrm{BP}}(s,t) \tag{7}$$

of s to a structure t randomly drawn from the ensemble. Alternatively, we may express the expected distance as a *per-nucleotide deviation* of the ensemble from s. This measure is known in the literature as *ensemble defect* $\mathrm{ED}(s)$ with respect to s and has been widely used in *in silico* structure design [23, 24]. $\mathrm{ED}(s)$ can be derived from $\langle d_{\mathrm{BP}}(s)\rangle$ by normalizing to the sequence length n and realizing that each base pair any structure t differs from s affects exactly two nucleotide positions, namely the pairing partners i and j of the respective base pair $(i,j)\in t$. Finally and most conveniently, the resulting value of $\mathrm{ED}(s)$ can be computed from the base pairing probabilities p_{ij} as

$$\begin{aligned} \mathrm{ED}(s) &= \frac{1}{n}\sum_t p(t) 2 d_{\mathrm{BP}}(s,t) \\ &= 1 - \frac{1}{n}\sum_{ij}\delta_{ij}(s) p_{ij} - \frac{1}{n}\sum_i \bar{\delta}_i(s) q_i. \end{aligned} \tag{8}$$

Here, we use the sequence length n and the indicator functions

$$\delta_{ij}(s) = \begin{cases} 1, & \text{if } (i,j)\in s \\ 0, & \text{otherwise} \end{cases} \quad \text{and} \quad \bar{\delta}_i(s) = \begin{cases} 1, & \text{if } \nexists j \mid (i,j)\in s \\ 0, & \text{otherwise} \end{cases} \tag{9}$$

which assume 1 only if base pair (i,j) is present in s ($\delta_{ij}(s)$), and if position i is unpaired ($\bar{\delta}_i(s)$), respectively. Variable $q_i = \sum_j 1 - p_{ij}$, again, is the probability of position i to stay unpaired. A complete derivation of Eq. (8) can be found in **Note 5**.

Implementations for both of the above measures are readily available in the `ViennaRNA Package`. Since their evaluation depends on base pair probabilities, the `Python` code below is slightly different to that of the previous MFE example. In particular, the corresponding evaluation functions are only available as methods attached to `fold_compounds`, so we need to resort to the object-oriented interface.

```
# create fold_compounds for both sequences
fc         = RNA.fold_compound(sequence)
fc_unmod   = RNA.fold_compound(unmodify(sequence))

# compute MFEs and corresponding structures
ss, mfe              = fc.mfe()
ss_unmod, mfe_unmod  = fc_unmod.mfe()

# set scaling for subsequent partition function computations
fc.exp_params_rescale(mfe)
fc_unmod.exp_params_rescale(mfe_unmod)

# compute partition function and pair probabilities
pp, dG               = fc.pf()
pp_unmod, dG_unmod   = fc_unmod.pf()

# compute ensemble defect wrt. reference structure
ensdef         = fc.ensemble_defect(structure_ref)
ensdef_unmod   = fc_unmod.ensemble_defect(structure_ref)

# compute positional entropies
posent         = fc.positional_entropy()
posent_unmod   = fc_unmod.positional_entropy()

print("A: " + structure_ref + " (reference)")
print('B: {} ({:6.2f}) d={:f}'.format(pp_unmod, dG_unmod, ensdef_unmod))
print('C: {} ({:6.2f}) d={:f}'.format(pp, dG, ensdef))

print(["{:.4g}".format(i) for i in posent_unmod])
print(["{:.4g}".format(i) for i in posent])
```

The first three `print()` functions yield

```
A: (((((((..((((........)))).(((((.......))))).....(((((.......))))))))))))).... (reference)
B: ((((((({,.{{((..,,{{{{||||....((({....,....)))),,,,|}}}..,))))))))))))).... (-24.86) d=0.462454
C: ((((((((..(((......}}).........................((((.......)))))))))))).... (-15.31) d=0.310662
```

followed by two more lines (not shown) with lists of the positional entropies for both sequences, respectively. In the above output, the reference structure (line starting with "`A:`") is followed by two lines that denote the results of the two partition function computations: the one for the unmodified sequence (line starting with "`B:`") and the other for the raw input sequence with annotated modified bases (line starting with "`C:`"). Here, the two dot-bracket-like strings indicate the pairing preference for each base (see the manpage of `RNAfold` for details). Each line is then followed by the respective free energy of the ensemble $\Delta G = -RTlnZ$ in kcal · mol^{-1}, and the ensemble defect wrt. to the reference structure. The positional

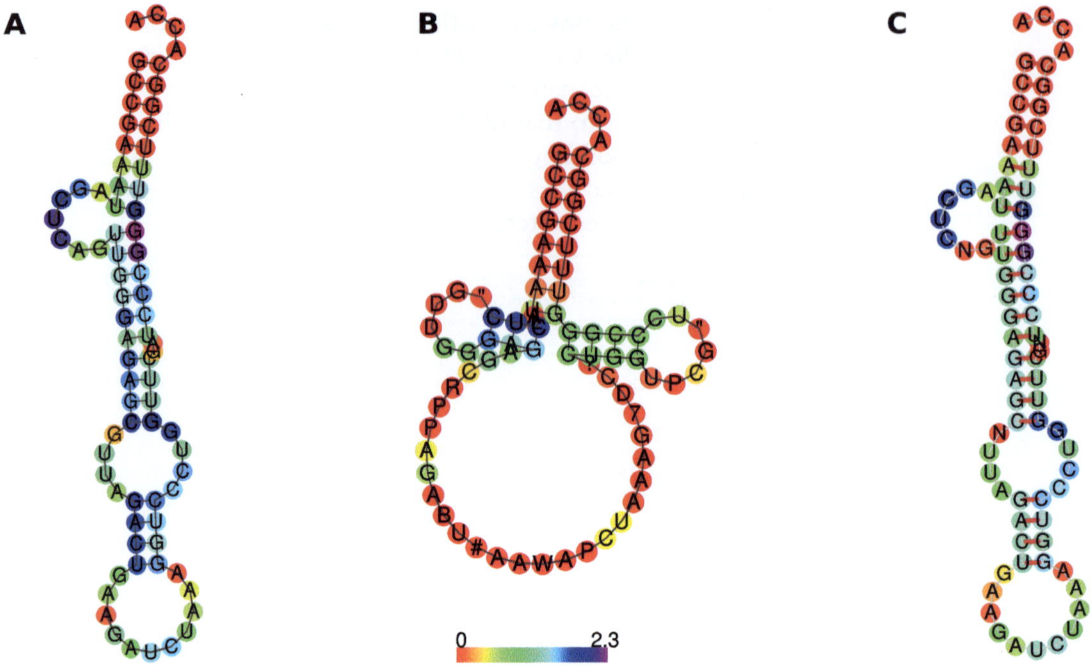

Fig. 6 Predicted MFE structures with positional entropy annotation. Shown here are the predicted MFE structures for human tRNA Phe (tdbR00000103). In (**a**) all modified bases have been replaced by their unmodifed counterpart, whereas in (**b**), the raw sequence with annotated modified bases has been taken for MFE prediction. In (**c**), modified bases known to block reverse transcriptase (RT) have been replaced by "N" and other modifications have been replaced by their unmodified canoncial nucleotide. All structures are annotated with positional entropies $S(i)$ shown in color. For (**a**) and (**c**), the most reliable positions reside in the terminal helix that connects the 5′ and 3′ end. This part of the predicted structure tallies with the acceptor stem of the tRNA reference structure. The remaining, mostly blue colored, bases show much higher position entropies and are therefore less reliable. The RT blocking sites in (**c**) have entropies $S(i) = 0$, since they appear in any structure. In (**b**), positional entropies are much lower in general. However, this is due to the extremely reduced ensemble of possible conformations that originates from the 17 formerly modified positions that in this prediction are forced to remain unpaired

entropies for both sequence variants are best visualized as color annotation within a structure layout plot as shown in Fig. 6. An example code to produce such color-annotated structure plots can be found in **Note** 4.

Still, the result of the above analysis suggests that a large proportion of structures within both ensembles, the one for the sequence without modified bases and the other where all modified bases are forced to stay unpaired, are quite dissimilar to the reference structure. According to the ensemble defect computation, almost half of the bases (0.46) of the unmodified sequence and about every third base (0.31) of the raw sequence input are paired differently. The positional entropy computations draw a very similar picture, *see* Fig. 6. For the unmodified sequence, everything but the acceptor stem can be relied on (mostly red), the remaining parts (mostly blue) have a large variety of alternative structural conformations. With the raw sequence input, the positional entropy is low for the majority of bases. This indicates that the computed MFE structure indeed seems a reliable representative of the underlying ensemble. This is not too much surprising, though, since all of the 17 modified bases are forced to stay unpaired, which considerably reduces not only the size of the entire ensemble but also the number of alternative structures.

Taken together, the prediction of the correct tRNA structure fails with both of the models introduced so far. On the one hand, simply using the raw sequence data with one-letter annotation of the modified bases turns out to be too stringent. While the natural function for *some* modifications may indeed be to prevent the formation of base pairs, others still seem essential to helix formation. The reference structure in our example shows that the anticodon stem consists of three Ψ-A pairs, and an additional *m2G*-C pair is closing the D-loop (*see* also Fig. 3a). Since all of them have been effectively rendered unpaired, the reference structure is not even among the solution set of the prediction. On the other hand, simply converting the modified bases into their unmodified counterpart yields a rod-like structure. This has almost nothing to do with the accepted functional form of a tRNA. Even more, the ensemble defect computations demonstrate that for any structure randomly drawn from the ensemble according to its Boltzmann probability, almost every second base will be paired differently compared to the reference.

3.8 Modifications Known to Prevent Base Pairing

As mentioned above, some modifications influence the pairing behavior of a base such that it effectively remains unpaired. This plays an important role in many sequencing experiments where RNA is reverse transcribed into cDNA. Here, modifications may block or stall the reverse transcriptase (RT). While that is quite inconvenient for experimentalists in most cases, it can also be taken advantage of. In particular, detecting RT blocking sites in experiments can actually reveal base modifications, a technique that became essential for the non-natural modifications induced by chemical RNA structure probing methods such as *SHAPE* and *DMS* [25]. In the comprehensive work of Motorin et al. [26], we can find a list of many naturally occurring modifications known to be RT blocking or stalling.

Several RT blocking modifications can be assumed to simply prevent base pair formation in the context of RNA secondary structures, as particular donors or acceptors for hydrogen bond formation are either missing or misplaced three-dimensionally. If we compare the modifications present in `tRNAdb` and those listed as RT blocking in Table 2.1 of Motorin et al. [26], we find that only four out of five listed modifications overlap: N^1-methyladenosine

(m^1A), N^1-methylguanosine (m^1G), N^2,N^2-dimethylguanosine (m_2^2G), and N^3-methylcytidine (m^3C).

Let us store their respective one-letter code in a list and create a new function that converts all bases in an RNA sequence into their unmodified counterpart, except for those listed as RT blocking.

```
# list of modifications known to block RT
mod_RT_blocking = [ '"', 'K', 'R', '\'' ]

# Returns the unmodified version of the input sequence
# but replaces modifications known to block RT with
# character 'N'
def unmodify_nonblocking(sequence):
    # split sequence into individual bases
    unmod_sequence = list(sequence)
    # replace nucleotide if necessary
    for i in range(0, len(unmod_sequence)):
        base = unmod_sequence[i]
        if base in mod_RT_blocking:
            unmod_sequence[i] = 'N'
        elif base in mod_A:
            unmod_sequence[i] = 'A'
        elif base in mod_C:
            unmod_sequence[i] = 'C'
        elif base in mod_G:
            unmod_sequence[i] = 'G'
        elif base in mod_U:
            unmod_sequence[i] = 'U'
    # return all bases glued together again
    return "".join(unmod_sequence)
```

The function `unmodify_nonblocking()` is almost identical to `unmodify()` introduced before. The former only differs in an additional first rule present that replaces the RT blocking sites with letter "`N`," the *unknown nucleotide*. This implicitly prevents these sites from forming base pairs in subsequent computations, while any other modifications are simply replaced by their unmodified counterpart. We can now proceed and perform MFE predictions on the corresponding sequence where we masked all RT blocking sites.

```
# create a fold compound for the sequence with RT blocking sites
fc_RTb  = RNA.fold_compound(unmodify_nonblocking(sequence))

# compute corresponding MFE and structure
ss_RTb, mfe_RTb = fc_RTb.mfe()

# compute base pair distance to reference structure
dist_RTb  = RNA.bp_distance(structure_ref, ss_RTb)

# set scaling for subsequent partition function computations
fc_RTb.exp_params_rescale(mfe_RTb)

# compute partition function and pair probabilities
pp_RTb, dG_RTb = fc_RTb.pf()

# compute ensemble defect wrt. reference structure
ensdef_RTb = fc_RTb.ensemble_defect(structure_ref)

# compute positional entropies
posent_RTb = fc_RTb.positional_entropy()

print("Ref: " + structure_ref)
print('MFE: {} ({:6.2f}) dist={:d}'.format(ss_RTb, mfe_RTb, dist_RTb))
print('PF:  {} ({:6.2f}) d={:f}'.format(pp_RTb, dG_RTb, ensdef_RTb))
```

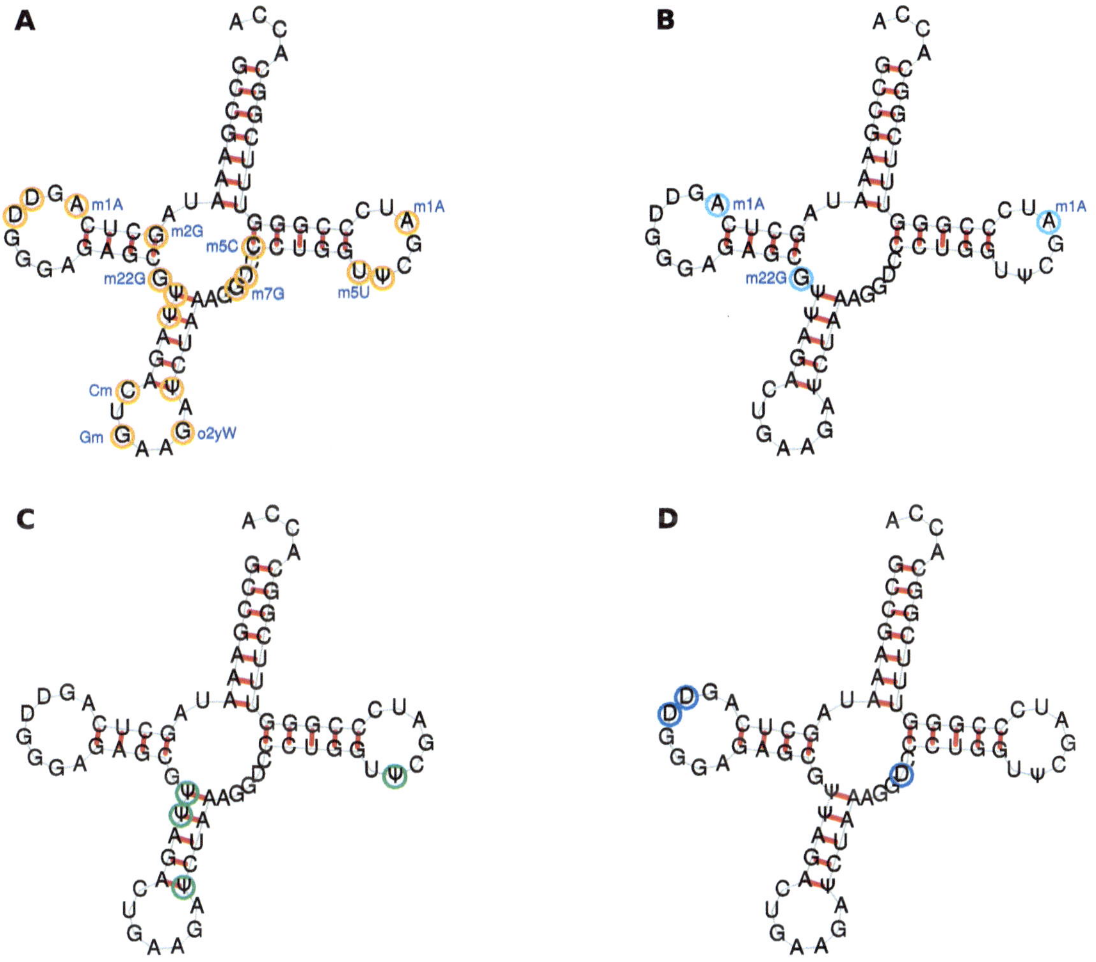

Fig. 7 Base modifications in human tRNA Phe that are handled explicitly. This panel shows the reference structure together with 17 known modifications, highlighted by yellow circles in (**a**). (**b**) shows the subset of 3 modifications that are known to block reverse transcriptase. (**c**) Highlights the 4 pseudouridines Ψ present in the sequence, while (**d**) denotes those positions where a uridine is modified to dihydro-uridine

As shown in Fig. 7b, just three of the 76 nucleotides of our tRNA example are affected by masking the RT blocking sites with N. We would therefore not expect too many differences if we compare the predictions against those where all bases are reverted to their unmodified version. This is indeed the case for the output of the above code:

```
Ref: (((((((..((((........)))).(((((.......))))).....(((((.......))))))))))))....
MFE: ((((((((.......((((((((((....((((...........))))....))))..))))))))))))))....  (-22.60) dist=29
PF:  (((((({,.{((({....{{{{||||....((((.....,....)))),,.,|}}}..}))))))))))))....  (-23.79) d=0.440332
```

Here, the predicted MFE structure is identical to that of the fully unmodified sequence. Only a slightly increased ΔG by about $1\ \mathrm{kcal} \cdot \mathrm{mol}^{-1}$ and subtle differences between the base pairing propensities can be seen. The ensemble defect wrt. the reference structure also decreased by just about 0.02 from previously 0.462454 to 0.440332. A comparison of the positional entropies against the previous predictions reveals the exact same pattern: there are only small differences between the fully unmodified sequence and the current one where we masked all RT blocking sites, *see* Fig. 6c.

Again, we investigate the 20 energetically best sub-optimal structures, to find out how structures close to the ground state in terms of free energy differ under the changes of our structure constraints. For this, we use the `subopt()` method attached to `fold_compound` objects. This is the equivalent of the above example for computing sub-optimal structures, except for the fact that we do not require to pass the input sequence as a parameter since it is already stored within the `fold_compound` object. The code below

```
print("20 Suboptimal Structures for sequence with known RT blocking sites")
for i, s in enumerate(fc_RTb.subopt(300), start=1):
    print('{:2d} {} ({:6.2f}) dist={:d}'.format(\
        i, \
        s.structure, \
        s.energy, \
        RNA.bp_distance(structure_ref, s.structure)))
    if i == 20:
        break
```

produces the output

```
20 Suboptimal Structures for sequence with known RT blocking sites
 1 (((((((.......(((((((((....((((...........))))....))))..))))))))))))).... (-22.60) dist=29
 2 ((((((...((((....((((((((....((((...........))))....))))..))))))))))))).... (-22.30) dist=31
 3 ((((...(((((....((((((((....((((...........))))....))))..))))))))))))).... (-22.00) dist=33
 4 (((((((..((((........))))....((((..........))))(((((.......)))))))))))).... (-21.90) dist=9
 5 (((((((..((((........)))).........((((...))))(((((.......)))))))))))).... (-21.60) dist=9
 6 ((((((((.((((........))))....((((..........))))((.((.......)).)))))))))).... (-21.50) dist=19
 7 ((((((...((((........))))....((((..........))))(((((.......))))).))))))).... (-21.40) dist=10
 8 ((((((((.((((........)))).........((((...))))((.((.......)).)))))))))).... (-21.20) dist=19
 9 ((((((((.......(((((.((((....((((...........))))....))))...)))))))))))))).... (-21.20) dist=28
10 (((((((..((((........)))).(((((.......))))).....(((.((.......)).)))))))))).... (-21.20) dist=10
11 (((((((...(((....((((((((....((((...........))))....))))..))))))))))))).... (-21.20) dist=29
12 ((((((...(((....(((((((((....((((...........))))....))))..))))))))))))).... (-21.20) dist=31
13 ((((((...((((........)))).........((((...))))(((((.......))))).))))))).... (-21.10) dist=10
14 ((((((((.((((........))))....((((..........)))).((((.......)))))))))))).... (-21.00) dist=11
15 (((((((..((((........)))).(((((.......))))).....(((((.......)))))))))))).... (-21.00) dist=0
16 ((((((((........((((((((....((((...........))))....))))..))))).))))))))).... (-20.90) dist=28
17 (((((((..((((........)))).((((((....).))))).....(((.((.......)).)))))))))).... (-20.90) dist=11
18 ((((((...((((....(((.((((....((((...........))))....))))...)))))))))))))).... (-20.90) dist=30
19 ((((...((((....((((((((....((((...........))))....))))..))))))))))))).... (-20.90) dist=33
20 ((((((((.........((((((((....((((...........))))....))))..))))..)))))))).... (-20.80) dist=27
```

Note that now the reference structure is located at position 15 in the list of possible solutions sorted by energy. Still, at least half of the 20 best solutions do not resemble the well-known cloverleaf structure, as also shown in Fig. 8.

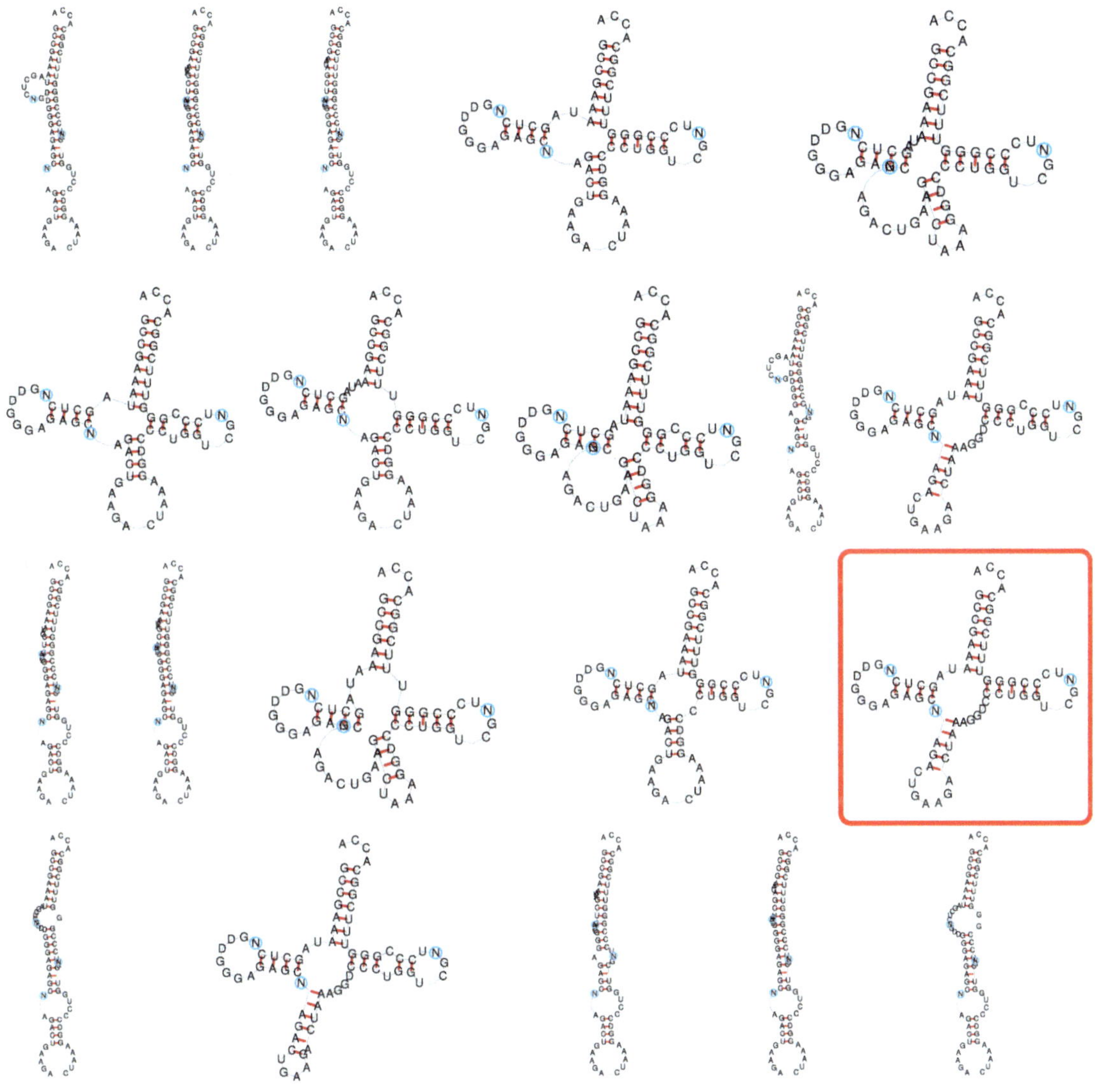

Fig. 8 Predicted suboptimal structures for the tRNA sequence where all bases known to block reverse transcriptase (RT) are prohibited to base pair. Shown here are the 20 energetically best secondary structures, sorted by free energy (left to right, top to bottom). RT blocking modifications in the source sequence are marked by blue circles. The correct reference structure (red rectangle) now resides at position 15 in the list of solutions, but still, half of the 20 solutions do not resemble the clover-leaf structure

It seems that, so far, we have exhausted our knowledge about strict pairing rules for the modified bases present in our tRNA example. The correct solution, i.e., the reference structure, is still too far from the optimal solution predicted according to the algorithm parameters. In particular, the predicted MFE structure is $-1.6\ \text{kcal}\cdot\text{mol}^{-1}$, more stable than the reference structure under the same parameters. Thus, the only obvious way that might increase the prediction performance is to specifically treat modified bases differently in terms of their induced structural stability. To do so, however, we require ***additional energy parameters*** for modified bases that are suitable for the underlying Nearest Neighbor (NN) model.

3.9 Stacking Parameters for Ψ-A Pairs

At the time this chapter was written, no complete NN parameter sets are available for any modified RNA bases. But parameters exist for base pair stacks that consist of at least one Ψ-A pair [27]. The data has been derived from UV-melting experiments of short oligos with a single Ψ-A pair and adjacent combinations of stacking canonical base pairs, except for the G-U wobble pair. In the last column of Table 3 of Hudson et al. [27], the authors list a $\Delta\Delta G_{37^\circ}$ U-A value that denotes the difference in free energy of a Ψ-A pair to that of an unmodified U-A pair in the same structural context at a temperature of 37 °C. This difference will become most useful in the following, as it can be readily used to augment the structure prediction with the help of the *constraint framework* of the `ViennaRNA Package` [28].

We have already been using *implicit constraints* in the above examples, when RT blocking bases have been masked by letter "`N`" or when we simply used the raw input sequence with non-standard RNA alphabet. Such constraints are considered *hard* since they forcefully remove any structure candidates from the set of solutions where the modified base would form a base pair. On the other hand, the so-called *soft constraints* can be used to apply individual changes to the free energy evaluation $E(s) = \sum_L E_L$ of secondary structures, s. In particular, each loop that is probed by the recursive decomposition scheme can be addressed separately to supply an additional energy contribution E_L^c. The *soft constrained* total energy of the structure then becomes $\hat{E}(s) = \sum_L E_L + E_L^c$, which enables a variety of applications [28]. Here we want to make use of the available stacking energy parameters and apply certain corrections whenever the algorithm evaluates base pair stacks that consist of at least one Ψ-A pair.

First, we will use integer values to encode the bases of the input sequence in the examples below, *see* **Note 6**. The `ViennaRNA Package` uses a default encoding that maps the canonical nucleotides `A`, `C`, `G`, and `U` to the numbers 1, 2, 3, and 4, respectively. Value 0 is reserved for any unknown nucleotides (`N`). We will extend this encoding for a fifth base, Ψ, which we encode as number 5. Utilizing a built-in function of the `ViennaRNA Package`, we can easily obtain the default encoding and write a wrapper function that simply replaces 4 by 5 for each Ψ in the input sequence.

```
import re

# standard sequence encoding in ViennaRNA is
# 0 ... unknown nucleotide
# 1 ... adenine (A)
# 2 ... cytosine (C)
# 3 ... guanine (G)
# 4 ... uridine (U)
#
# We will further use
# 5 ... pseudouridine (P)
def get_encoding(sequence):
    # obtain encoding of unmodified sequence with masked blocking bases
    enc = list(RNA.seq_encode(unmodify_nonblocking(sequence)))

    # go through list of pseudo-uridine positions in our sequence
    for p in [ m.start() + 1 for m in re.finditer('P', sequence) ]:
        enc[p] = 5

    return enc
```

Note that here we trust our previous hard constraint: modifications that are known to block RT will remain unpaired.

Next, we store the $\Delta\Delta G_{37^\circ}$ U-A values in two 2-dimensional lists that can be addressed by the sequence encoding: one list to store the data for Ψ-A pairs and second for A-Ψ enclosing a base pair stack. Both variants have to be handled separately due to the directionality $(5' - 3')$ of the parameters. Similar to the sub-optimal structure predictions above, the soft constraints API requires us to provide the contributions E_L^c in units of $\text{dacal} \cdot \text{mol}^{-1}$, which conveniently turn into integer numbers:

```
# energy parameters for Psi-A pairs
# taken from Hudson et al. 2013
stacking_psi_A = [
  # N   A     C    G     U     P
  [ 0,  0,    0,   0,    0,    0 ], # N
  [ 0,  0,    0,   0,    -77,  0 ], # A
  [ 0,  0,    0,   -14,  0,    0 ], # C
  [ 0,  0,    -9,  0,    0,    0 ], # G
  [ 0,  -181, 0,   0,    0,    0 ], # U
  [ 0,  0,    0,   0,    0,    0 ], # P
]

stacking_A_psi = [
  # N   A     C    G     U     P
  [ 0,  0,    0,   0,    0,    0 ], # N
  [ 0,  0,    0,   0,    -69,  0 ], # A
  [ 0,  0,    0,   -105, 0,    0 ], # C
  [ 0,  0,    -69, 0,    0,    0 ], # G
  [ 0,  -170, 0,   0,    0,    0 ], # U
  [ 0,  0,    0,   0,    0,    0 ], # P
]
```

Now, the $\Delta\Delta G_{37^\circ}$ U-A value for a G-C pair stacking on top of a Ψ-A pair can be accessed as

```
print(stacking_psi_A[3][2])
```

where the two list indices are the corresponding integer encodings for nucleotides $G = 3$ and $C = 2$. Executing the above `print()` function yields a value of $-9\ \mathrm{dacal} \cdot \mathrm{mol}^{-1} = -0.09\ \mathrm{kcal} \cdot \mathrm{mol}^{-1}$.

Using the generalized soft constraints framework of the `ViennaRNA Package` is straight-forward: We only need to write and bind a *callback* function of the form

```
def callback(i, j, k, l, loop, data):
    return 0 # energy correction in dacal/mol
```

that returns free energy corrections E_L^c in units of $\mathrm{dacal} \cdot \mathrm{mol}^{-1}$. The implemented prediction algorithms will automatically execute the callback in each of their recursive decomposition steps. When the callback is executed, it is passed through the type of loop that is currently being evaluated (via parameter `loop`) and the corresponding delimiting sequence coordinates `i`, `j`, `k`, and `l`. The last parameter, `data`, is an arbitrary data structure that can be specified by the user and will be passed through each time the callback is executed. Usually, this data structure simply encapsulates variables required to properly execute the callback function.

In our example, `data` will be the encoded sequence, assuming that it is not present as a global variable. Since we attempt to correct stacking energies, the callback must return corrections only when stacks are actually being evaluated. This can be accomplished easily because two stacking base pairs (i,j) and (k,l) resemble a special case of ***internal loops***, where the `loop` variable takes on the value `RNA.DECOMP_PAIR_IL`, and the loop-delimiting positions are $k = i + 1$ and $l = j - 1$. Consequently, a callback to correct for Ψ-A containing stacks is:

```
# soft constraints callback for P-A stacks
def psi_A_stack_correction(i, j, k, l, loop, data):
    if (loop == RNA.DECOMP_PAIR_IL):
        # corrections for stacking interactions
        if i + 1 == k and \
           j - 1 == l:
            # correct for known stacks that involve at
            # least one pseudouridine
            # 1. P-A pair enclosing other pair (k,l)
            if data[i] == 5 and data[j] == 1:
                return stacking_psi_A[data[k]][data[l]]
            # 2. A-P pair enclosed by other pair (i,j)
            elif data[k] == 1 and data[l] == 5:
                return stacking_psi_A[data[j]][data[i]]
            # 3. A-P pair enclosing other pair (k, l)
            elif data[i] == 1 and data[j] == 5:
                return stacking_A_psi[data[k]][data[l]]
            # 4. P-A pair enclosed by other pair (i,j)
            elif data[k] == 5 and data[l] == 1:
                return stacking_A_psi[data[j]][data[i]]

    return 0 # return 0 correction by default
```

While the above function is sufficient for MFE (or sub-optimal structure) predictions, it can not directly be used in any partition function computations. Here, the corresponding Boltzmann factors $exp(-E^c_L/RT)$ are required and we need to implement an additional callback. However, in our case, we can simply recycle the `psi_A_stack_correction()` function as shown in the following code block:

```
import math

# get value of RT for current temperature
RT = RNA.exp_param().kT

# soft constraint callback for P-A stacks,
# partition function variant returning
# Boltzmann factor with energy in cal/mol
def psi_A_stack_correction_exp(i, j, k, l, loop, data):
    # get energy correction in dacal/mol
    e = psi_A_stack_correction(i, j, k, l, loop, data)
    if e != 0:
        e = e * 10. # convert to cal/mol
        return math.exp(-e / RT) # return Boltzmann factor
    else:
        return 1.0 # return exp(0) = 1.0 by default
```

Note here that callbacks returning Boltzmann factors $exp(-E/RT)$ must use units of $\mathrm{cal} \cdot \mathrm{mol}^{-1}$ for the respective energies.

Put everything together and proceed with the structure prediction including the soft-constrained Ψ-A stack energy corrections:

```
# create a fold compound for the sequence with RT blocking sites
fc_psi  = RNA.fold_compound(unmodify_nonblocking(sequence))

# bind P-A correction callbacks and sequence encoding
fc_psi.sc_add_f(psi_A_stack_correction)
fc_psi.sc_add_exp_f(psi_A_stack_correction_exp)
fc_psi.sc_add_data(get_encoding(sequence))

# compute corresponding MFE and structure
ss_psi, mfe_psi = fc_psi.mfe()

# compute base pair distance to reference structure
dist_psi  = RNA.bp_distance(structure_ref, ss_psi)

# set scaling for subsequent partition function computations
fc_psi.exp_params_rescale(mfe_psi)

# compute partition function and pair probabilities
pp_psi, dG_psi = fc_psi.pf()

# compute ensemble defect wrt. reference structure
ensdef_psi = fc_psi.ensemble_defect(structure_ref)

# compute positional entropies
posent_psi = fc_psi.positional_entropy()

print("Ref: " + structure_ref)
print('MFE: {} ({:6.2f}) dist={:d}'.format(ss_psi, mfe_psi, dist_psi))
print('PF:  {} ({:6.2f}) d={:f}'.format(pp_psi, dG_psi, ensdef_psi))
```

In the above code, we first bind the Ψ-A energy correction callbacks and the required sequence encoding list. The remaining part for structure prediction and distance computations stays the same as before. No further function calls are necessary, since the soft-constraint framework of the `ViennaRNA Package` now automatically takes care of correcting energy contributions whenever necessary.

Finally, running the above code produces

```
GCCGAAAUALCUC"GDDGGGAGAGCRPPAGABU#AAWAPCUAAAG7DC?CUGGTPCG"UCCCGGGUUUCGGCACCA
Ref: (((((((..((((........)))).(((((.......))))).....(((((.......))))))))))))....
MFE: ((((((((.......((((((((((....((((...........))))....))))..))))))))))))).... (-22.74) dist=29
PF:  ((((((({,.((((....{{{,||||.,,,(({{.....}},,,.}}}}{{,{|,,,..,)))))))))))))).... (-24.16) d=0.360202
```

but, compared to the previous prediction without corrections for stacking Ψ-A base pairs, the MFE structure remained the same, with slightly less free energy, though. The miniscule decrease in free energy can be attributed to the predicted pair between A23 and Ψ55 that resides stacked within a helix. However, this time, the ensemble defect wrt. the reference structure dropped by almost 0.1 to about 0.36, indicating that the ensemble is shifted more towards the reference structure. Further analyzing the energetically best 20 sub-optimal structures again, i.e., executing the following code,

```python
print("20 Suboptimals for sequence with RT blocking sites and P-A pairs")
for i, s in enumerate(fc_psi.subopt(300), start=1):
    print('{:2d} {} ({:6.2f}) dist={:d}'.format(\
        i, \
        s.structure, \
        s.energy, \
        RNA.bp_distance(structure_ref, s.structure)))
    if i == 20:
        break
```

reveals more evidence of this shift: the correct solution is found at position 5 within the list of solutions sorted by energy

```
20 Suboptimals for sequence with RT blocking sites and P-A pairs
 1 ((((((((.......(((((((((....((((...........))))....))))..))))))))))))).... (-22.74) dist=29
 2 ((((((...((((....((((((((....((((...........))))....))))..))))))))))))).... (-22.44) dist=31
 3 (((((...(((((....((((((((....((((...........))))....))))..))))))))))))).... (-22.14) dist=33
 4 (((((((..((((........)))).(((((.......)))))....(((.((......)).)))))))))).... (-22.11) dist=10
 5 (((((((..((((........)))).(((((.......))))).....(((((.......))))))))))))).... (-21.91) dist=0
 6 (((((((..((((........))))....((((...........))))(((((.......))))))))))))).... (-21.90) dist=9
 7 (((((((..((((........)))).((((((....).)))))....(((.((......)).)))))))))).... (-21.81) dist=11
 8 (((((((..((((........)))).((((((....).))))).....(((((.......))))))))))))).... (-21.61) dist=1
 9 ((((((...((((........)))).(((((.......)))))....(((.((......)).))).))))))).... (-21.61) dist=11
10 (((((((..((((........)))).........((((....))))(((((.......))))))))))))).... (-21.60) dist=9
11 ((((((((.((((........)))).(((((.......))))).....((.((......)).)))))))))).... (-21.51) dist=10
12 ((((((((.((((........))))....((((...........))))((.((......)).)))))))))).... (-21.50) dist=19
13 (((((((..((((........))))..((((.......)))).....(((.((......)).)))))))))).... (-21.41) dist=11
14 ((((((...((((........)))).(((((.......))))).....(((((.......))))).))))))).... (-21.41) dist=1
15 ((((((...((((........))))....((((...........))))(((((.......))))).))))))).... (-21.40) dist=10
16 ((((((((.......(((((.((((....((((...........))))....))))...)))))))))))))).... (-21.34) dist=28
17 ((((((...(((....((((((((....((((...........))))....))))..))))))))))))).... (-21.34) dist=29
18 ((((((...(((....(((((((((....((((...........))))....))))..))))))))))))).... (-21.34) dist=31
19 ((((((...((((........)))).((((((....).)))))....(((.((......)).))).))))))).... (-21.31) dist=12
20 ((((((((.((((........)))).((((((....).))))).....((.((......)).)))))))))).... (-21.21) dist=11
```

Moreover, the difference in free energy between the MFE structure and the reference structure is just 0.83 $\text{kcal} \cdot \text{mol}^{-1}$. A graphical representation of this result is shown in Fig. 9. We can

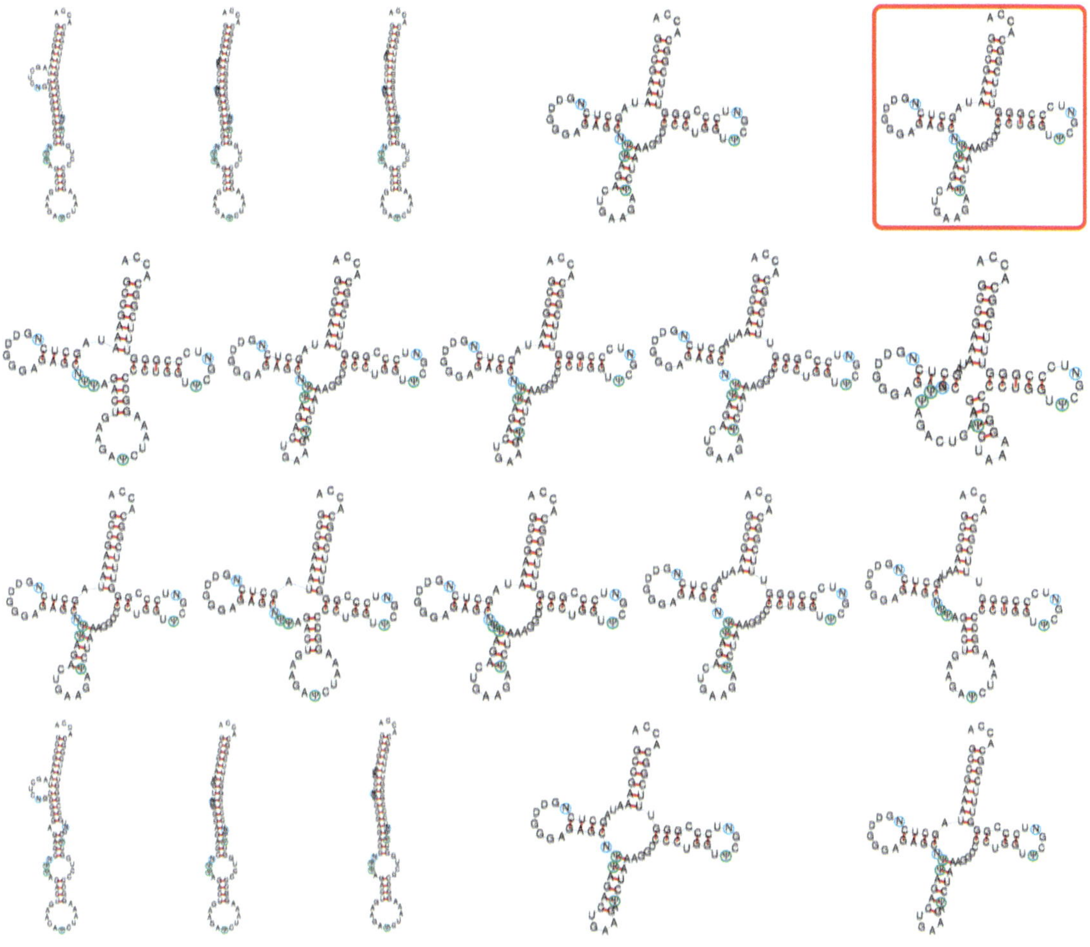

Fig. 9 Predicted suboptimal structures for the tRNA sequence where all bases known to block reverse transcriptase (RT) are prohibited to base pair and stacking energies for Ψ-A pairs are corrected for. Shown are the 20 energetically best secondary structures, sorted by free energy (left to right, top to bottom). RT blocking modifications in the source sequence are marked by blue circles and pseudouridines are highlighted in cyan. Now, many more structures resemble the expected clover-leaf shape and the correct reference structure at position 5 (red rectangle) is much closer to the MFE

conclude that, indeed, adding dedicated stacking energies for Ψ-A pairs increases the prediction performance for this particular example. Whether or not the prediction would be perfect if more free energy parameters for other types of loops with Ψ were available needs to be shown. At the time this chapter is written, no more parameters have been available and this question remains unanswered.

3.10 De-stabilization by Dihydrouridine

As demonstrated above, even an incomplete set of energy parameters for base pairs with modified nucleotides can enhance the quality of structure prediction algorithms. The addition of Ψ-A stacking energies already resulted in a substantial increase in prediction performance. Nevertheless, MFE prediction still did not yield the correct reference structure, i.e., the clover-leaf shape.

Mature tRNAs are known to harbor many different kinds of modified bases, some of which are essential for them to adopt the well-known clover-leaf shape [5]. However, when searching the literature for energy parameters with modified bases other than Ψ, only very few experimental studies can be found, where the (de-)stabilizing effect of particular modifications on their structural context has been explored. An outstanding example for such a study is one on the probably most frequent base modification found in tRNAs: the dihydrouridines (D) that give rise to the name of corresponding subdomain, the D-arm. In [29], the authors performed NMR experiments on mono- and oligonucleotides containing D to infer the destabilizing effect of the C3'-endo sugar conformation in favor of the C2'-endo conformation. Their findings suggest that the C3'-endo form is destabilized by about 1.5–5.3 $\text{kcal} \cdot \text{mol}^{-1}$. Hence, the de-stacking effect of base pairs that involve at least one D can be assumed to be in the same order. We therefore, as a conservative guess, can assume that whenever D is involved in base pair stacking, it contributes with a destabilizing penalty of ≈ 1.5 $\text{kcal} \cdot \text{mol}^{-1}$. Similar to Ψ-A pairs, this effect can be accommodated for with a correction term in the form of a soft constraint.

As a prerequisite, let us first extend the sequence encoding to explicitly account for dihydrouridines by encoding them as number 6:

```
# standard sequence encoding in ViennaRNA is
# 0 ... unknown nucleotide
# 1 ... adenine (A)
# 2 ... cytosine (C)
# 3 ... guanine (G)
# 4 ... uridine (U)
#
# We will further use
# 5 ... pseudouridine (P)
# 6 ... dihydrouridine (D)
def get_encoding_psi_D(sequence):
    # obtain encoding of unmodified sequence with masked blocking bases
    enc = list(RNA.seq_encode(unmodify_nonblocking(sequence)))

    # go through list of pseudo-uridine positions in our sequence
    for p in [ m.start() + 1 for m in re.finditer('P', sequence) ]:
        enc[p] = 5

    for d in [ m.start() + 1 for m in re.finditer('D', sequence) ]:
        enc[d] = 6

    return enc
```

Each D in the input sequence can now be easily identified. What remains now is to adjust the callback function that already takes care to account for Ψ-A stacks:

```python
# soft constraints callback for P-A stacks and destabilization of D
def psi_A_D_stack_correction(i, j, k, l, loop, data):
    correction = 0 # no correction (0) by default

    if (loop == RNA.DECOMP_PAIR_IL):
        # corrections for stacking interactions
        if  i + 1 == k and \
            j - 1 == l:
            # correct for known stacks that involve at
            # least one pseudouridine
            # 1. P-A pair enclosing other pair (k,l)
            if data[i] == 5 and data[j] == 1:
                correction += stacking_psi_A[data[k]][data[l]]
            # 2. A-P pair enclosed by other pair (i,j)
            elif data[k] == 1 and data[l] == 5:
                correction += stacking_psi_A[data[j]][data[i]]
            # 3. A-P pair enclosing other pair (k, l)
            elif data[i] == 1 and data[j] == 5:
                correction += stacking_A_psi[data[k]][data[l]]
            # 4. P-A pair enclosed by other pair (i,j)
            elif data[k] == 5 and data[l] == 1:
                correction += stacking_A_psi[data[j]][data[i]]

            if  data[i] == 6 or \
                data[j] == 6 or \
                data[k] == 6 or \
                data[l] == 6:
                # dihydrouridines de-stabilize stacking by at least
                # 1.5kcal/mol, Dalluge et al. 1996
                correction += 150

    return correction


# soft constraints callback for P-A stacks and destabilization of D
# partition function variant returning  Boltzmann factor
# with energy in cal/mol
def psi_A_D_stack_correction_exp(i, j, k, l, loop, data):
    # get energy correction in dacal/mol
    e = psi_A_D_stack_correction(i, j, k, l, loop, data)
    if e != 0:
        e = e * 10. # convert to cal/mol
        return math.exp(-e / RT) # return Boltzmann factor
    else:
        return 1.0 # return exp(0) = 1.0 by default
```

Note here that we added only one more condition that checks for any encoded dihydrouridines (encoded by integer 6). Whenever at least one D is present while evaluating a base pair stack the additional destabilizing contribution of 150 dacal · $\mathrm{mol}^{-1} = 1.5\ \mathrm{kcal} \cdot \mathrm{mol}^{-1}$ is added to the overall correction energy. Let us repeat the prediction part where we now use different variable names and the new callback function to not be confused with any previous predictions:

```
# create a fold compound for the sequence with RT blocking sites
fc_psi_D  = RNA.fold_compound(unmodify_nonblocking(sequence))

# bind P-A and D correction callbacks and sequence encoding
fc_psi_D.sc_add_f(psi_A_D_stack_correction)
fc_psi_D.sc_add_exp_f(psi_A_D_stack_correction_exp)
fc_psi_D.sc_add_data(get_encoding_psi_D(sequence))

# compute corresponding MFE and structure
ss_psi_D, mfe_psi_D = fc_psi_D.mfe()

# compute base pair distance to reference structure
dist_psi_D  = RNA.bp_distance(structure_ref, ss_psi_D)

# set scaling for subsequent partition function computations
fc_psi_D.exp_params_rescale(mfe_psi_D)

# compute partition function and pair probabilities
pp_psi_D, dG_psi_D = fc_psi_D.pf()

# compute ensemble defect wrt. reference structure
ensdef_psi_D = fc_psi_D.ensemble_defect(structure_ref)

# compute positional entropies
posent_psi_D = fc_psi_D.positional_entropy()

print("Ref: " + structure_ref)
print('MFE: {} ({:6.2f}) dist={:d}'.format(ss_psi_D, mfe_psi_D, dist_psi_D))
print('PF:  {} ({:6.2f}) d={:f}'.format(pp_psi_D, dG_psi_D, ensdef_psi_D))
```

The corresponding predictions are as follows

```
Ref: (((((((..((((........)))).(((((.......))))).....(((((.......))))))))))))....
MFE: (((((((..((((........)))).(((((.......)))))....(((.((.....)).))))))))))).... (-22.11) dist=10
PF:  (((((((..((((........)))).(((((.......)))))....,((,((.....}),))))))))))).... (-23.54) d=0.144076
```

Compared to the previous prediction where we explicitly handled Ψ-A pairs, the additional correction for the destabilization by D yields an MFE structure that is much closer to the correct clover-leaf shape of the reference structure in terms of base pair distance. Also, the ensemble defect decreased substantially to 0.144. This means that in equilibrium, most of the structures are actually quite similar to the reference structure. Nevertheless, the predictions still do not seem perfect, as the difference between the reference and the MFE structure is still 10 base pairs. To further investigate this discrepancy, let us investigate the 20 most energetically stable structures again:

```
print("20 Suboptimals with RT blocking, P-A pairs and destabilizing D")
for i, s in enumerate(fc_psi_D.subopt(300), start=1):
    print('{:2d} {} ({:6.2f}) dist={:d}'.format(\
        i, \
        s.structure, \
        s.energy, \
        RNA.bp_distance(structure_ref, s.structure)))
    if i == 20:
        break
```

which results in

```
20 Suboptimals with RT blocking, P-A pairs and destabilizing D
 1 (((((((..((((........)))).(((((.......)))))....(((.((.......)).)))))))))))).... (-22.11) dist=10
 2 (((((((..((((........)))).(((((.......))))).....(((((.......)))))))))))).... (-21.91) dist=0
 3 (((((((..((((........)))).((((((....).)))))....(((.((.......)).)))))))))))).... (-21.81) dist=11
 4 (((((((..((((........)))).((((((....).))))).....(((((.......)))))))))))).... (-21.61) dist=1
 5 ((((((...((((........)))).(((((.......)))))....(((.((.......)).))).)))))).... (-21.61) dist=11
 6 ((((((((.((((........)))).(((((.......))))).....((.((.......)).)))))))))).... (-21.51) dist=10
 7 (((((((..((((........))))..((((.......)))).....(((.((.......)).)))))))))))).... (-21.41) dist=11
 8 ((((((...((((........)))).(((((.......))))).....(((((.......))))).)))))).... (-21.41) dist=1
 9 ((((((...((((........)))).((((((....).)))))....(((.((.......)).))).)))))).... (-21.31) dist=12
10 ((((((((.((((........)))).((((((....).))))).....((.((.......)).)))))))))).... (-21.21) dist=11
11 (((((((..((((........)))).(((((.(....))))))....(((.((.......)).)))))))))))).... (-21.21) dist=11
12 (((((((..((((........))))..((((.......)))).....(((((.......)))))))))))).... (-21.21) dist=1
13 (((((((..((((........))))..(((((....).)))).....(((.((.......)).)))))))))))).... (-21.11) dist=12
14 ((((((...((((........)))).((((((....).))))).....(((((.......))))).)))))).... (-21.11) dist=2
15 ((((((((.((((........)))).(((((.......))))).....((((.......)))))))))))).... (-21.01) dist=2
16 (((((((..((((........)))).(((((.(....)))))).....(((((.......)))))))))))).... (-21.01) dist=1
17 (((((((..((((........))))..(((((....).)))).....(((((.......)))))))))))).... (-20.91) dist=2
18 ((((((...((((........))))..((((.......)))).....(((.((.......)).))).)))))).... (-20.91) dist=12
19 ((((((((.((((........))))..((((.......)))).....((.((.......)).)))))))))).... (-20.81) dist=11
20 ((((((((.((((........)))).((((((....).))))).....((((.......)))))))))))).... (-20.71) dist=3
```

and corresponding graphical visualization in Fig. 10.

Now, all of the investigated sub-optimal solutions resemble a clover-leaf shape. This is somewhat surprising, since we only moderately destabilized stacks containing dihydrouridines. Moreover, the correct reference structure is already at position 2 of the list, with just 0.2 $\text{kcal} \cdot \text{mol}^{-1}$ difference in energy compared to the predicted MFE. This is only about $1/3$ of thermal fluctuation ($RT \approx 0.6$ $\text{kcal} \cdot \text{mol}^{-1}$), so both solutions can be accepted optimal. A closer look at the difference between the two reveals that they only differ in the base pairing pattern of the T-arm. In particular, the T-arm helix of the predicted MFE structure is shifted by one nucleotide toward the $5'$ end. It might be possible that the m5C at position 49, or the tertiary interaction between m5U/T at position 54 and m1A at position 58 in the TΨC-arm (T-arm), or even both further shift the ensemble towards the reference structure [5, 6, 30].

3.11 Concluding Remarks

In this chapter, we have shown that accounting for the structural effects of post-transcriptional nucleotide modifications can drastically increase the prediction performance for RNA secondary structures. Those effects can be incorporated in predictions as *hard* constraints that effectively prevent particular modified bases from pairing, and *soft* constraints that change the stability of loop types, e.g., stacked base pairs.

However, in many cases, only a limited amount of data on such structural effects is known. In particular, physics-based secondary structure prediction requires accurate energy parameters for modified bases that, to-date, are only available for a very small number of bases and loop types. Moreover, even if some data is available, there is no guarantee that such auxiliary information actually increases the prediction performance. Some modifications may act cooperatively, i.e., the RNA only changes the structure toward the functional conformation if multiple modified bases are accounted for properly [5].

In fact, the adaptations to the structure prediction algorithm within this chapter yield better predictions for the Human tRNA

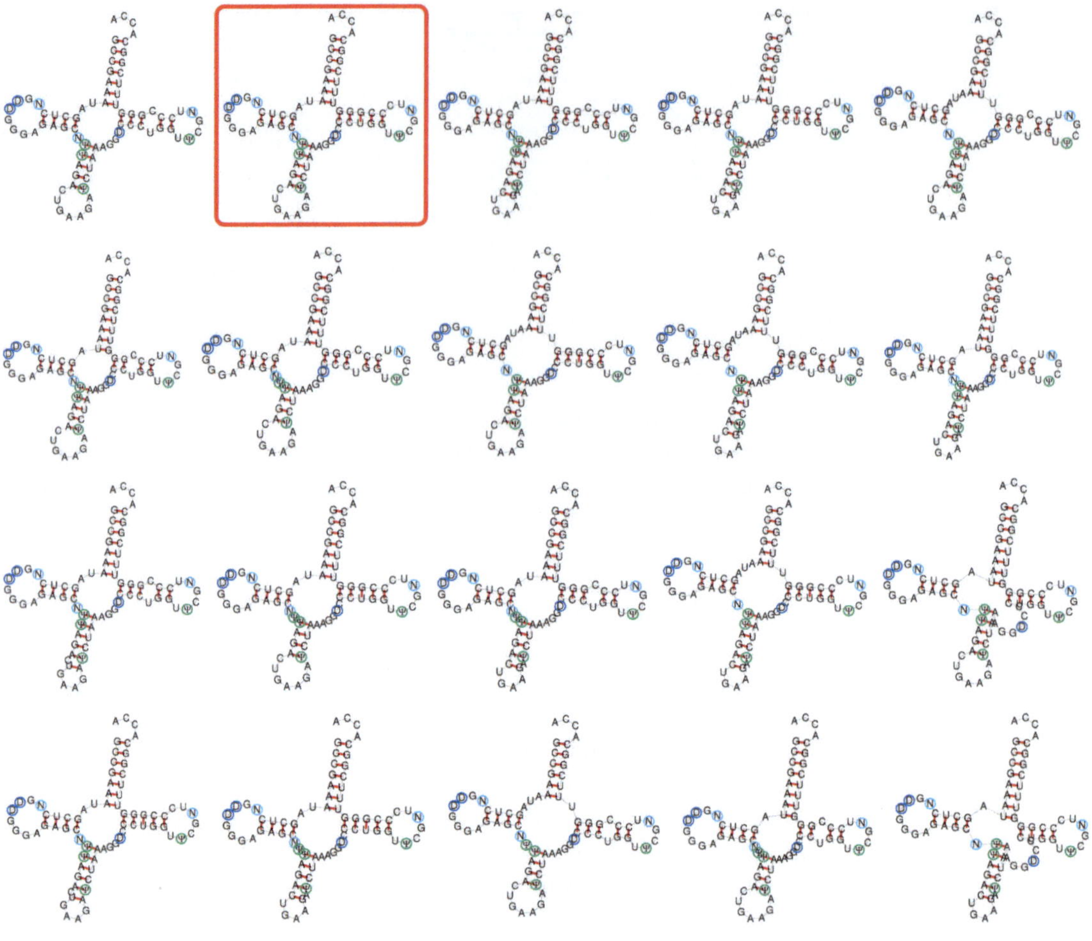

Fig. 10 Predicted suboptimal structures for the tRNA sequence where all bases known to block reverse transcriptase (RT) are prohibited to base pair and energy correction for stacks consisting of D and/or Ψ-A pairs. Shown are the 20 energetically best secondary structures, sorted by free energy (left to right, top to bottom). Specially treated modified bases are marked by circles of different blue colors. Two of the three dihydrouridines are located in the D-loop (left arm) and one in the four-way junction between anticodon arm and T-arm (lower right). All of the 20 suboptimal solutions resemble the clover-leaf shape, and only differ in a few (shifted) base pairs. The correct reference structure is at position 2 (red rectangle)

Phe example. But a more general view on all 623 tRNAs within the full `tRNAdb` data set draws a more intricate picture. To express and compare the prediction performance in terms of correctly predicted base pairs, one usually computes the *true positive rate* (TPR), also known as *sensitivity*, as well as the *positive predictive value* (PPV) with

$$\mathrm{TPR} = \frac{\mathrm{TP}}{\mathrm{TP} + \mathrm{FN}} \qquad \mathrm{PPV} = \frac{\mathrm{TP}}{\mathrm{TP} + \mathrm{FP}}. \tag{10}$$

These values can be obtained from (i) the number of true positives (TP), i.e., the actual base pairs within the reference structure,

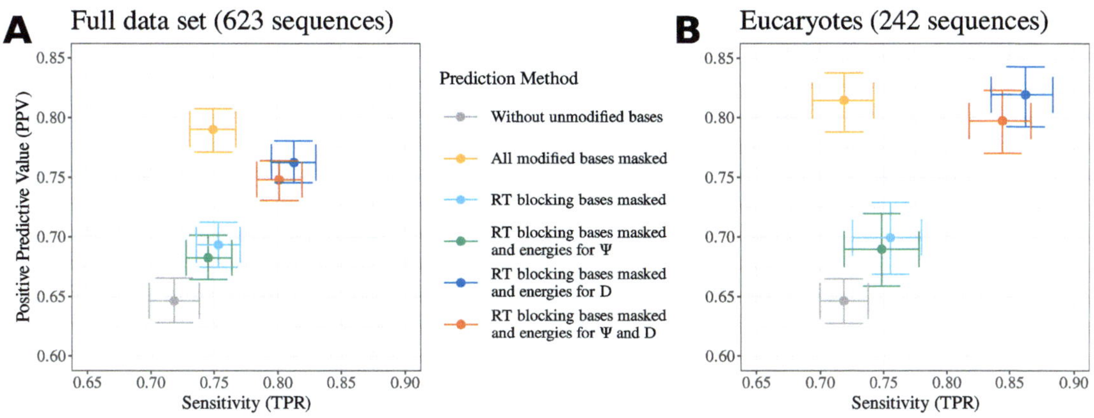

Fig. 11 Prediction performance on `tRNAdb` data set. Shown here are the average sensitivity/true positive rate (TPR) and positive predictive value (PPV) for secondary structure predictions with respect to the reference structures available at `tRNAdb` for (**a**) the full data set and (**b**) the subset of eukaryotic tRNAs. Error bars indicate 95% confidence intervals as obtained by bootstrapping with 1,000 iterations. Here, we distinguish six different MFE prediction settings, highlighted by color: (i) all modified bases are replaced by their unmodified counterpart (grey), (ii) all modified bases are masked, i.e., constrained to stay unpaired (dark yellow), (iii) modified bases are masked only if they are known to block reverse transcriptase (RT) (light blue), (iv) modified bases are masked if they block RT plus adjusted stacking energies for Ψ-A pairs (green), (v) modified bases are masked if they block RT plus adjusted destabilization of base pair stacks involving dihydrouridine (D) (dark blue), and (vi) modified bases are masked if they block RT plus energy corrections for both, Ψ-A pairs and stacks with D

(ii) the number of false positives (FP), i.e., the predicted base pairs that are not part of the reference structure, and (iii) the number of false negatives (FN), i.e., the base pairs that are part of the reference structure but missing in the prediction.

As shown in Fig. 11, a special treatment of modified bases during structure prediction, on average, indeed increases the prediction performance. However, compared to only incorporating the destabilization of D, predictions are slightly worse when both modifications, pseudouridine and dihydrouridine, are accounted for. It seems that for many tRNAs within the data set, the extension with Ψ-A stacking energies decreases the TPR and PPV. A deeper investigation of this effect for the different types of tRNAs and the corresponding domains of life (eukaryotes, bacteria, archaea, and viruses) draws a quite diverse set of pictures (*see* Fig. 11b for the subset of eukaryotic tRNAs, other subsets not shown). These discrepancies may be attributed to cooperativity effects as well as to different usage of the repertoire of base modifications within the kingdoms of life. Note here that, on average, PPV for the entire `tRNAdb` data set is highest when all modified bases are treated as "*must not pair*," without any further constraints. This behavior is an artefact of the underlying constraints: the more potential base pairs are forbidden, the less FP base pairs can be predicted. Hence, the higher the PPV.

4 Notes

1. **Installation into a non-standard directory**. Linux or Mac OS X users who do not possess *sudo* privileges or simply want to install the programs and Python bindings of the ViennaRNA Package into a different location can specify the installation target directory at configure-time. The target directory path then simply needs to be passed to the additional --prefix flag when running the configure script:

```
$ ./configure --prefix=/your/path/
```

Here /your/path/ is the target path, e.g., your home directory. All programs and libraries will then be installed into directories /your/path/bin, /your/path/lib, /your/path/share, etc. To make the installed executable programs and more importantly the Python interface available, we need to include the installation path in PATH and PYTHONPATH environment variables, e.g., by adding the lines

```
PATH=/your/path/bin:${PATH}
PYTHONPATH=/your/path/lib/python3.8/site-packages:${PYTHONPATH}
```

at the end of the *bash* configuration file ~/.bashrc. Note that the above PYTHONPATH contains the version number of the Python interpreter used at the time of installation and may need to be adjusted according to your current setup. Also note that on Linux systems, the most convenient installation target directory may be $HOME/.local, since Python interpreters usually by default include $HOME/.local/lib/python3.x/site-packages in their search path for external packages and, thus, explicit setting of the PYTHONPATH variable is not required.

2. In principle, BioPython already provides a rich source of high-level functions for bioinformatics file format parsing and writing. However, while the corresponding SeqIO.parse() function is able to read FASTA formatted input files, it only expects sequence input, not structure information in dot-parenthesis format stored within the record. Therefore, SeqIO.parse() treats any secondary structure information in the input file as if it were part of the sequence. Nevertheless, in cases where one can be absolutely sure that the input file contains dot-bracket-formatted secondary structure information after the sequence block, the following code may be of help to separate the two:

```python
from Bio import SeqIO

with open("files/tRNA_phe.fasta", "r") as handle:
    for record in SeqIO.parse(handle, "fasta"):
        # get FASTA header
        fasta_header  = str(record.id)
        # extract actual sequence part (first half)
        sequence      = str(record.seq[:int(len(record.seq)/2)])
        # extract structure part (second half)
        structure_ref = str(record.seq[int(len(record.seq)/2):])
```

3. The default output file format for secondary structure layout plots in the `ViennaRNA Package` is *Encapsulated PostScript* (EPS). EPS files are simple text-files with printer instructions to render the final plot. The `Python` interface of the `ViennaRNA Package` provides the `file_ps_rnaplot()` function that allows for creating an RNA structure layout in EPS format given a sequence and corresponding dot-parenthesis structure. For instance, the following lines of code produce a simple black and white structure layout file named *ref_struct_simple.eps* for the tRNA reference used in this chapter:

```python
import RNA
seq = "GCCGAAAUALCUC\"GDDGGGAGAGCRPPAGABU#AAWAPCUAAAG\
       7DC?CUGGTPCG\"UCCCGGGUUUCGGCACCA"
ss  = "(((((((..((((........)))).(((((.......)))))...\
       ..(((((.......))))))))))))...."

RNA.file_PS_rnaplot(seq, ss, "ref_struct_simple.eps", None)
```

To annotate the layout, one can use `file_ps_rnaplot_a()` function and pass additional PostScript code. The files generated with the `file_PS_rnaplot_a()` function already provide some convenience PostScript functions suitable for annotation, but any PostScript code may be passed. For instance, to mark all modified bases with a dark-yellow circle, one may extend the above example as follows:

```python
import re
# create a list of sequence positions of the modified bases
all_mods = [m.start() + 1 for m in re.finditer('N', seq)]
# change color to dark-yellow and linewidth to 1.5
annot = "0.9 0.62 0 setrgbcolor 1.5 setlinewidth "
# create a string with cmark convenience commands
annot += " ".join(["{:d} cmark".format(i) for i in all_mods])
# wrap the annotation string into gsave/grestore to keep color
# and line width changes local
annot = "gsave " + annot + " grestore"
# create EPS file
RNA.file_PS_rnaplot_a(seq, ss, "ref_struct_mod.eps", "", annot)
```

Have a look at the two EPS files using an EPS viewer such as `gv` or `evince` in a `Linux` environment, or `preview` in Mac OSX.

More comprehensive examples can be constructed by changing the `annot` string accordingly. The complete set of convenience PostScript functions can be found in the resulting EPS text file below the line

```
% extra definitions for standard annotations
```

Use a regular text editor to open the EPS file, not an EPS viewer or word processor. For further information, you may also want to consult a PostScript programming language manual.

4. Similar to **Note** 3, a structure layout can be annotated by the *positional entropy*. For that purpose, one may use the `Perl 5` script "*relplot.pl*" that comes with the `ViennaRNA Package` installation. However, this script requires an existing structure layout EPS file as well as the corresponding dot-plot with the base pair probabilities. In Subheading 3.7, we already computed all required data but did not create any output files yet. To prevent us from creating unnecessary output files and running additional `Perl` code, one can write some `Python` code that mimics the function of the *relplot.pl* script.

   Assuming that the positional entropies are stored in variable `posent` (*see* Subheading 3.7), the following code produces the required annotation PostScript function, creates a PostScript array that consists of the positional entropy data, and finally produces an RNA structure layout file "*ref_struct_-posent.eps*" that is annotated by the positional entropies:

```python
# PostScript function to draw colored circles (taken from relplot.pl)
relplot_fun = """
/range 0.8 def
/drawreliability {
  /Smax 2.3 def
  0
  coor {
    aload pop
    S 3 index get
    Smax div range mul
    invert {range exch sub} if
    dup 0 ge
    {1 1 sethsbcolor} {pop 1 1 1 setrgbcolor} ifelse
    newpath
    fsize 2 div 0 360 arc
    fill
    1 add
  } forall
} bind def
"""
# positional entropy data (one per row)
relplot_dat = "\n".join(["{:6f}".format(i) for i in posent[1:]])
# PostScript array for the positional entropies
relplot_def = "/S [" + relplot_dat + " ] def"
# merge all PostScript blocks
PS_code = [
    relplot_fun,
    relplot_def,
    "/invert false def",
    "drawreliability"
]
annot = "\n".join(PS_code)
# create annotated output file
RNA.file_PS_rnaplot_a(seq, ss, "ref_struct_posent.eps", annot, "")
```

5. In Eq. (8), we state that it is possible to compute the ensemble distance $ED(s)$ without exhaustive summation over all structures t. Here, we provide a complete derivation for this claim.

The transformations make use of the pairing probabilities p_{ij}, the alternative form of the base pair distance

$$d_{\mathrm{BP}}(s,t) = \sum_{i<j} \left(\delta_{ij}(s) + \delta_{ij}(t) - 2\delta_{ij}(s)\delta_{ij}(t)\right) \tag{11}$$

and the two indicator functions

$$\delta_{ij}(s) = \begin{cases} 1, & \text{if } (i,j) \in s \\ 0, & \text{otherwise} \end{cases} \tag{12}$$

$$\bar{\delta}_i(s) = \begin{cases} 1, & \text{if } \nexists j \mid (i,j) \in s \\ 0, & \text{otherwise} \end{cases} \tag{13}$$

that denote (12) whether or not base pair (i,j) is present in s and (13) whether position i is unpaired in s, respectively. Let $q_i = \sum_j (1 - p_{ij})$ further denote the probability that position i is unpaired, then the full derivation becomes

$$\begin{aligned}
\mathrm{ED}(s) &= \frac{1}{n}\sum_t p(t)2d_{\mathrm{BP}}(s,t) \\
&= \frac{1}{n}\sum_t p(t)2\sum_{i<j}\left(\delta_{ij}(s) + \delta_{ij}(t) - 2\delta_{ij}(s)\delta_{ij}(t)\right) \\
&= \frac{1}{n}\sum_t p(t)\sum_{ij}\left(\delta_{ij}(s) + \delta_{ij}(t) - 2\delta_{ij}(s)\delta_{ij}(t)\right) \\
&= \frac{1}{n}\sum_{ij}\left(\delta_{ij}(s)\sum_t p(t) + \sum_t p(t)\delta_{ij}(t) - 2\delta_{ij}(s)\sum_t p(t)\delta_{ij}(t)\right) \\
&= \frac{1}{n}\sum_{ij}\left(\delta_{ij}(s) + p_{ij} - 2\delta_{ij}(s)p_{ij}\right) \\
&= \frac{1}{n}\sum_{ij}(1 - \delta_{ij}(s))p_{ij} + \frac{1}{n}\sum_{ij}(1 - p_{ij})\delta_{ij}(s) \\
&= \frac{1}{n}\sum_i \bar{\delta}_i(s)\sum_j p_{ij} + \frac{1}{n}\sum_{ij}\delta_{ij}(s) - \frac{1}{n}\sum_{ij} p_{ij}\delta_{ij}(s) \\
&= \frac{1}{n}\sum_i \bar{\delta}_i(s)(n - q_i) + \frac{1}{n}\sum_{ij}\delta_{ij}(s) - \frac{1}{n}\sum_{ij} p_{ij}\delta_{ij}(s) \\
&= \frac{1}{n}\left(\sum_i \bar{\delta}_i(s)n + \sum_{ij}\delta_{ij}(s)\right) - \frac{1}{n}\sum_i \bar{\delta}_i(s)q_i - \frac{1}{n}\sum_{ij}\delta_{ij}(s)p_{ij} \\
&= 1 - \frac{1}{n}\sum_{ij}\delta_{ij}(s)p_{ij} - \frac{1}{n}\sum_i \bar{\delta}_i(s)q_i
\end{aligned} \tag{14}$$

6. The `ViennaRNA` Package internally uses an integer encoding for the RNA sequences it operates on. This allows one to easily address positions within a list (or array) through the encoding instead of using a dictionary (or hash) with the sequence letter as key. At first sight, both variants seem similar, both exhibit an asymptotic time complexity of $O(1)$ to retrieve a particular value. However, when the number of retrievals approaches a few millions or more, the burden of the hash function can quickly become a bottle neck.

References

1. Hopfinger MC, Kirkpatrick CC, Znosko BM (2020) Predictions and analyses of RNA nearest neighbor parameters for modified nucleotides. Nucleic Acids Res 48:8901–8913. https://doi.org/10.1093/nar/gkaa654.
2. Agris PF (1996) The importance of being modified: roles of modified nucleosides and Mg2+ in RNA structure and function. Progr Nucleic Acid Res Mol Biol 53:79–129. https://doi.org/10.1016/S0079-6603(08)60143-9
3. Denmon AP, Wang J, Nikonowicz EP (2011) Conformation effects of base modification on the anticodon stem-loop of Bacillus subtilis tRNATyr. J Mol Biol 412:285–303. https://doi.org/10.1016/j.jmb.2011.07.010
4. Seelam PP, Sharma P, Mitra A (2017) Structural landscape of base pairs containing post-transcriptional modifications in RNA. RNA 23(6):847–859. https://doi.org/10.1261/rna.060749.117
5. Lorenz C, Lünse CE, Mörl M (2017) tRNA modifications: impact on structure and thermal adaptation. Biomolecules 7(2):35. https://doi.org/10.3390/biom7020035
6. Helm M (2006) Post-transcriptional nucleotide modification and alternative folding of RNA. Nucl Acids Res 34:721–733. https://doi.org/10.1093/nar/gkj471
7. Harcourt EM, Kietrys AM, Kool ET (2017) Chemical and structural effects of base modifications in messenger RNA. Nature 541:339–346. https://doi.org/10.1038/nature21351
8. Machnicka MA, Olchowik A, Grosjean H, Bujnicki JM (2014) Distribution and frequencies of post-transcriptional modifications in tRNAs. RNA Biol 11:1619–1629. https://doi.org/10.4161/15476286.2014.992273
9. Tanzer A, Hofacker IL, Lorenz R (2018) RNA modifications in structure prediction - status quo and future challenges. Methods 156:32–39. https://doi.org/10.1016/j.ymeth.2018.10.019
10. Reuter JS, Mathews DH (2010) RNAstructure: software for RNA secondary structure prediction and analysis. BMC Bioinformat 11:3. https://doi.org/10.1186/1471-2105-11-129
11. Lorenz R, Bernhart SH, Höner zu Siederdissen C, Tafer H, Flamm C, Stadler PF, Hofacker IL (2011) ViennaRNA package 2.0. Algorithms Mol Biol 6(1):26. https://doi.org/10.1186/1748-7188-6-26
12. Pan T (2018) Modifications and functional genomics of human transfer RNA. Cell Res 28:395–404. https://doi.org/10.1038/s41422-018-0013-y
13. Lyons SM, Fay MM, Ivanov P (2018) The role of RNA modifications in the regulation of tRNA cleavage. FEBS Lett 592:2828–2844. https://doi.org/10.1002/1873-3468.13205
14. Jühling F, Mörl M, Hartmann RK, Sprinzl M, Stadler PF, Pütz J (2009) tRNAdb 2009: compilation of tRNA sequences and tRNA genes. Nucl Acids Res 37(suppl_1):D159–D162. https://doi.org/10.1093/nar/gkn772
15. Dunin-Horkawicz S, Czerwoniec A, Gajda MJ, Feder M, Grosjean H, Bujnicki JM (2006) MODOMICS: a database of RNA modification pathways. Nucl Acids Res 34. https://doi.org/10.1093/nar/gkj084
16. Boccaletto P, Machnicka MA, Purta E, Piątkowski P, Bagiński B, Wirecki TK, de Crécy-Lagard V, Ross R, Limbach PA, Kotter A et al (2018) MODOMICS: a database of RNA modification pathways. 2017 update. Nucl Acids Res 46:D303–D307. https://doi.org/10.1093/nar/gkx1030
17. Cantara WA, Crain PF, Rozenski J, McCloskey JA, Harris KA, Zhang X, Vendeix FAP, Fabris D, Agris PF (2011) The RNA, modification database, RNAMDB, (2011) update. Nucl Acids Res 39. https://doi.org/10.1093/nar/gkq1028
18. McCloskey JA, Rozenski J (2005) The small subunit rRNA modification database. Nucl

Acids Res 33. https://doi.org/10.1093/nar/gki015

19. Grüning B, Dale R, Sjödin A, Chapman BA, Rowe J, Tomkins-Tinch CH, Valieris R, Köster J (2018) Bioconda: sustainable and comprehensive software distribution for the life sciences. Nat Methods 15(7):475–476. https://doi.org/10.1038/s41592-018-0046-7
20. Zuker M (1989) On finding all suboptimal foldings of an RNA molecule. Science 244(4900):48–52. https://doi.org/10.1126/science.2468181
21. Wuchty S, Fontana W, Hofacker IL, Schuster P (1999) Complete suboptimal folding of RNA and the stability of secondary structures. Biopolymers 49(2):145–165. https://doi.org/10.1002/(SICI)1097-0282(199902)49:2%3C145::AID-BIP4%3E3.0.CO;2-G
22. McCaskill JS (1990) The equilibrium partition function and base pair binding probabilities for RNA secondary structure. Biopolymers 29(6–7):1105–1119. https://doi.org/10.1002/bip.360290621
23. Zadeh JN, Wolfe BR, Pierce NA (2011) Nucleic acid sequence design via efficient ensemble defect optimization. J Comput Chem 32(3):439–452. https://doi.org/10.1002/jcc.21633
24. Wolfe BR, Pierce NA (2015) Sequence design for a test tube of interacting nucleic acid strands. ACS Synth Biol 4(10):1086–1100. https://doi.org/10.1021/sb5002196
25. Bevilacqua PC, Ritchey LE, Su Z, Assmann SM (2016) Genome-wide analysis of RNA secondary structure. Ann Rev Genet 50:235–266. https://doi.org/10.1146/annurev-genet-120215-035034
26. Motorin Y, Muller S, Behm-Ansmant I, Branlant C (2007) Identification of modified residues in RNAs by reverse transcription-based methods. Methods Enzymol 425:21–53. https://doi.org/10.1016/S0076-6879(07)25002-5
27. Hudson GA, Bloomingdale RJ, Znosko BM (2013) Thermodynamic contribution and nearest-neighbor parameters of pseudouridine-adenosine base pairs in oligoribonucleotides. RNA 19(11):1474–1482. https://doi.org/10.1261/rna.039610.113
28. Lorenz R, Hofacker IL, Stadler PF (2016) RNA folding with hard and soft constraints. Algor Mol Biol 11(1):1–13. https://doi.org/10.1186/s13015-016-0070-z
29. Dalluge JJ, Hashizume T, Sopchik AE, McCloskey JA, Davis DR (1996) Conformational flexibility in RNA: the role of dihydrouridine. Nucl Acids Res 24(6):1073–1079. https://doi.org/10.1093/nar/24.6.1073
30. Motorin Y, Helm M (2010) tRNA stabilization by modified nucleotides. Biochemistry 49(24):4934–4944. https://doi.org/10.1021/bi100408z

Chapter 9

How to do RNA-RNA Interaction Prediction? A Use-Case Driven Handbook Using IntaRNA

Martin Raden and Milad Miladi

Abstract

Computational prediction of RNA-RNA interactions (RRI) is a central methodology for the specific investigation of inter-molecular RNA interactions and regulatory effects of non-coding RNAs like eukaryotic microRNAs or prokaryotic small RNAs. Available methods can be classified according to their underlying prediction strategies, each implicating specific capabilities and restrictions often not transparent to the non-expert user. Within this work, we review seven classes of RRI prediction strategies and discuss the advantages and limitations of respective tools, since such knowledge is essential for selecting the right tool in the first place.

Among the RRI prediction strategies, accessibility-based approaches have been shown to provide the most reliable predictions. Here, we describe how IntaRNA, as one of the state-of-the-art accessibility-based tools, can be applied in various use cases for the task of computational RRI prediction. Detailed hands-on examples for individual RRI predictions as well as large-scale target prediction scenarios are provided. We illustrate the flexibility and capabilities of IntaRNA through the examples. Each example is designed using real-life data from the literature and is accompanied by instructions on interpreting the respective results from IntaRNA output. Our use-case driven instructions enable non-expert users to comprehensively understand and utilize IntaRNA's features for effective RRI predictions.

Key words RNA-RNA interaction, Computational methods, Review, IntaRNA, Handbook, Application

1 Introduction

Inter-molecular base pairing between RNA molecules, called RNA-RNA interaction (RRI), is key to the regulatory functions of non-coding RNAs (ncRNAs) [23]. For instance, many prokaryotic small RNAs (sRNAs) bind to the ribosomal binding site of the target messenger RNAs (mRNAs) [65, 73], as a functional mechanism to hinder the mRNA translation. In eukaryotes, the embedding of micro RNAs (miRNAs) into ribonucleoprotein complexes enables precise mRNA target recognition that is based on miRNA-mRNA interaction formation [26, 28, 50]. Similarly, small nuclear

Ronny Lorenz (ed.), *RNA Folding: Methods and Protocols*, Methods in Molecular Biology, vol. 2726,
https://doi.org/10.1007/978-1-0716-3519-3_9,

RNAs (snRNAs) utilize the RRI principle to recognize the exon-intron boundary regions for splicing the precursor mRNAs [42]. Thus, the reliable identification of RRIs is essential for studying the complex interrelations of many cellular regulatory networks.

In the last decade, various experimental approaches for the large scale transcriptome-wide identification of RRIs have been introduced (*see* [23] for a review). Typically, the information from such experiments is restricted to the identification of the interacting RNA pairs that are accompanied by hints where the RRI occurs over the RNA transcripts. These experiments are however sensitive to transcript expression levels, in a way e.g., the functional RRIs can be missed if the expression level of interacting partners is low. Importantly, the experimental outcome cannot be easily generalized for the other tissues that tend to have different expression dynamics. Therefore, despite the usability of experimental approaches, for the detailed and comprehensive study of the regulatory mechanisms of RRIs and to explore custom sets of RNAs for their interaction potential, computational RRI prediction methods are still a crucial asset and are widely applied. In other words, the concerted application of computational and experimental methods is key to reduce respective biases and to complement each other.

1.1 Objectives of this Handbook

Within this work, we provide a handbook on how to do RNA-RNA interaction prediction using the tool INTARNA [11, 40], which is one of the leading tools in terms of accuracy and applicability for RRI prediction [34, 36, 70]. To better understand the possibilities and limitations of INTARNA and other computational RRI prediction, we first briefly review and classify existing RRI prediction strategies and respective available tools.

Subsequently, we provide a brief introduction to INTARNA and the underlying algorithms to highlight and discuss its advantages and limitations. Application scenarios of INTARNA are exemplified by several use cases from literature accompanied with discussions that guide the reader through the interpretation of the results. The latter will enable non-expert users to correctly use and utilize INTARNA in their work. To gain the most from the flexibility and comprehensive feature set of INTARNA, the hands-on examples are making use of the command-line interface rather than INTARNA's webserver [51, 78].

1.2 Computational RRI Prediction

In the most basic form, an RRI prediction tool gets as input two RNA sequences that have a putative interaction potential. The prediction result would be the optimal or a set of suboptimal pairs of interaction sites within the two sequences. The interaction sites are typically subsequences of the original input that form stable structures via complementary base pairs.

The in silico prediction of RNA-RNA interactions is typically based on thermodynamic models with many commonalities with

Table 1
Classes of RNA-RNA interaction (RRI) prediction approaches, their main advantages and limitations

RRI prediction strategy	Main advantages	Main limitations
Hybridization-only	Fast	Ignores intra pairing, inflated length & energy
Simultaneous RRI & folding	Comprehensive interaction model	Extremely slow
Concatenation-based	Effective for short RNAs	No interaction of loops
Accessibility-based *exact*	Accurate prediction	Slow, no intra-mol. structure in RRI
Accessibility-based *heuristics*	Fast, accurate prediction	Sensitive to quality of accessibility
Two-step RRI	Very fast, scalable	Compromising sensitive
Comparative	Evolutionary conserved interaction	Requires homology info, only conserved RRI identified
Meta-tool	Improved quality	Difficult to balance contributions

intra-molecular RNA structure prediction. RRI prediction approaches can be classified by the algorithmic strategy, which implies how RRIs are modeled and which interaction patterns are considered. Table 1 provides a first overview of major classes. Respective tools are introduced in the following. For details on the algorithms and the theoretical concepts, please refer to [52]. The performance evaluation and benchmarks of the tools are available e.g., in [36, 48, 70].

1.2.1 Hybridization-Only RRI Prediction

A straightforward approach to predict RRIs is to identify two interacting subsequences that form the most stable inter-molecular helix (or sequence of loop-separated helices). This can be done analogously to sequence alignment strategies while summing up energy contributions due to inter-molecular base-pairings. Thus, the algorithms have a time and space complexity of $O(nm)$, where n, m denote sequence lengths of the interacting RNAs. Representatives are e.g., RNAHYBRID [56], MIRANDA [18], RNADUPLEX [37], DUPLEXFOLD [57], RISEARCH [74], or TARGETRNA [69]. While known to be fast, such hybridization-focused models ignore the competition between inter- and intra-molecular base pairing, i.e., that interaction formation is competing with the intra-molecular pairings of already folded RNAs. Thus, respective tools often overestimate interaction length and strength by ignoring the RNAs' structures.

1.2.2 Simultaneous RRI and Structure Prediction

A more accurate model of RRI prediction thus requires the simultaneous prediction of inter- and intra-molecular base pairs. One such general RRI model was introduced by Dmitri D. Pervouchine in the tool IRIS [49], which models combinations of inter-molecular helices that link two RNA molecules forming non-crossing structures. While this provides a powerful framework, the high computational time complexity of $O(n^3m^3)$ limits its applicability [72]. The same restrictions applies to similar tools like PIRNA [58] or RACTIP [32].

1.2.3 Concatenation-Based RRI Prediction

An alternative way to optimize the combination of both inter- and intra-molecular base pairs is to concatenate both sequences via a dummy linker sequence. The resulting pseudo sequence is then used as input to a modified RNA structure prediction approach, which is constrained to have the linker region unpaired and its enclosing loop properly scored. Eventually, a joint structure of the concatenated sequences can be predicted with similar time complexity to RNA folding, i.e., $O((n+m)^3)$. Examples of tools implementing this strategy are UNAFOLD [41] (former MFOLD [82]), PAIRFOLD [4], RNACOFOLD [10, 37], and BIFOLD [57].

While this works well if one of the RNAs is very short (e.g., miRNAs [45]), a central problem results from the used RNA structure prediction approach. More specifically, all "Zuker-like" approaches [83] predict only non-crossing joint structures. This limits the scope of considered interaction patterns by this approach. For instance, no interactions between loops can be predicted, since these would result in a crossing (pseudo-knot) pattern within the joint structure. One way to overcome this problem is to use and extend a structure prediction approach that considers crossing structures to the cost of increased time complexities, as e.g., done in the NUPACK package [16].

1.2.4 Accessibility-Based RRI Prediction

Another strategy to incorporate intra-molecular structure formation into RRI prediction is by adding penalty terms if an interaction spans structured subsequences [46]. To this end, first sequence-specific unpaired probabilities $P^u(i..j)$ for all subsequences $i..j$ (from position i to position j) of each RNA are precomputed. These terms represent how likely a subsequence is ***not involved*** in intra-molecular base pairing, among the whole structural ensemble that can be formed by the sequence. This *accessibility* measure P^u is converted into penalty terms $\Delta(i..j) = -RT\log(P^u(i..j))$, where R denotes the gas constant and T the temperature. The Δ terms reflect the amount of energy needed to make the respective subsequence unpaired (and thus ready for inter-molecular base-pair formation). Eventually, the energy of all possible interactions are computed by its hybridization energy (similar to hybridization-only approaches) plus the respective Δ-penalty terms of the subsequences that cover the inter-molecular base pairs. The interaction with minimal energy is reported as the optimal interaction. Further details are discussed below in Subheading 3.1.1 and are further elaborated in [52].

This approach was first implemented in RNAUP [46]. It enables highly accurate interaction prediction [70]. However, the computational complexity is high with an $O(n^2m^2)$ time complexity. To overcome the high complexity, heuristics have been introduced to speedup the accessibility-based RRI prediction. One way to speedup the computations is by reducing the number of considered interactions. For instance, INTARNA [11, 40] considers for RRI extension only optimal subinteractions, which excludes the majority of possibilities that are not part of stable RRIs. That way, time complexity can be reduced by two orders of magnitude from $O(n^2m^2)$ (e.g., in RNAUP) to $O(nm)$ (e.g., in INTARNA), while keeping the prediction quality on a near optimal level. INTARNA – like RNAUP – still makes use of accessibility terms for explicit subsequences as computed by RNAPLFOLD [9, 27]. An alternative heuristic to simplify the model was introduced with RNAPLEX and its extended web interface RNAPREDATOR [17, 66, 67] such that the accessibility computations are simplified. That is, instead of explicitly computing and considering Δ terms for subsequences, they are approximated by combining single-positing accessibilities. This scheme enables the integration of position-wise accessibility penalties into hybridization-only approaches, resulting in the same lower time complexity of $O(nm)$. Similar strategies have also been applied in ACCESSFOLD [15] and RIBLAST [19].

1.2.5 Two-Step RRI Prediction

To further speedup RRI prediction, some recent works introduced a two-step scheme. Using this scheme, first a fast and simple method is applied to identify promising (sub)interactions that are subsequently refined in the second step with more accurate models e.g., by using an accessibility-based method like INTARNA [68]. Possible pre-filters for the first step are local alignment methods like BLASTN [3] or GUUGLE [22], which align one RNA to the reverse complement of the other RNA and identify nearly perfect inter-molecular (sub)helices. These inter-molecular helices serve as anchors for subsequent RRI extension on both ends, as e.g., done in ASSA [5] or PSRNATARGET [14].

To increase prediction quality, many tools incorporate additional features derived from biological observations. For instance, eukaryotic sRNAs and prokaryotic miRNAs are known to form stable subinteractions of about seven consecutive base pairs with their regulated target RNAs, a so called seed interaction. Thus, tools like TARGETSCAN [1], RISEARCH2 [2], STARPICKER [81], RIBLAST [19], and the recent version of INTARNA (version 3.0+) identify and extend putative seeds and provide powerful and high-performance instruments for large scale genome-wide RRI analyses.

1.2.6 Comparative RRI Prediction

RRI prediction tools tend to have high false-positive prediction rates when they are applied in genome-wide target screens. To overcome this, comparative approaches have been introduced. Such methods combine target predictions from homologous regions/genes from evolutionary related organisms to identify conserved regulatory pathways among species. To this end, TARGETRNA2 incorporates the conservation of sRNA queries among species [33] to better identify common target RNAs. As an alternative, tools like PETCOFOLD [62], RNAALIDUPLEX [37], and a variant of RNAPLEX utilize sequence alignment input for RRI prediction. This allows for instance the automatic identification of conserved subsequences that are central for RRI formation and thus can provide an important constraint for prediction. A more comprehensive approach, namely COPRARNA [77], combines statistics on phylogenetically related INTARNA target predictions to identify common and thus evolutionary conserved targets. While COPRARNA was shown to provide much fewer false-positive predictions [48, 77], such methods are restricted to evolutionary conserved interactions and are thus blind to species-specific RNAs or RRIs [31].

1.2.7 Meta-tool RRI Prediction

A final class of RRI prediction approaches tries to amend for the limitations of the individual tools by combining their results, as e.g., manually done in [64]. One representative is SPOT [34], which uses INTARNA, TARGETRNA2, STARPICKER, and COPRARNA. It was shown that predictions supported by more than one tool can improve the overall prediction quality, while the overall performance of SPOT is dominated by INTARNA.

2 Materials

2.1 System Requirements

To reproduce the examples of this handbook, only a standard PC is needed. While in general all major operating systems like Microsoft Windows, Linux, and Apple macOS are supported, we will assume and use in the following a Linux-based system using a `bash` shell, which is available also for Windows 10 and macOS. Other command-line interfaces can be also used. Eventually, the examples and respective INTARNA calls are independent of the used shell and operating system.

We recommend the use of a basic text editor, with monospaced font without formatting, to create and edit the input text files in FASTA format. Some examples are based on files from freely available online repositories. They have to be downloaded using the provided links.

2.2 Availability and Installation of INTARNA

INTARNA is an open-source project that is continuously developed and extended [11, 21, 40, 44]. Originally built for command-line usage and embedding in high-throughput pipelines, it is also available via its web interface within the Freiburg RNA tools webserver [51, 78]. The webserver is useful for easy access and usage of INTARNA but has restrictions for input and parameters to cap the usage of computational resources. Therefore, the subsequent use-case scenarios will be based on INTARNA's command-line interface and thus require a local installation. INTARNA's software project page provides detailed installation and usage instructions, source code, and for major releases precompiled INTARNA programs for Microsoft Windows. A convenient way to install INTARNA over different platforms is by using the CONDA[1] package manager [24], which is shown in Example 2.1. This handbook is based on INTARNA version 3.2.0 compiled with Vienna RNA package version 2.4.14. Slight deviations from the mentioned results can be expected using more recent versions. For getting an overview of available options, a compact overview of INTARNA's parameters can be obtained with "`IntaRNA --help`" while "`IntaRNA --fullhelp`" lists all available options.

```
IntaRNA software:
https://github.com/BackofenLab/IntaRNA
IntaRNA webserver:
http://rna.informatik.uni-freiburg.de/IntaRNA
```

Example 2.1: Local installation via conda (Linux system with bash shell)

Installing conda (if not already available)

```
# download Miniconda installer
wget https://repo.continuum.io/miniconda/Miniconda3-latest-Linux-x86_64.sh -O ~/miniconda.sh
# note: '-b' runs the installer in non-interactive silent mode (agreeing to licence etc.)
bash ~/miniconda.sh -b -p $HOME/miniconda
```

Everytime when starting a new shell (before first running IntaRNA, see Note 2)

```
# add conda to your program search path
export PATH="$HOME/miniconda/bin:$PATH"
```

Installing IntaRNA via conda

```
conda install -y -c conda-forge -c bioconda intarna
```

Checking installed IntaRNA version

```
IntaRNA --version
```

[1] Detailed conda information and installation guidelines on https://conda.io

3 Methods

3.1 IntaRNA

INTARNA [11, 40] is a generic, accessibility-based RRI prediction tool, which incorporates seeds and many other constraints to achieve high accuracy results in low runtime. In its recent version (3.0+), it implements an accessibility-based two-step approach by first identifying putative seed interactions that are subsequently locally extended to optimal hybridizations. As shown by various benchmarks [36, 48, 70] and its incorporation in various other prediction tools and pipelines [20, 34, 77] as discussed above, it is one of the state-of-the-art RRI prediction tools.

While INTARNA performs already well using default parameters, its performance and output can be optimized for different RRI prediction tasks and scenarios using dedicated parameter sets. To reduce the need for manual adjustments, INTARNA comes with a set of predefined parameter setups, so called *personalities*. A personality is eventually an alternative to the default values of the parameters. Specific parameters of a personality can still be overwritten by the user by manually passing the respective parameter. Some of the personalities are used in the following. A complete list as well as detailed information on available parameters and examples are provided via INTARNA's online documentation (*see* **Note 1**).

3.1.1 Formalities in a Nutshell

For two given RNA molecules, an interaction I is defined by a set of inter-molecular base pairs. Within the following, we assume the base pairs to be ordered, non-overlapping, and non-crossing, i.e., they can be drawn side-by-side as shown in Fig. 1. Furthermore, we assume that every nucleotide can form at most one base pair. The latter implies that nucleotides involved in inter-molecular base pairing cannot form intra-molecular base pairs within the individual molecules. Accessibility-based methods like INTARNA consider the subsequences that contain inter-molecular base pairs to be unpaired (see resp. "acc. penalty" boxes in Fig. 1) and incorporate energy

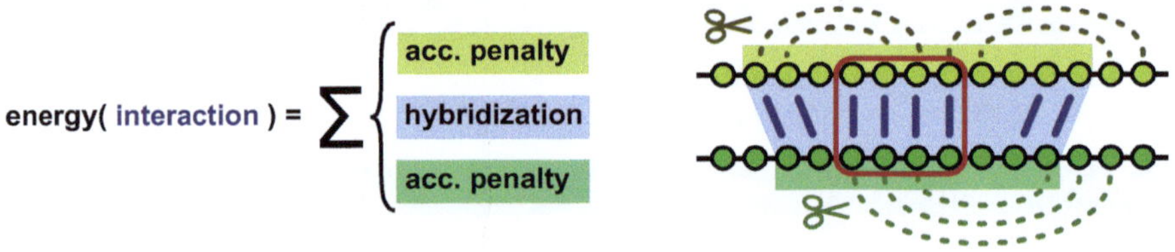

Fig. 1 For a given interaction, i.e., a set of base pairs (middle, depicted in dark blue) formed by two RNA molecules (top and buttom, olive and dark-green), an energy is derived by correcting the hybridization energy (light-blue middle box; defined by summing Nearest-Neighbor loop terms) with two accessibility penalties (top and bottom, yellow and light-green boxes), which incorporate how much energy is needed to break all intra-molecular base pairs (dotted lines) within the interacting subsequences (depicted by scissors). The red box highlights a seed subinteraction that consists of four base pairs. Seeds of a specified length are a requisite for any interaction considered by INTARNA

penalty terms to make the interacting subsequences accessible, i.e., free of any intra-molecular base pairing. The probability P^u of a subsequence to be unpaired is transformed into an accessibility penalty $\Delta E = -RT\log(P^u)$, where R denotes the gas constant and T the temperature.

To assess the stability of an interaction, its hybridization energy is estimated using the thermodynamic Nearest-Neighbor (NN) model that was initially introduced for RNA structure prediction [80]. The NN model provides local energy terms for sub-interaction formed by consecutive base pairs, so called loops. Loops of consecutive, i.e., stacked, base pairs provide the strongest contributions, while loops covering unpaired bases are considered less stable and less favorable. Given this, accessibility-based models identify the most stable interaction that shows the minimal sum of hybridization energy plus respective accessibility penalties [11, 46].

In addition, prediction quality and computation time can be improved when additional knowledge is incorporated with respective constraints. One such constraint is to consider only interactions that contain at least one putative seed subinteraction formed by a fixed number of e.g., four to seven base pairs, which is depicted by the red box in Fig. 1. An interpretation for this seed requirement is to ensure that an initial contact of the two RNAs can indeed expand to a full interaction and is not directly break up due to local instability, since such kinetic effects are not covered by the prediction model.

For further details on the energy model and computational details, we refer to these dedicated reviews [52, 79], and the original work on RNAUP [46] and INTARNA [11].

3.1.2 Advantages and Limitations

The most powerful advantage of accessibility-based RRI prediction is the implicit incorporation of intra-molecular structure that can be formed by the interacting RNAs. This allows, in contrast to most concatenation-based models, the prediction of interaction patterns formed by loops, i.e., by unpaired regions within hairpin or helical structures depicted in Fig. 2a. Furthermore, the used accessibility

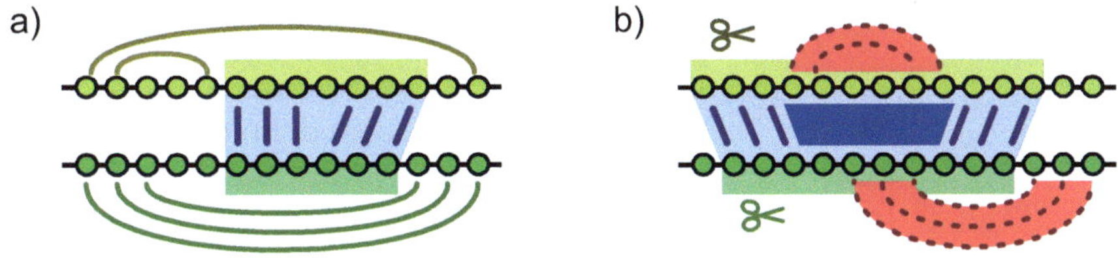

Fig. 2 (**a**) Accessibility-based models can predict interactions in any structural context, e.g., within helices (top) or hairpin loops (bottom). (**b**) But as they assume the *whole* interacting subsequences (lime/green boxes) to be *without* intra-molecular base pairing (red arcs), large inter-molecular loops (dark blue box) are considered fully unstructured. Since large unstructured regions are rated unstable, accessibility-based approaches typically provide only one consecutive block of inter-molecular base pairing with smaller bulges

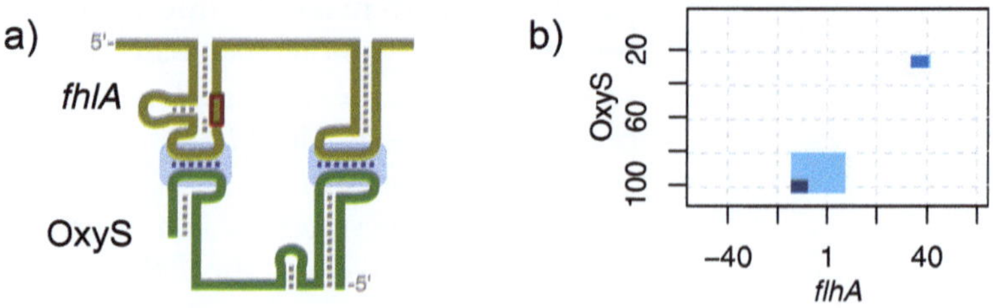

Fig. 3 (**a**) Double-kissing-hairpin interaction of *E. coli* OxyS sRNA with the *fhlA* mRNA as reported in [6] where the *fhlA* start codon is highlighted in red and the two concurrent RRI are depicted in blue. Base pairs are abstracted by dotted lines. (**b**) Respective interaction potential predicted by INTARNA which highlights any index pair covered by a putative interaction in blue. Position 1 of *fhlA* refers to the first position of the start codon. Darker shades of blue represent more stable (lower energy) RRI while white areas are not predicted to interact. The left kissing hairpin from a) can form a more stable interaction compared to the right one (lighter blue of upper right box)

penalties are not based on a single intra-molecular structure (as considered by concatenation-based approaches), but incorporate all possible structures that form base pairs in the subsequence of interest. These penalties are a much more general and reliable estimate of the likelihood of subsequence be involved in an RRI, which provide eventually a more accurate prediction.

Exact computation of accessibility terms are time consuming since it is similar to individual structure prediction [46]. Thus, approximate methods based on local folding have been introduced [9], which are employed by INTARNA. Furthermore, INTARNA can take precomputed accessibility terms from file, which speeds up RRI prediction.

But explicit accessibility consideration induces also the strongest limitation of current accessibility-based methods. Since the whole interaction-spanning subsequences are considered free of intra-molecular base pairing, no mixed base pairing pattern can be modeled. This is illustrated in Fig. 2b, that is, while the RRI contains long non-interacting subsequences, possible intra-molecular base pairing is still penalized and treated as if it has to be broken.

This problem also translates to the prediction of multiple (distant) interactions that can be formed simultaneously as e.g., known for the OxyS-*fhlA* RRI [6] depicted in Fig. 3. Since the intra-molecular structure formation at individual sites is not independent from each other, respective unpaired probabilities (accessibility) of single sites cannot be combined into a joint probability that both sites are unpaired simultaneously. INTARNA can still be used in a hierarchical approach to predict concurrent interactions, which will be discussed in Subheading 3.2.3.

Accessibility estimates computed by INTARNA are based on non-crossing secondary structure model. While these can be enhanced e.g., with structural constraints or structure probing data (Subheading 3.3.4), they are not modeling crossing base pairing pattern present in three-dimensional functional fold. Nevertheless, if such accessibility profiles (in terms of unpaired probabilities per subsequence) are available, they can be directly and correctly incorporated into INTARNA's RRI prediction.

As for individual RNA structure prediction, the consideration of the most stable (energy minimal) RRI can be misleading if alternative but slightly less stable interactions can be formed [36]. Thus, we advise to investigate suboptimal RRI to amend for artifacts of the used parameters. Furthermore, since INTARNA is using the Nearest-Neighbor energy model, except the temperature, no other environmental conditions like ion concentrations can be altered.

Runtime and prediction quality of INTARNA depend on the applied parameters. For instance, the seed-extension strategy works best for strong seed constraints such that only few putative seeds are to be considered. Too strong constraints (e.g., enforcing too many base pairs within seeds) exclude many interactions from the prediction. Thus, we encourage the user to reflect on the objective of the study at hand, which will help to pick either one of INTARNA's predefined personalities or to setup a study-specific parameter set. The latter can be stored and loaded from file to ensure reproducibility and to simplify INTARNA's usage. That way, the whole flexibility of INTARNA's input/output user interface can be exploited to enable an efficient and exact prediction of RNA-RNA interactions.

3.2 Specific RNA-RNA Interaction Prediction

A standard application of INTARNA is to predict RRIs for two (or more) given RNAs. This is done (i) to reveal putative RRI, (ii) to estimate their stability, and (iii) to investigate regulatory mechanisms induced by such RRIs.

A few preliminary considerations are important to ensure the best prediction results:

Two crucial questions to start with:

- *What temperature should INTARNA use for the RRI model?*
- *What preliminary knowledge (e.g. interaction region) can be supplied?*

The first question, i.e., the correct setup of the model temperature (`--temperature` parameter, default 37° C), heavily influences the thermodynamic model and thus energy estimation, while RNA-specific constraints can significantly improve predictions. The latter covers for instance structure probing data (*see* Subheading

3.3.4) or hard constraints on individual RNA structures to guide accessibility assessment or approximate pre-location of putative seed or even RRI regions. All such side constraints can be provided as additional input to INTARNA to guide its output.

Within the following, we will illustrate how to use INTARNA for specific RRI prediction based on a well studied interaction between Escherichia coli's (*E. coli*) OxyS sRNA with its target mRNA *fhlA* [6]. To this end, we prepare files with the respective subsequences in FASTA format and call INTARNA by providing the filenames as arguments for the query "`-q`" and target "`-t`" RNA parameter (*see* Example 3.1). We could also directly input the sequences (instead of files) and provide name tags via `--qId` and `--tId`, resp., but the FASTA format simplifies the call and supports multi-RNA input for query and/or target. In the latter case, all pairwise combinations would be considered for prediction. Since we are only working on subsequences, we also provide the index of the first sequence position of each RNA (otherwise indexing starts with 1). For mRNAs like *fhlA*, often a start-codon-related indexing is wanted. That is, indexing starts with 1 at the beginning of the start codon AUG. All preceding positions are indexed by their (negated) distance to the start. INTARNA supports both indexing methods, which is automatically selected based on the first nucleotide's index provided via `--tIdxPos0` or `--qIdxPos0` for target and query, respectively. If not provided, a 1-based indexing is used.

3.2.1 Interpreting the Basic Results

INTARNA reports RRIs with negative free energies, since values are in relation to a non-interacting model with an interaction energy of 0. A standard invocation call of INTARNA providing only the respective RNA sequences outputs a depiction of the most stable RRI with a respective overall energy term (*see* Example 3.1). For our OxyS-*fhlA* example, this reveals one of two kissing-hairpin RRIs reported in [6] (depicted in Fig. 3a). The reported interaction matches the experimentally verified interaction from the literature, which forms a stable helix of 7bps that blocks the Shine-Dalgarno of *fhlA* (GGA at mRNA positions -11..-9 [6]). As a rule of thumb, one should be cautious about predicted RRIs that have an energy above − 2 kcal/mol or even − 3 since they are unlikely to be stable enough within the solution. On the other hand, the interaction from Example 3.1 with an energy of about − 5 kcal/mol can be considered thermodynamically stable. This shows already one of the main problems of computational RRI prediction: it is likely that two RNAs can form *at least some* RRI, which are identified and consequently reported. Eventually, the user has to interpret and assess the sanity of any RRI prediction given his/her expert knowledge about the biological/experimental context that is not be covered by the prediction model.

Example 3.1: Standard input and output

Minimal call

```
IntaRNA -t fhlA.fasta -q OxyS.fasta
```

with explicit *fhlA* index information

```
--tIdxPos0=-53
```

Standard output

```
fhlA|NC000913|-53..+60|doi:10.1006/jmbi.2000.3942:Fig7
           -15     -9
             |      |
5'-AGU...GCUU       ACAA...ACA-3'
             UCCUGGA
             +++++++
             AGGACCU
    3'-CGCCU        CUAG...AAG-5'
             |      |
           104     98
OxyS|NC_000913|56..164|doi:10.1006/jmbi.2000.3942:Fig7

interaction energy = -4.94 kcal/mol
```

Input File fhlA.fasta + OxyS.fasta: Subsequences of *E. coli fhlA* mRNA and OxyS sRNA in FASTA format as shown in [6]-Fig.7:

```
>fhlA|NC000913|-53..+60|doi:10.1006/jmbi.2000.3942:Fig7
AGUUAGUCAAUGACCUUUUGCACCGCUUUGCGGUGCUUUCCUGGAACAACAAAAUGU
CAUAUACACCGAUGAGUGAUCUCGGACAACAAGGGUUGUUCGACAUCACUCGGACA
```

```
>OxyS|NC_000913|56..164|doi:10.1006/jmbi.2000.3942:Fig7
GAAACGGAGCGGCACCUCUUUUAACCCUUGAAGUCACUGCCCGUUUCGAGAGUUUCUC
AACUCGAAUAACUAAAGCCAACGUGAACUUUUGCGGAUCUCCAGGAUCCGC
```

3.2.2 Detailed Investigation of the Output

The standard call and output is useful for a first check but comes with two limitations: (i) per default INTARNA applies a fast and approximate prediction model and (ii) no detailed assessment of the RRI is possible. Thus, we recommend an adapted parameter set for detailed analyses that is preset via the `IntaRNAexact` *personality*. The latter can either be set via parameter (`--personality=IntaRNAexact`) or by calling the dedicated binary wrapper (`IntaRNAexact`). This personality runs exact (i.e., non-heuristic) accessibility and interaction prediction, which results in longer runtimes but could provide better predictions. Note, to reduce runtime, it is still useful to restrict the maximal subsequence length that can be spanned by an RRI (*see* **Note 3**). Within the following example, we use a threshold of 60 nucleotides (`--intLenMax=60`) that is recommended for sRNA prediction [54].

Example 3.2 shows the extended output, which shows:

- details of the energy terms (see "`E(..energy-type..)`")
- overall hybridization energy (i.e., "`E(hybrid) = -11.58`")
- unpaired probabilities of interacting subsequences (*fhlA* ~ 0.37, OxyS ~ 0.00005)
- details on putative seeds ("+" base pairs in plot and bottom information)

The overall energy difference of 0.02 when comparing to Example 3.1 derives from the non-constraint accessibility computation. The latter terms reveal that the interacting region is not necessarily unpaired in OxyS ("`Pu(seq2)=0.00005`") while the hairpin, which encloses the *fhlA* binding site ("`Pu(seq1)=0.37`"), seems to be very stable.

Example 3.2: Detailed output of exact prediction

```
IntaRNAexact --intLenMax=60 -t fhlA.fasta --tIdxPos0=-53 -q OxyS.fasta
```

Detailed visualization (see Note 4)

```
--outMode=D
```

```
fhlA|NC000913|-53..+60|doi:10.1006/jmbi.2000.3942:Fig7
            -15     -9
              |      |
5'-AGU...GCUU        ACAA...ACA-3'
             UCCUGGA
             +++++++
             AGGACCU
     3'-CGCCU        CUAG...AAG-5'
              |      |
            104     98
OxyS|NC_000913|56..164|doi:10.1006/jmbi.2000.3942:Fig7

interaction seq1   = -15..-9
interaction seq2   = 98..104

interaction energy = -4.92 kcal/mol
  = E(init)        = 4.1
  + E(loops)       = -15.6
  + E(dangleLeft)  = -0.1
  + E(dangleRight) = -0.98
  + E(endLeft)     = 0.5
  + E(endRight)    = 0.5
    : E(hybrid)    = -11.58
  + ED(seq1)       = 0.62
    : Pu(seq1)     = 0.36569
  + ED(seq2)       = 6.04
    : Pu(seq2)     = 5.54465e-05

seed seq1   = -15..-9
seed seq2   = 98..104
seed energy = -4.92
seed ED1    = 0.62
seed ED2    = 6.04
seed Pu1    = 0.36569
seed Pu2    = 5.54465e-05
```

CSV output of top-2 predictions

```
-n 2 --outMode=C --outCsvSort=E
--outCsvCols="E,Pu2,subseqDB,hybridDB"
```

```
E;Pu2;subseqDB;hybridDB
-4.92;5.54465e-05;-15UCCUGGA&98UCCAGGA;-15|||||||&98|||||||
-3.62;0.0167597;34CAAGGGU&24ACCCUUG;34|||||||&24|||||||
```

In case an RRI contains multiple putative seed interactions, all are shown and highlighted within the detailed output mode. Furthermore, additional output useful for detailed investigations can be obtained by setting the optional parameters for generating outputs such as:

- Position-(pair)wise energy profiles [40] (discussed next) or
- Probabilities that index pairs are covered by an RRI [44] (termed as spot probabilities)

Energy profiles provide for each RNA position or intermolecular index pair the minimal energy of any interaction spanning the index (pair). While computationally extensive, this allows to easily assess whether the most stable RRI is exclusive or only one of the comparable alternatives. When extending the call of Example 3.2 with "`--out=pMinE:mfeAllIdxPairs.csv`," we derive the RRI potential for all index combinations of OxyS-*fhlA*. Figure 3b visualizes the respective data and reveals a second putative interaction with low energy in the coding region of *fhlA* that is not overlapping with the RRI with lowest energy. This highlights the importance of investigating non-optimal RRIs, which are reported by INTARNA by setting the "number-of-suboptimal-RRI" parameter "-n" to a value larger than one.

The latter is illustrated in Example 3.2-right, which makes use of the tabularized output format. This CSV output enables a flexible framework to assess any interaction detail of interest in a computationally accessible standard table format. For a complete list and explanation of available features, please refer to INTARNA's online documentation (**Note 1**). The example shows that the alternative interaction of OxyS within the coding sequence of *fhlA* reported in [6] (right RRI in Fig. 3a) is indeed correctly predicted as the second-best interaction at the experimentally verified position. Furthermore, while being less stable, the second site shows a lower unpaired probability for OxyS (`Pu2`), which might hint at a more likely interaction initiation of this site.

3.2.3 Prediction of Multiple Concurrent RRIs

As already discussed and depicted in Fig. 3a, OxyS and *fhlA* can form two RRIs simultaneously. While both RRIs are predicted correctly, single-site accessibility-based approaches as INTARNA are per se not able to predict the correct overall energy term for a concurrent model where both RRIs are formed together. This results from the fact that the unpaired probabilities of the individual RRIs are computed independently and thus cannot be fused into a joint probability that both sites are unpaired at the same time. For the latter, a conditional unpaired probability computation is needed, which is discussed in detail in [59].

In contrast to most other tools, INTARNA is able to compute conditional probabilities of concurrent sites once a given RRI is fixed. That is, the user can specify additional constraints that mark the fixed RRI's subsequences as not available for intra- and intermolecular base pairing. The latter results in the computation of corrected conditional unpaired probabilities (and thus accessibility terms) as well as enforcing the prediction of alternative RRIs that can be formed simultaneously. For our OxyS-*fhlA* example, we fix the most stable RRI that blocks the Shine-Delgarno sequence (left subinteraction in Fig. 3a) and provide its overall energy of -4.92 kcal/mol as additional input. The latter enables the correct computation of the overall energy of the concurrent model when both subinteractions are formed concurrently. Example 3.3 shows the respective call and the correct prediction of the second RRI of the two-site interaction along with the joint energy of $-8,54$ kcal/mol when both RRI are formed. If desired, the described approach can be iterated to model even more concurrent RRIs.

When comparing the joint energy of the concurrent formation of both interactions possible for OxyS-*fhlA* from Example 3.3-right with the sum of individual energies shown in Example 3.2-right, no difference can be seen. That is, the joint energy equals the sum of the first two suboptimal interactions. From this, we can deduce that the formation of the individual subinteractions is independent for OxyS-*fhlA*, i.e., binding at one site is not effecting the other binding site's accessibility (even if minor local changes of the intra-molecular structure are possible).

Example 3.3: Corrected prediction of 2nd RRI in two-site concurrent model

```
IntaRNAexact --intLenMax=60 -t fhlA.fasta --tIdxPos0=-53 -q OxyS.fasta
```

First RRI via standard call

```
fhlA|NC000913|-53..+60|doi:10.1006/jmbi.2000.3942:Fig7
          -15     -9
            |      |
5'-AGU...GCUU      ACAA...ACA-3'
             UCCUGGA
             +++++++
             AGGACCU
    3'-CGCCU       CUAG...AAG-5'
            |      |
          104     98
OxyS|NC_000913|56..164|doi:10.1006/jmbi.2000.3942:Fig7

interaction energy = -4.92 kcal/mol
```

Second concurrent RRI via constraints

```
--tAccConstr="b:-15--9"
--qAccConstr="b:98-104"
--energyAdd="-4.92"
```

```
fhlA|NC000913|-53..+60|doi:10.1006/jmbi.2000.3942:Fig7
          +34     +40
            |      |
5'-AGU...ACAA      UGUU...ACA-3'
             CAAGGGU
             +++++++
             GUUCCCA
3'-CGC...UGAA      AUUU...AAG-5'
            |      |
           30     24
OxyS|NC_000913|56..164|doi:10.1006/jmbi.2000.3942:Fig7

interaction energy = -8.54 kcal/mol
```

Note, such an iterative approach makes full use of the speedup from single-site prediction methods while predicting complex multi-site interaction patterns that are typically only computed via very slow simultaneous RRI and structure prediction approaches. Since individual sites are predicted independently, the joint concurrent RRI patter might not be optimal compared to explicit multi-site models. On the other hand, no limitations concerning the relative placement of the individual sites are present but steric constraints should be considered when contemplating a concurrent model. To conclude, INTARNA provides a powerful framework to predict concurrent multi-site interactions via single-site RRI optimization.

3.3 Incorporating Experimental Information

To guide and enhance RRI prediction, it is useful to incorporate as much knowledge from literature or experiments as possible. In the following, we illustrate some scenarios that can be easily incorporated into INTARNA predictions and which can be combined at will.

3.3.1 Modeling of Already Bound Molecules

A standard scenario is the incorporation of concurrent binding of other molecules previous to the RRI to be predicted. One use case, the prediction of multiple concurrent RRI of the same RNA, was already discussed above. Similarly, one can incorporate non-RNA interaction information. For instance, Hfq is a protein, which is known to support and stabilize sRNA-mRNA interactions [7, 61]. In the presence of experimental evidence of Hfq-binding or predictions based on respective sequence motifs [29, 30], respective subsequences can be constrained e.g., to be unpaired using INTARNA's accessibility constraint encoding. For instance, following [60], we could constrain the putative Hfq-binding site `AAUAAC`, which complies with its $\texttt{ARN}_n$ binding motif [71], using the constraint `--qAccConstr=b:65-70`. This marks the region as single stranded and blocked by some (not further modeled) molecule. The same way, additional knowledge about the individual RNA structures can be incorporated to ensure a correct(ed) accessibility assessment and thus RRI prediction.

3.3.2 Known Inter-Molecular Base Pairs from Mutation Studies

When investigating the mechanistic details of RRI-induced regulatory, dedicated predictions incorporating compensatory mutation experiments are useful (*see* **Note 5** on how INTARNA can be used for their design). As an example, we reevaluate the GcvB-*phoB* interaction reported in [13], which was experimentally verified via compensatory mutation experiments. The latter showed that the subsequences 154–158 of GcvBs interaction site R3 pair with a full complementary subsequence in *phoB* at positions -17 to -13 upstream of the start codon.

Example 3.4 shows that standard prediction reports the core region of the known RRI of R3 from [13] as second-best interaction. The predicted mfe interaction of GcvB regions 66–85 covers GcvB's interaction site R1. While [13] has a focus on R3, also, evidence for R1 interaction with *phoB* is found via *phoB*-R3 mutants from the compensatory mutation experiment (*see* [13]—Fig. 3).

When incorporating the hypothesis from the compensatory mutation experiment covering R3, namely the explicit pairing of the mutated subsequences, all reported RRIs are showing the provided subinteraction. This reveals that (i) INTARNA's predictions are (intrinsically) confined to the short unpaired region of GcvB's R3, which is in contrast to the longer interaction reported in [13] using TARGETRNA, and (ii) the mutated helix provides the most stable part of the GcvB-*phoB* interaction.

Example 3.4: Anchored prediction

```
IntaRNAexact -q GcvB.fasta -t phoB.fasta --tIdxPos0=-200 --intLenMax=60 -n 2
```

Standard prediction

```
phoB|NC_000913|b1130|-200..+100|genom-subsequence
           -56                     -34
             |                     |
5'-GAG...CCCC      C     AUC        CUAU...CCA-3'
             CAUAAC ACAUA   GCGUUACA
             ||:||| |||:|   ++++++++
             GUGUUG UGUGU   UGUAGUGU
3'-UUU...GUUU      U     ---        UGGC...UCA-5'
             |                     |
            85                     66
GcvB|NC_000913

interaction energy = -9.44 kcal/mol

phoB|NC_000913|b1130|-200..+100|genom-subsequence
           -18      -11
             |      |
5'-GAG...AUUA        AGAA...CCA-3'
             AGACAGGG
             ++++++++
             UCUGUCCC
3'-UUU...CCUG        AUUU...UCA-5'
             |      |
           159      152
GcvB|NC_000913

interaction energy = -9.29 kcal/mol
```

Anchored prediction

```
--seedTQ="-17|||||&154|||||"
```

```
phoB|NC_000913|b1130|-200..+100|genom-subsequence
           -18      -11
             |      |
5'-GAG...AUUA        AGAA...CCA-3'
             AGACAGGG
             |+++++||
             UCUGUCCC
3'-UUU...CCUG        AUUU...UCA-5'
             |      |
           159      152
GcvB|NC_000913

interaction energy = -9.29 kcal/mol

phoB|NC_000913|b1130|-200..+100|genom-subsequence
           -17     -11
             |     |
5'-GAG...UUAA       AGAA...CCA-3'
             GACAGGG
             +++++||
             CUGUCCC
3'-UUU...CUGU       AUUU...UCA-5'
             |     |
           158     152
GcvB|NC_000913

interaction energy = -8.48 kcal/mol
```

Input File GcvB.fasta + phoB.fasta: *E. coli* GcvB sRNA sequence and genomic context of *phoB* gene 200 up- and 100 down-stream of the start codon

```
>GcvB|NC_000913
ACUUCCUGAGCCGGAACGAAAAGUUUUAUCGGAAUGCGUGUUCUGGUGAACUUUUGGCUUACGGUUGUGAUGUUGUGUUGUUGUGUUUGCAAUUGGUCUGCGAUUCAGACCAUGGUAGCA
AAGCUACCUUUUUUCACUUCCUGUACAUUUACCCUGUCUGUCCAUAGUGAUUAAUGUAGCACCGCCUAAUUGCGGUGCUUU
```

```
>phoB|NC_000913|b1130|-200..+100|genom-subsequence
GAGCTATCACGATGGTTGATGAGCTGAAATAAACCTCGTATCAGTGCCGGATGGCGATGCTGTCCGGCCTGCTTATTAAGATTATCCGCTTTTTATTTTTTCACTTTACCTCCCCTCCCC
GCTGGTTTATTTAATGTTTACCCCCATAACCACATAATCGCGTTACACTATTTTAATAATTAAGACAGGGAGAAATAAAAATGCGCGTACTGGTTGTTGAAGACAATGCGTTGTTACGTC
ACCACCTTAAAGTTCAGATTCAGGATGCTGGTCATCAGGTCGATGACGCAGAAGATGCCA
```

3.3.3 RNA-Specific Regions, Motifs, etc.

Another source of knowledge is the identification of conserved interaction sites. Simple examples are eukaryotic miRNAs, which are known to form a stable seed interaction between 5′-end and their regulated target mRNA [50]. Thus, constraining the seed prediction towards the miRNAs' 5′-ends will enhance the prediction model with this miRNA-specific information. Another example is the evolutionary conservation of binding motifs. The GcvB sRNA, for instance, shows high structure conservation and almost 100% sequence conservation in two longer unpaired regions named R1 and R2 [25]. Both R1 and R2 are known from verified GcvB-target RRIs and are hypothesized to be involved in seed interactions.

We have extracted the positions 75–93 and 129–145 for R1 and R2, resp., in *S. Typhimurium* GcvB from [25]. Given this, we compare in Example 3.5 a standard IntaRNA call with a prediction constraint to form a seed either in R1 or R2 of GcvB for RRI prediction with *ilvE*. Without seed constraint, IntaRNA reports an RRI similar to the putative RRI reported in [63]. Notably, while covering R2, the RRI does not form any stable interaction within the R2 site but instead forms a long inter-molecular helix incorporating stem loop SL4. When using a seed-region constraint, INTARNA reports a slightly less stable RRI with seed in R2 with the same CA-rich target region in *ilvE*, which was hypothesized in [63] to be important for target recognition. We therefore conclude that especially seed region constraints are a powerful tool to improve detailed RRI investigations.

Example 3.5: Seed-region-constraint prediction

```
IntaRNAexact -q GcvB.ST.fasta -t ilvE.fasta --tIdxPos0=-200 --intLenMax=60
```

Standard prediction

```
ilvE|NC_003197|STM3903|-200..+100|genom-subsequence
          -56                         -32
            |                           |
5'-GGU...GAUC   ---  G    C   C          AAAU...GUA-3'
            UGC   CA AGCG UGC ACAUCACAAC
            |||   || |:|: ::| ++++++++++
            ACG   GU UUGU GUG UGUAGUGUUG
3'-UUU...GUUA   UUU  G    U   U          GCAU...UCA-5'
            |                           |
           91                          64
GcvB|NC_003197

interaction energy = -8.16 kcal/mol
```

GcvB-seed-constraint prediction

```
--seedQRange="75-93,129-145"
```

```
ilvE|NC_003197|STM3903|-200..+100|genom-subsequence
          -37      -30
            |        |
5'-GGU...ACAU        AUCC...GUA-3'
             CACAACAA
             ++++++++
             GUGUUGUU
3'-UUU...GUUU        GUGU...UCA-5'
            |        |
           85       78
GcvB|NC_003197

interaction energy = -6.32 kcal/mol
```

Input File GcvB.ST.fasta + ilvE.fasta: *S. Typhimurium* GcvB sRNA sequence and genomic context of *ilvE* gene 200 up- and 100 down-stream of the start codon

```
>GcvB|NC_003197
ACUUCCUGAGCCGGAACGAAAAGUUUUAUCGGAAUGCGUGUUCUGAUGGGCUUUUGGCUUACGGUUGUGAUGUUGUGUUGUUGUGUUUGCAAUUGGUCUGCGAUUCAGACCACGGUAGCG
AGACUACCCUUUUUCACUUCCUGUACAUUUACCCUGUCUGUCCAUAGUGAUUAAUGUAGCACCGCCAUAUUGCGGUGCUUU
```

```
>ilvE|NC_003197|STM3903|-200..+100|genom-subsequence
GGTTTTCAGGTGTGCTCCATGAATATGGAAGCCGCGACCGATGCGCAGAATATAAATATTGAATTGACCGTTGCCAGTCCCCGGTCGGTCGACTTACTGTTTAGTCAGTTAAGTAAACTG
GTAGATGTTGCGCATGTCGCGATCTGCCAGAGCGCTGCCACATCACAACAAATCCGCGCCTGAGCGCAAAAGGAAGAAAAATGACGACGAAAAAAGCTGATTATATTTGGTTCAATGGCG
AGATGGTGCGCTGGGAAGACGCGAAGGTTCACGTAATGTCTCACGCGCTGCACTACGGTA
```

3.3.4 Structure Probing SHAPE Data

In the recent years, various experimental protocols have been introduced to probe the structure of an RNA in vitro and in vivo. Respective data are also publicly often available [12]. For instance, SHAPE[2] [75] enables the experimental assessment of an RNA's accessibility on a nucleotide resolution even on a transcriptome-wide scale [35]. Structure probing data are efficiently incorporated into INTARNA predictions, to guide its accessibility computation [44]. To do so, it is sufficient to provide the SHAPE reactivity profiles via the respective "`--qShape`" or "`--tShape`" parameters. For details on the integration and an illustrative example, please refer to [44] and INTARNA's online documentation (**Note 1**). Recently, we have shown that even computationally simulated SHAPE profiles, produced by the SHAKER approach [43], can improve RRI prediction [54].

[2] SHAPE = Selective 2′-hydroxyl acylation analyzed by primer extension.

3.4 Large-Scale Target Prediction

Another common use case of RRI prediction methods is the screen for putative regulatory targets based on RNA-RNA interaction potential [8, 50, 79]. One can distinguish into (i) the more common target screen within a predefined set of target RNAs, which typically are compiled from transcriptomic or genomic subsequences, and (ii) target prediction on the whole genome sequence. For such tasks, typically fast methods like two-step approaches or heuristic tools, which make heavy use of additional constraints and distributed (parallel) computation, are applied. In the following, we will show how INTARNA can be used for such tasks.

General points to speedup large-scale predictions are (i) to parallelize computation (using the "`--threads`" argument) and (ii) to reuse accessibility information if multiple INTARNA calls are done using the same sequence(s). For the latter, one either stores the accessibility penalties from the first run using e.g., "`--out=tED:targetAccessibility.ed`" or calls INTARNA with a dummy sequence to trigger only accessibility computation (*see* Example 3.6). Afterwards, one can provide the stored data to subsequent INTARNA runs using "`--tAcc=E`" (to load ED from file) and "`--tAccFile=targetAccessibility.ed`" (ED input file selection), which is also used in Example 3.6.

3.4.1 Transcriptomic Data or Genomic Regions

In the following, we reproduce the genomic sRNA-target-screen pipeline of the INTARNA webserver [51, 78] to exemplify a large-scale target search. Since sRNAs are known to block the ribosomal binding site of mRNAs via RRI [8], genomic subsequences 200 up- and 100 down-stream of each gene's start codon are extracted[3] and stored with the respective NCBI locus tag. Example 3.6 shows the top-5 predictions (lowest energy) of predicted RRIs of ChiX with all 4,319 genomic start regions of *E. coli*. Note, to speedup, this example makes maximal use of parallelized RRI prediction by setting `--threads=0`, which runs one prediction thread per available CPU. The first RRI with target `b0481` shows extremely low energy and is an artifact of the data set, since the upstream region of `b0481` contains parts of the anti-sense sequence of the *chiX* gene that is consequently fully complementary for positions 1–79. The subsequent predictions correspond to verified RRIs of ChiX with *chbC* [47], *chiP* [55], and *dpiB/citA* [39], respectively. To exclude artifacts from the fast heuristic prediction mode used for target screens, we do an exact prediction for the top-ranked targets 2–4 using the "`--tSet`" parameter. All three predicted RRIs match the reported interaction site from the literature, and all top predictions involve ChiX subsequences 39–56, which is an unpaired region directly downstream of the proximal binding site of Hfq [61].

---

[3] Using `get_refseq_from_refid.pl` and `parse_region_from_genome.pl` from the COPRARNA package [77] available at https://github.com/PatrickRWright/CopraRNA

Example 3.6: Target screen on genomic regions

Precomputation of target accessibility via dummy call (pseudo query + no output)

```
IntaRNA -t NC_000913.fa --out=tAcc:NC_000913.ed -q AAA -n 0 --out=/dev/null --noSeed
```

Multi-threaded target screen using precomputed target accessibility data

```
IntaRNA -t NC_000913.fa -q ChiX_NC_000913.fasta --out=NC_000913.ChiX.csv \
  --outMode=C --outCsvCols="E,id1,start1,end1,id2,start2,end2" --outCsvSort=E \
  --tAcc=E --tAccFile=NC_000913.ed \
  --threads=0
```

top-5 from NC_000913.ChiX.csv output

```
E;id1;start1;end1;id2;start2;end2
-91.38;b0481;1;79;ChiX;1;79
-18.36;b1737;134;152;ChiX;38;56
-16.41;b0681;182;193;ChiX;45;56
-15.45;b0619;164;175;ChiX;46;57
-14.5;b1427;123;136;ChiX;43;56
```

Exact prediction for target 2-4

```
IntaRNAexact [...] \
 --tSet="1726-1726,667-667,1414-1414"
```

```
E;id1;start1;end1;id2;start2;end2
-17.99;b1737;134;152;ChiX;38;56
-16.1;b0681;182;193;ChiX;45;56
-14.92;b0619;164;175;ChiX;46;57
```

Input File NC_000913.fa + ChiX_NC_000913.fasta: *E. coli* genomic subsequences 200 up- and 100 down-stream of the start codon of each gene and ChiX sRNA sequence

available from https://github.com/BackofenLab/IntaRNA-benchmark in sub-directories of 'input/NC_000913'

While these target screen results for ChiX are extremely good, not all target screens provide as good true target rates. For instance, the verified interactions of GcvB (as e.g., listed in the INTARNA benchmark data set [21]) are often not among the top-ranked predictions. This does not necessarily imply that all top-ranked targets are false predictions. Nevertheless, the low rank of verified targets might be a result from a strong impact of the Hfq protein in the RRI formation of GcvB or other effects not modeled in the prediction, which shows the limitation of single-organism screens. Here, comparative approaches like COPRARNA are successful to improve the reliability of sRNA target screen [38, 77]. COPRARNA combines target screens from different organisms into a single prediction of evolutionary conserved targets. For details how to employ COPRARNA, we refer to a dedicated handbook [76].

3.4.2 Screening a Full Genome

If the user is interested in a full genome-wide screen for possible RRIs of a single or multiple RNAs versus a full genome or a collection of very long sequences, further considerations are useful. For instance, a recommended option would be to restrict the predictions to regions of high accessibility to reduce the computational workload while heeding the central assumption of accessibility-based RRI prediction. This can either be done using a study-specific fixed threshold, as has been done in [68] and applicable via INTARNA's "--outMinPu" constraint, or via some dynamic heuristic as has been applied in MECHRNA [20], which employs INTARNA using the --tRegionLenMax' option.

While successful, such heuristics might compromise some of the important functional RRIs depending on the customized parameters used. Thus, INTARNA provides an automatic decomposition of long sequences into overlapping windows of a given length ("--windowWidth") and overlap ("--windowOverlap"). This does not only enable exhaustive predictions with low memory consumption but also allows vast parallelized processing of the individual windows, which significantly accelerates genome-wide RRI prediction. Furthermore, INTARNA's window-based processing is fully compatible with all other constraints like the discussed restriction to accessible regions. Finally, especially for genome-based prediction, the already discussed precomputation of accessibility penalties is recommended (*see* Example 3.6).

3.5 Conclusions

RNA-RNA interaction (RRI) prediction tools are an important complement to experimental RRI identification methods. Available tools and methods can be classified based on the underlying prediction strategy. We identify and review seven distinct classes and discuss the specific strengths and weaknesses of the respective strategy. This knowledge is essential to select a suitable RRI prediction tool for the task at hand.

INTARNA is a fast and accurate RRI prediction tool that incorporates seed, accessibility and interaction constraints. Continuous maintenance and high performance make it one of the state-of-the-art RRI prediction tools. It provides a powerful and flexible interface, which enables use-case-specific RRI investigation via customized and predefined parameter sets. Within this handbook, we have exemplified different application scenarios of INTARNA using known interactions from the literature. The provided discussion on parameter selection and result interpretation will help non-expert users to comprehensively use INTARNA and assess reported RRIs.

4 Notes

INTARNA online documentation: https://backofenlab. github.io/IntaRNA/

1. INTARNA's online documentation provides detailed information on available parameters with respective examples. It is freely available at the link from above. You should also refer to the online documentation if you are using a different version of INTARNA ($\neq 3.2.0$), due to the continuous development of the tool.

2. If you don't extend the program search path, INTARNA and its dependencies will not be found when called. Alternatively, you can also add the given `export` command to your shell startup script, e.g., the `.bashrc` file when using the `bash` shell.
3. The selection of reasonable bounds on the maximal interaction length depends on the RNA class, since very short miRNAs (length about 20–25nt) will form much shorter RRI compared to e.g., anti-sense RNAs that might form a full length duplex.
4. INTARNA provides sequence-specific information within its output. Therein, target-related information is always annexed with a `1` while query-specific information is annotated by an annex of `2`. For instance, within the CSV output, the index range within the target is provided via `start1` and `end1` while the query's unpaired probability of the interacting subsequence is given by `Pu2` (*see* Example 3.2).
5. In compensatory mutation experiments, subsequences in both interacting RNAs are mutated at the putative RRI site. Within in vitro experiments, it is tested that (i) the mutated sequences still show a similar sRNA-induced effect on translation as observed for wildtype sequences and that (ii) both mutant-wildtype combinations show a reduced impact. This experimental design can be applied also to in silico experiments, i.e., the effect is measured in terms of reduced interaction potential. COPOMUS [53] implements such a framework using INTARNA to assess the RRI potential. Eventually, it reports a classification and ranking of all potential compensatory mutations within INTARNA's mfe RRI to help experimenters in the design of the in vitro verification. COPOMUS is part of the INTARNA package, and its webserver is freely available at http://rna.informatik.uni-freiburg.de/CopomuS.

References

1. Agarwal V, Bell GW, Nam JW, Bartel DP (2015) Predicting effective microRNA target sites in mammalian mRNAs. eLife 4:e05005
2. Alkan F, Wenzel A, Palasca O, Kerpedjiev P, Rudebeck A, Stadler PF, Hofacker IL, Gorodkin J (2017) RIsearch2: suffix array-based large-scale prediction of RNA-RNA interactions and siRNA off-targets. Nucleic Acids Res 45(8):e60–e60
3. Altschul SF, Gish W, Miller W, Myers EW, Lipman DJ (1990) Basic local alignment search tool. J Mol Biol 215(3):403–410
4. Andronescu M, Zhang ZC, Condon A (2005) Secondary structure prediction of interacting RNA molecules. Journal of Molecular Biology 345(5):987–1001
5. Antonov I, Marakhonov A, Zamkova M, Medvedeva Y (2018) ASSA: fast identification of statistically significant interactions between long RNAs. J Bioinform Comput Biol 16(01):1840001
6. Argaman L, Altuvia S (2000) fhlA repression by OxyS RNA: kissing complex formation at two sites results in a stable antisense-target RNA complex. J Mol Biol 300(5):1101–1112
7. Arluison V, Hohng S, Roy R, Pellegrini O, Règnier P, Ha T (2007) Spectroscopic observation of RNA chaperone activities of Hfq in post-transcriptional regulation by a small non-coding RNA. Nucleic Acids Res 35(3):999–1006

8. Backofen R, Hess WR (2010) Computational prediction of sRNAs and their targets in bacteria. RNA Biol 7(1):33–42
9. Bernhart SH, Hofacker IL, Stadler PF (2005) Local RNA base pairing probabilities in large sequences. Bioinformatics 22(5):614–615
10. Bernhart SH, Tafer H, Mückstein U, Flamm C, Stadler PF, Hofacker IL (2006) Partition function and base pairing probabilities of RNA heterodimers. Algorithms Mol Biol 1(1):3
11. Busch A, Richter AS, Backofen R (2008) IntaRNA: efficient prediction of bacterial sRNA targets incorporating target site accessibility and seed regions. Bioinformatics 24(24): 2849–2856
12. Choudhary K, Deng F, Aviran S (2017) Comparative and integrative analysis of RNA structural profiling data: current practices and emerging questions. Quantit Biol 5(1):3–24
13. Coornaert A, Chiaruttini C, Springer M, Guillier M (2013) Post-transcriptional control of the Escherichia coli PhoQ-PhoP two-component system by multiple sRNAs involves a novel pairing region of GcvB. PLOS Genet 9(1): 1–16
14. Dai X, Zhuang Z, Zhao PX (2018) psRNATarget: a plant small RNA target analysis server (2017 release). Nucleic Acids Res 46(W1): W49–W54
15. DiChiacchio L, Sloma MF, Mathews DH (2015) AccessFold: predicting RNA-RNA interactions with consideration for competing self-structure. Bioinformatics 32(7): 1033–1039
16. Dirks R, Bois J, Schaeffer J, Winfree E, Pierce N (2007) Thermodynamic analysis of interacting nucleic acid strands. SIAM Rev 49(1): 65–88
17. Eggenhofer F, Tafer H, Stadler PF, Hofacker IL (2011) RNApredator: fast accessibility-based prediction of sRNA targets. Nucleic Acids Res 39(suppl 2):W149–W154
18. Enright A, John B, Gaul U, Tuschl T, Sander C, Marks D (2003) MicroRNA targets in Drosophila. Genome Biol 5:R1
19. Fukunaga T, Hamada M (2017) RIblast: an ultrafast RNA-RNA interaction prediction system based on a seed-and-extension approach. Bioinformatics 33(17):2666–2674
20. Gawronski AR, Uhl M, Zhang Y, Lin YY, Niknafs YS, Ramnarine VR, Malik R, Feng F, Chinnaiyan AM, Collins CC, Sahinalp SC, Backofen R (2018) MechRNA: prediction of lncRNA mechanisms from RNA-RNA and RNA-protein interactions. Bioinformatics 34(18):3101–3110
21. Gelhausen R, Will S, Hofacker IL, Backofen R, Raden M (2019) IntaRNAhelix—composing RNA-RNA interactions from stable intermolecular helices boosts bacterial sRNA target prediction. J Bioinform Comput Biol 17(5): 1940009
22. Gerlach W, Giegerich R (2006) GUUGle: a utility for fast exact matching under RNA complementary rules including G-U base pairing. Bioinformatics 22(6):762–764
23. Gong J, Ju Y, Shao D, Zhang QC (2018) Advances and challenges towards the study of RNA-RNA interactions in a transcriptome-wide scale. Quant Biol 6(3):239–252
24. Grüning B, Dale A Ryan Sjödin, Chapman BA, Rowe J, Tomkins-Tinch CH, Valieris R, Bioconda Team, Köster J (2018) Bioconda: sustainable and comprehensive software distribution for the life sciences. Nat Methods 15(7):475–476
25. Gulliver EL, Wright A, Lucas DD, Mégroz M, Kleifeld O, Schittenhelm RB, Powell DR, Seemann T, Bulitta JB, Harper M, Boyce JD (2018) Determination of the small RNA GcvB regulon in the Gram-negative bacterial pathogen Pasteurella multocida and identification of the GcvB seed binding region. RNA 24(5): 704–720
26. Guo H, Ingolia NT, Weissman JS, Bartel DP (2010) Mammalian microRNAs predominantly act to decrease target mRNA levels. Nature 466:835–840
27. H Bernhart S, Mückstein U, Hofacker I (2011) RNA accessibility in cubic time. Algorithms Mol Biol 6:3
28. He L, J Hannon G (2004) MicroRNAs: Small RNAs with a big role in gene regulation. Nat Rev Genet 5:522–31
29. Hoekzema M, Romilly C, Holmqvist E, Wagner EGH (2019) Hfq-dependent mRNA unfolding promotes sRNA-based inhibition of translation. EMBO J 38(7):e101199
30. Holmqvist E, Wright PR, Li L, Bischler T, Barquist L, Reinhardt R, Backofen R, Vogel J (2016) Global RNA recognition patterns of post-transcriptional regulators Hfq and CsrA revealed by UV crosslinking in vivo. EMBO J 35(9):991–1011
31. Hör J, Gorski SA, Vogel J (2018) Bacterial RNA biology on a genome scale. Mol Cell 70(5):785–799
32. Kato Y, Sato K, Hamada M, Watanabe Y, Asai K, Akutsu T (2010) RactIP: fast and accurate prediction of RNA-RNA interaction using integer programming. Bioinformatics 26(18): i460–i466

33. Kery MB, Feldman M, Livny J, Tjaden B (2014) TargetRNA2: identifying targets of small regulatory RNAs in bacteria. Nucleic Acids Res 42(W1):W124–W129
34. King AM, Vanderpool CK, Degnan PH (2019) sRNA target prediction organizing tool (SPOT) integrates computational and experimental data to facilitate functional characterization of bacterial small RNAs. mSphere 4(1): 10–1128.
35. Kutchko KM, Laederach A (2017) Transcending the prediction paradigm: novel applications of SHAPE to RNA function and evolution. Wiley Interdiscip Rev RNA 8(1)
36. Lai D, Meyer IM (2015) A comprehensive comparison of general RNA-RNA interaction prediction methods. Nucleic Acids Res 44(7): e61–e61
37. Lorenz R, Bernhart SH, Höner zu Siederdissen C, Tafer H, Flamm C, Stadler PF, Hofacker IL (2011) ViennaRNA package 2.0. Algorithms Mol Biol 6(1):26
38. Lott SC, Schäfer RA, Mann M, Backofen R, Hess WR, Voss B, Georg J (2018) GLASSgo—automated and reliable detection of sRNA homologs from a single input sequence. Front Genet 9:124
39. Mandin P, Gottesman S (2009) A genetic approach for finding small RNAs regulators of genes of interest identifies RybC as regulating the DpiA/DpiB two-component system. Mol Microbiol 72(3):551–565
40. Mann M, Wright PR, Backofen R (2017) IntaRNA 2.0: enhanced and customizable prediction of RNA-RNA interactions. Nucleic Acids Res 45(W1):W435–W439
41. Markham NR, Zuker M (2008) UNAFold. Humana Press, Totowa, pp 3–31
42. Matlin AJ, Clark F, Smith CWJ (2005) Understanding alternative splicing: towards a cellular code. Nat Rev Mol Cell Biol 6:386–398
43. Mautner S, Montaseri S, Miladi M, Raden M, Costa F, Backofen R (2019) ShaKer: RNA SHAPE prediction using graph kernel. Bioinformatics 35(14):i354–i359
44. Miladi M, Montaseri S, Backofen R, Raden M (2019) Integration of accessibility data from structure probing into RNA-RNA interaction prediction. Bioinformatics 35(16):2862–2864
45. Min H, Yoon S (2010) Got target? Computational methods for microRNA target prediction and their extension. Exp Mol Med 42: 233–244
46. Mückstein U, Tafer H, Hackermüller J, Bernhart SH, Stadler PF, Hofacker IL (2006) Thermodynamics of RNA-RNA binding. Bioinformatics 22(10):1177–1182
47. Overgaard M, Johansen J, Møller-Jensen J, Valentin-Hansen P (2009) Switching off small RNA regulation with trap-mRNA. Mol Microbiol 73(5):790–800
48. Pain A, Ott A, Amine H, Rochat T, Bouloc P, Gautheret D (2015) An assessment of bacterial small RNA target prediction programs. RNA Biol 12(5):509–513
49. Pervouchine D (2004) IRIS: intermolecular RNA interaction search. Genome informatics Int Conf Genome Inform 15:92–101
50. Peterson S, Thompson J, Ufkin M, Sathyanarayana P, Liaw L, Congdon CB (2014) Common features of microRNA target prediction tools. Front Genet 5:23
51. Raden M, Ali SM, Alkhnbashi OS, Busch A, Costa F, Davis JA, Eggenhofer F, Gelhausen R, Georg J, Heyne S, Hiller M, Kundu K, Kleinkauf R, Lott SC, Mohamed MM, Mattheis A, Miladi M, Richter AS, Will S, Wolff J, Wright PR, Backofen R (2018) Freiburg RNA tools: a central online resource for RNA-focused research and teaching. Nucleic Acids Res 46(W1):W25–W29
52. Raden M, Mohamed MM, Ali SM, Backofen R (2018) Interactive implementations of thermodynamics-based RNA structure and RNA-RNA interaction prediction approaches for example-driven teaching. PLOS Comput Biol 14(8):e1006341
53. Raden M, Gutmann F, Uhl M, Backofen R (2020) CopomuS-ranking compensatory mutations to guide RNA-RNA interaction verification experiments. Int J Mol Sci 21(11): 3852
54. Raden M, Müller T, Mautner S, Gelhausen R, Backofen R (2020) The impact of various seed, accessibility and interaction constraints on sRNA target prediction—a systematic assessment. BMC Bioinform 21:15
55. Rasmussen AA, Johansen J, Nielsen JS, Overgaard M, Kallipolitis B, Valentin-Hansen P (2009) A conserved small RNA promotes silencing of the outer membrane protein YbfM. Mol Microbiol 72(3):566–77
56. Rehmsmeier M, Steffen P, Höchsmann M, Giegerich R (2004) Fast and effective prediction of microRNA/target duplexes. RNA 10(10):1507–1517
57. Reuter JS, Mathews DH (2010) RNAstructure: software for RNA secondary structure prediction and analysis. BMC Bioinform 11(1):129
58. Salari R, Backofen R, Sahinalp SC (2009) Fast Prediction of RNA-RNA Interaction. In: Salzberg SL, Warnow T (eds) Algorithms in Bioinformatics. Springer, Berlin, pp 261–272

59. Salari R, Backofen R, Sahinalp SC (2010) Fast prediction of RNA-RNA interaction. Algorithms Mol Biol 5:5
60. Salim NN, Feig AL (2010) An upstream Hfq binding site in the fhlA mrna leader region facilitates the OxyS-fhlA interaction. PLOS ONE 5(9):1–11
61. Santiago-Frangos A, Kavita K, Schu DJ, Gottesman S, Woodson SA (2016) C-terminal domain of the RNA chaperone Hfq drives sRNA competition and release of target RNA. Proc Natl Acad Sci 113(41):E6089–E6096
62. Seemann SE, Richter AS, Gesell T, Backofen R, Gorodkin J (2011) PETcofold: predicting conserved interactions and structures of two multiple alignments of RNA sequences. Bioinformatics 27(2):211–219
63. Sharma CM, Papenfort K, Pernitzsch SR, Mollenkopf HJ, Hinton JCD, Vogel J (2011) Pervasive post-transcriptional control of genes involved in amino acid metabolism by the Hfq-dependent GcvB small RNA. Mol Microbiol 81(5):1144–1165
64. Soutourina OA, Monot M, Boudry P, Saujet L, Pichon C, Sismeiro O, Semenova E, Severinov K, Le Bouguenec C, Coppée JY, Dupuy B, Martin-Verstraete I (2013) Genome-wide identification of regulatory RNAs in the human pathogen clostridium difficile. PLOS Genet 9(5):1–18
65. Storz G, Vogel Prof Dr J, Wassarman K (2011) Regulation by small RNAs in bacteria: expanding Frontiers. Mol Cell 43:880–91
66. Tafer H, Hofacker IL (2008) RNAplex: a fast tool for RNA-RNA interaction search. Bioinformatics 24(22):2657–2663
67. Tafer H, Amman F, Eggenhofer F, Stadler PF, Hofacker IL (2011) Fast accessibility-based prediction of RNA-RNA interactions. Bioinformatics 27(14):1934–1940
68. Terai G, Iwakiri J, Kameda T, Hamada M, Asai K (2016) Comprehensive prediction of lncRNA–RNA interactions in human transcriptome. BMC Genomics 17(1):12
69. Tjaden B (2008) TargetRNA: a tool for predicting targets of small RNA action in bacteria. Nucleic Acids Res 36(suppl 2):W109–W113
70. Umu SU, Gardner PP (2016) A comprehensive benchmark of RNA-RNA interaction prediction tools for all domains of life. Bioinformatics 33(7):988–996
71. Updegrove TB, Zhang A, Storz G (2016) Hfq: the flexible RNA matchmaker. Curr Opin Microbiol 30:133–138
72. Varadarajan S (2019) Optimizing RNA-RNA interaction computations. In: 2019 IEEE/ACM International Symposium on Code Generation and Optimization (CGO), pp 269–270
73. Wagner E, Gerhart H, Romby P (2015) Small RNAs in bacteria and archaea: who they are, what they do, and how they do it. Adv Genet 90:133–208
74. Wenzel A, Akbasli E, Gorodkin J (2012) RIsearch: fast RNA-RNA interaction search using a simplified nearest-neighbor energy model. Bioinformatics 28(21):2738–2746
75. Wilkinson KA, Merino EJ, Weeks KM (2006) Selective 2′-hydroxyl acylation analyzed by primer extension (SHAPE): quantitative RNA structure analysis at single nucleotide resolution. Nat Protoc 1(3):1610
76. Wright PR, Georg J (2018) Workflow for a computational analysis of an sRNA candidate in bacteria. Methods Mol Biol 1737:3–30
77. Wright PR, Richter AS, Papenfort K, Mann M, Vogel J, Hess WR, Backofen R, Georg J (2013) Comparative genomics boosts target prediction for bacterial small RNAs. Proc Natl Acad Sci USA 110(37):E3487–96
78. Wright PR, Georg J, Mann M, Sorescu DA, Richter AS, Lott S, Kleinkauf R, Hess WR, Backofen R (2014) CopraRNA and IntaRNA: predicting small RNA targets, networks and interaction domains. Nucleic Acids Res 42 (Web Server issue):W119–23
79. Wright PR, Mann M, Backofen R (2018) Structure and interaction prediction in prokaryotic RNA biology. Microbiol Spectr 6(2)
80. Xia T, SantaLucia Jr J, Burkard ME, Kierzek R, Schroeder SJ, Jiao X, Cox C, Turner DH (1998) Thermodynamic parameters for an expanded nearest-neighbor model for formation of RNA duplexes with Watson-Crick base pairs. Biochemistry 37(42):14719–14735
81. Ying X, Cao Y, Wu J, Liu Q, Cha L, Li W (2011) sTarPicker: a method for efficient prediction of bacterial sRNA targets based on a two-step model for hybridization. PLOS ONE 6(7):1–12
82. Zuker M (2003) Mfold web server for nucleic acid folding and hybridization prediction. Nucleic Acids Res 31(13):3406–3415
83. Zuker M, Stiegler P (1981) Optimal computer folding of large RNA sequences using thermodynamics and auxiliary information. Nucleic Acids Res 9(1):133–148

Chapter 10

LocARNA 2.0: Versatile Simultaneous Alignment and Folding of RNAs

Sebastian Will

Abstract

Generating accurate alignments of non-coding RNA sequences is indispensable in the quest for understanding RNA function. Nevertheless, aligning RNAs remains a challenging computational task. In the twilight-zone of RNA sequences with low sequence similarity, sequence homologies and compatible, favorable (a priori unknown) structures can be inferred only in dependency of each other. Thus, simultaneous alignment and folding (SA&F) remains the gold-standard of comparative RNA analysis, even if this method is computationally highly demanding. This text introduces to the recent release 2.0 of the software package LocARNA, focusing on its practical application. The package enables versatile, fast and accurate analysis of multiple RNAs. For this purpose, it implements SA&F algorithms in a specific, lightweight flavor that makes them routinely applicable in large scale. Its high performance is achieved by combining ensemble-based sparsification of the structure space and banding strategies. Probabilistic banding strongly improves the performance of LocARNA 2.0 even over previous releases, while simplifying its effective use. Enabling flexible application to various use cases, LocARNA provides tools to globally and locally compare, cluster, and multiply aligned RNAs based on optimization and probabilistic variants of SA&F, which optionally integrate prior knowledge, expressible by anchor and structure constraints.

Key words Simultaneous alignment and folding, RNA alignment, RNA secondary structure prediction, Multiple sequence alignment, Comparative analysis

1 Introduction

RNA molecules are known as multifaceted players in biological systems, which perform a variety of specific functions through their complex and individual spatial structures. A key technology for unraveling the biological function of RNA molecules is their computational comparison. Such methods are e.g., used to identify of functionally similar RNAs based on evolutionary conserved sequence and structure, find conserved sequence and structure patterns, characterize RNA families, and predict reliable structure models.

Ronny Lorenz (ed.), *RNA Folding: Methods and Protocols*, Methods in Molecular Biology, vol. 2726,
https://doi.org/10.1007/978-1-0716-3519-3_10,

Since structure can be conserved for RNAs of even quite remote sequences, comparing RNAs cannot solely rely on standard sequence alignment methods but requires to take structure information into account. This poses specific challenges, in the typical case where the RNAs' structures are a priori unknown.

Focusing on this case of unaligned RNAs with unknown structure, three main strategies have been identified to infer alignments and likely structure models [5]. Crucially, such methods often rely on RNA secondary structure prediction (in the sense of free energy minimization [26] and tools like RNAfold of the Vienna RNA package [11]) to complement the structural information. The first strategy aligns the RNA sequences by solely relying on "one-dimensional" sequence information; RNA-specific structure prediction-based methods are then applied to infer a common two-dimensional secondary structure from the alignment, typically called the *consensus structure*. Such inference is efficiently performed by tools like RNAalifold [1] in an extension of standard RNA energy minimization. The second reverse strategy starts by predicting structure models of the single RNAs (e.g., using efficient energy minimization as implemented in RNAfold); then, it aligns the structure-annotated RNAs by specialized tree alignment or tree editing-like algorithms. This scheme is directly implemented by tools like RNAforester [10] and MARNA [18].

Both strategies are limited in theory and practice by their separation of alignment and structure prediction. In the first strategy, the alignment has to be constructed without structural information; this works well only for highly similar RNAs, but was demonstrated to plainly fail for RNAs below 60% sequence similarity [6]. The second strategy requires reliably predicted structures, whereas the accuracy of prediction from single sequences is limited. In principle, this cyclic dependence of the alignment and structure prediction phase is best resolved by optimizing the structure model and the alignment at the same time, i.e., by simultaneous alignment and folding (SA&F). Already in 1985, Sankoff presented an efficient solution [17] to this problem; however, the time and space complexity of this original SA&F algorithm prohibit its direct practical application.

Several later implementations of the Sankoff algorithm made moderate applications of SA&F feasible due to strong heuristic restrictions of the search space [7, 8]. Only much later, developments paved the way for large-scale, routine application of the SA&F-paradigm for RNA structure analysis. Notably, PMcomp [9] introduced the idea of light-weight SA&F that strongly reduces the computational overhead of SA&F (at first, by a constant factor), which was then picked up by a whole family of SA&F methods, e.g., [3, 20, 22, 23].

1.1 LocARNA 2.0

Here, we provide an introduction to the practical application of LocARNA, a software package that implements a series of RNA alignment algorithms following the general line of light-weight SA&F. By applying additional strategies that overcome the efficiency problems of SA&F, LocARNA combines the potential for very fast RNA comparisons with the advantages of SA&F. Moreover, it implements variants and extensions that support flexible use in routine as well as many non-standard scenarios.

Specifically, the text introduces for the first time specific characteristics and features of the latest release LocARNA 2.0. Among the new features, in-detail improvements and under-the-hood changes, the single most notable benefit of this release comes from a newly implemented sequence similarity-based probabilistic banding strategy, which transparently speeds up all alignment variants. This technique neither requires alignment-mode specific parameter settings nor is likely to compromise the alignment accuracy, since it automatically adapts to the alignment mode and relevant alignment space; thereby, it significantly improves the performance and usability of LocARNA.

1.2 Scope and Overview of the Text

The text discusses the following topics concerning the practical use of the software in more depth:

- Installation and setup of the software
- General usage for pairwise and multiple alignment
- The general mechanism of pairwise SA&F by LocARNA
- The construction of multiple alignments
- Implemented heuristics that balance speed and accuracy
- Structure constraints, anchor constraints, and realignment

This detailed discussion is necessarily limited to a selection of the possibilities available due to the software package. Nevertheless, it covers its elementary as well as advanced usage in specialized scenarios. We are going to touch the theoretical background of the software in terms of computational models, heuristics, and combinatorial optimization algorithms where this is directly beneficial for the successful, advanced use of the software. For further technical details, the reader is referred to the corresponding original research publications [14–16, 21–24].

1.3 Noteworthy Omissions from This Text

Many advanced or specialized features of the LocARNA package had to be omitted from this introductory book chapter. Nevertheless, to provide at least a more comprehensive overview, the most important ones shall be mentioned in a cursory summary for further reference:

Local and Semi-global Alignment While the chapter focuses on global alignment of entire input sequences, LocARNA supported various modes of local alignment from its beginnings (this circumstance causing the acronym). Such, the package provides local and semi-local alignment, where it either finds the best alignment of sub-sequences or allows free end gaps to accommodate scenarios like finding the occurrence of a small sequence in a larger one, loosely defined 3'-ends, etc. Local alignment even comes with a special flavor that can leave out dissimilar substructure (*structure local alignment*; [16]).

Clustering of RNAs One of the earliest applications of LocARNA was the clustering of non-coding RNAs by their (local) sequence and structure similarity. Clustering moderately sized sets of RNAs (several hundreds to thousands) using SA&F is sufficiently fast and directly possible using a special mode of the multiple alignment tool `mlocarna`. For this purpose, it is recommended to use multi-threading (`mlocarna` option `--threads`) and limit the maximum alignment size (option `--max-alignment-size`). For this use case, `mlocarna` computes a distance-annotated tree of the input RNAs and intermediary "cluster" alignments up to the given maximum size.

Probabilistic Alignment, Alignment Reliabilities, and Consistency-Based Multiple Alignment LocARNA implements the computation of partition functions over SA&F, which is used to derive match probabilities and column-wise reliability score [21]. Match probabilities are moreover the basis for improvements to the construction of multiple alignments. This is achieved by computing probabilistic consistency-based multiple alignment (similar to ProbCons [4], which was introduced for pure sequence alignment). Moreover, probabilistic alignment is used in LocARNA to score improvements due to (experimental) iterative refinement of multiple alignment.

Exact Matches of Local Sequence-Structure Patterns The package implements the method ExpaRNA-P [15], which finds exact matches in RNA structure ensembles. These can be useful to understand similarities of large RNA molecules (without *a priori* known structure). We suggested a method to use such matches as anchor speed up alignments of large structural RNAs.

Improved Pairwise Alignment Methods The package supports to exchange the pairwise alignment method in its construction of multiple alignments. It implements the pairwise SA&F algorithm SPARSE [22], which extends the alignment model of PMcomp (by loop insertions and deletions comparable to the original

Sankoff algorithm) and speeds up pairwise RNA alignment using strong structure-ensemble based sparsification. Other supported pairwise alignment methods are CARNA [2] and Pankov [14].

2 Material: Installing LocARNA

It is recommended to install the software on Linux or MacOS using the package manager Conda as described below. Windows users can install Ubuntu with Windows Subsystem for Linux (WSL) and then follow the installation instructions for Linux as below from the Ubuntu terminal.

As alternative to the Conda installation, the software can be compiled and installed from a source tar-ball or directly from its repository on GitHub https://github/s-will/locarna. While compiling from source should be limited to special cases, instructions for the installation from source are provided with the LocARNA online documentation at https://s-will.github.io/LocARNA/. Recall that this chapter describes the release 2.0 of LocARNA.

2.1 Installation via Conda

It's recommended to install Conda from https://conda.io/en/latest/miniconda.html following the given instructions for your system. LocARNA is best installed in a dedicated Conda environment. First, this environment has to be created (once) and activated (once per session) by running

```
conda create -n locarna
conda activate locarna
```

in a terminal. Then, LocARNA is installed in this environment by

```
conda install -c conda-forge -c bioconda locarna
```

As a main benefit of using the package manager Conda, this single command installs the most recent LocARNA package without requiring compilation and together with its dependencies, i.e., first of all, the Vienna RNA package [11] (Conda package `viennarna`). The option `-c bioconda` tells Conda to find the package on the Bioconda channel.

After successful installation, the following call should report the LocARNA version (at least 2.0).

```
locarna --version
```

3 Methods

3.1 Elementary Usage

The central functionality of the package is the computation of (multiple) alignments and consensus structures via the tool mlocarna, starting from a set of RNA sequences. As a running example, we consider five tRNAs. We assume that their sequences are specified in a fasta file:

File tRNA_5.fa:

```
>X03715.1_288-361
GCGCCCAUAGAUCAAUUGGAUAGAUCGUUUGACUACGGAUCAAAAGGUUGAGGGUUCGAUUCCUUCUGGGCGCG
>J01435.1_264-194
UAGAUUGAAGCCAGUAAGUAGGGUAUUUAGUUGUUAACUAAAUUUUCGUAGGUUUGAAUCCUUCCAAUCUA
>X16886.1_781-711
AGUAAAGUAAGCUAAUAAAAAGCUUUUGGGUUCAUACCUCAAAAAUGGAAGGAUAAAUACCUCCCUUUAUU
>V00654.1_12038-12108
ACUUUUAAAGGAUAGUAGUUUAUCCGUUGGUCUUAGGAACCAAAAAAUUGGUGCAACUCCAAAUAAAAGUA
>AE004237.1_2976-2903
GCGUCCGUAGCUCAGUUGGUUAGAGCACCACCUUGACAUGGUGGGGUCGGUGGUUCGAGUCCACUCGGACGCA
```

To run the examples, first create (or obtain) this file. Running mlocarna (again, from the command line) without any further parameters like

```
mlocarna tRNA_5.fa
```

computes a multiple alignment that aims to explain both the sequence and the structure homology of the sequences. The command prints the alignment and the inferred common structure as text output to the screen.[1] (At the same time, mlocarna writes several result and intermediary files to a subdirectory tRNA_5.out.)

[1] For this presentation, we added the formatting option --width 60, which breaks the sequence lines after 60 characters, instead of 120 per default. In consequence, this alignment is printed in two blocks.

```
J01435.1_264-194          UAGAUUGAAGCCAGUAAG--UAGGGUAUUUAGUUGUUAACUAAAUUUUCGUAGGUUUGAA
X16886.1_781-711          AGUAAAGUAAGCUAAUAA--AAAGCUUUUGGGUUCAUACCUCAAAAAUGGAAGGAUAAAU
X03715.1_288-361          GCGCCCAUAGAUCAAUUGGAUAGAUCGUUUGACUACGGAUCAAAAGGUUGAGGGUUCGAU
AE004237.1_2976-2903 GCGUCCGUAGCUCAGUUGGUUAGAGCACCACCUUGACAUGGUGGGGGUCGGUGGUUCGAG
V00654.1_12038-12108 ACUUUUAAAGGAUAGUAGUUU-AUCCGUUGGUCUUAGGAACCAAAAA--AUUGGUGCAAC

J01435.1_264-194          UCCUUCCAAUCUA-
X16886.1_781-711          ACCUCCCUUUAUU-
X03715.1_288-361          UCCUUCUGGGCGCG
AE004237.1_2976-2903 UCCACUCGGACGCA
V00654.1_12038-12108 UCCAAAUAAAAGUA

alifold                   (((((((..((((.........)))).(((((((....).)))))).....(((((......
                          .)))))))))))). (-39.72 = -16.98 + -22.74)
```

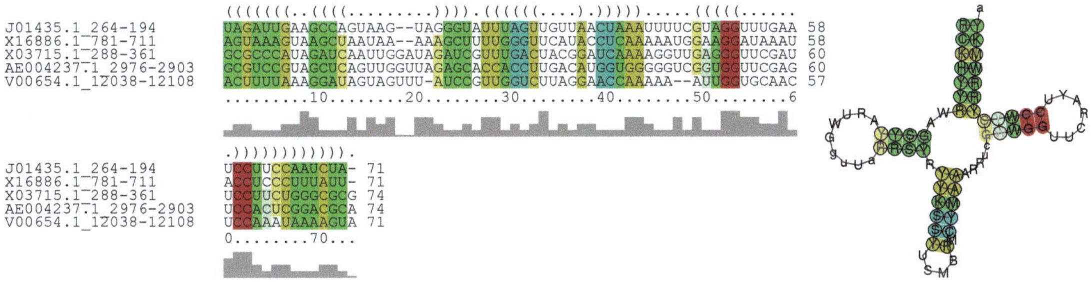

Fig. 1 The alignment and consensus structure for the running example of five tRNAs. These figures are generated using RNAalifold from the alignment computed by `mlocarna` with default parameters. The color annotation encodes evolutionary support of base pairs in the specific way of RNAalifold; the bars below of the alignment show column-wise sequence conservation; finally, RNAalifold reports "most informative" IUPAC code for each column in the 2D structure plot (RNAalifold option `--mis`)

The last line of the output reports the consensus structure of the alignment; this structure is computed—based on LocARNA's alignment—by the tool RNAalifold (with command line arguments `--mis --cfactor=0.6 --nfactor=0.5`). The output line includes the total energy calculated by this tool, as well as its composition from the average Turner energy and conservation score. RNAalifold as well produces graphical output of the alignment and consensus structure as shown in Fig. 1. The RNAalifold-specific color annotation of base-paired columns and base pairs in these figures reflects the diversity of consensus base pairs—related to the idea of compensatory mutations. (Please find more details of the graphical representation in RNAalifold's documentation.)

Input and Output Files of `mlocarna` The tool writes it results to a target directory, whose name is (by default; otherwise use `--tgtdir`) derived from the input file; e.g., for input file

Examples/tRNA.fa, the target directory would be Examples/ tRNA.out. The fasta input file is preserved in the target directory as input/input.fa. Results are written to the subdirectory results; e.g., mlocarna writes the result alignment to results/result. aln in the format of ClustalW. Given the option --stockholm, mlocarna additionally writes the result alignment and it's consensus structure in Stockholm format to results/result.stk. Along with these files, mlocarna writes several other files, containing final and intermediary results, to the target directory. For example, it writes figures as the ones of Fig. 1 of the result alignment and consensus structure as produced by RNAalifold in PostScript format to subdirectory results. As an alternative to fasta input, mlocarna can read its input sequences from a file in ClustalW format. This requires the option --realign, since it serves to realign already aligned sequences. By default, mlocarna ignores the gaps in the input alignment.

Getting Specific Help for mlocarna The command line tool mlocarna can be finely controlled by a rich set of command line parameters. Note that only some of them will be immediately obvious and are therefore easily applicable in standard applications of the tool. Others will become clear only in the course of this text and/or after more experience with the tool. A reference of the command line parameters is provided with the manual page of mlocarna. It is found online at https://s-will.github.io/LocARNA/md_src_Utils_mlocarna.html and can be displayed from the command line using mlocarna's option --man or by the command

```
man mlocarna
```

3.2 Remarks on the Nature of Default Results and Advanced Usage of LocARNA

For users of the tool (even occasional ones), it is helpful to understand the nature of LocARNA's main output in more detail. In its default mode, given unaligned input sequences without *a priori* known structure, the tool predicts a *good* alignment of the *entire* input sequences (*global* alignment). In the case of RNAs, a good alignment must not only align similar nucleotides with little insertions and deletions but also possess a consensus structure that is thermodynamically favorable for each of the single RNAs. LocARNA therefore optimizes a combined evaluation of the alignment and the predicted consensus structure.

On the one hand, thanks to this ability of considering sequence similarity and quality of predicted structure at the same time, which is the core idea of simultaneous alignment and folding, LocARNA can produce good alignments of RNAs in the twilight or even midnight zone, where strong sequence similarity signals are lacking and thus sequence-only alignment methods fail. This is as well the case of our running example from the deep midnight zone (with only 40% average pairwise sequence identity.)

On the other hand, alignment of RNAs with low sequence identity poses specific problems that are not all known from simpler sequence alignment tools. Consequently, LocARNA could struggle to find a good alignment for various reasons. While some issues are fundamental and subject of ongoing research, many of these problems can be dealt with by specific parametrization of the tool (e.g., relative focus on sequence or structure similarity, gap cost), specifying a priori knowledge in the form of constraints (anchor constraints, specifying aligned positions and/or structure constraints, providing knowledge on the structure space), and specific alignment modes (e.g., local and semi-local alignment, but as well, probabilistic alignment to learn about local alignment reliability). An important aspect is the high computational complexity of the method, which requires to make trade-offs (which in LocARNA are interpretable and controlled) between accuracy and execution speed.

In summary, while LocARNA responds to these challenges of RNA alignment by offering a high degree of flexibility (different alignment modes, different SA&F variants, various heuristics, parametrization, constraints, ...), using these possibilities effectively requires some insight into its computational machinery (as discussed in the next subsection).

3.3 The Construction of Multiple and Pairwise Alignments by LocARNA

The tool `mlocarna` constructs multiple alignments by the progressive method as it is well-known from classic alignment tools like ClustalW [19].

Note that the LocARNA package implements several extensions and refinement of the progressive strategy, like probabilistic consistency-transformations and iterative refinement, which are available with specific alignment modes. Moreover, alignment construction based on T-Coffee is available via the tool `locarnate` of the package. To limit the scope of this text, we discuss only the purely progressive method, the default strategy of `mlocarna`, in more detail.

The main idea of the progressive method is to construct multiple alignments based on the repeated computation of pairwise alignments. This has the main advantage that in the case of sequence alignment and as well of simultaneous alignment and folding, the pairwise problem is precisely defined and can be solved efficiently. In contrast, the multiple alignment problem is computationally hard (NP hard); therefore, progressive construction does not guarantee to find the very best multiple alignment.

Pairwise Alignment by the Tool locarna By default, `mlocarna` employs the low level tool `locarna` for computing the pairwise alignments (illustrated in Fig. 2). Applying a dynamic programming algorithm [23], this tool optimizes over pairwise alignments and their simultaneously foldings into a common RNA secondary structure. It follows the general idea of PMcomp [9] to derive the optimal alignment and common structure from the two RNA sequences and the base pair probabilities of the both single sequences. The latter are predicted based on the thermodynamic model by the McCaskill algorithm [13] (as implemented in the Vienna RNA package). Together with the sequences, we consider the base pair probabilities input of the pairwise alignment tool.

Figure 3 visualizes the base pair probabilities of each of two example sequences as heatmaps (equivalent to the dot plot representation of the Vienna RNA package). The figure also shows that the probable base pairs are typically very sparse and highlights the more probable base pairs. The LocARNA algorithm was designed to exploit this sparsity, which in LocARNA fundamentally speeds up computations and reduces the memory footprint over traditional SA&F. The effect is controlled by a probability threshold (argument `-p` of the tools `mlocarna` and `locarna`; the threshold for the minimum base pair probability defaults to 10^{-3}).

Banding Strategies In addition to controlling the structure space using a minimum base pair probability, LocARNA controls the alignment space. A common simple strategy to speed up pairwise alignment algorithm is to restrict the maximum difference between

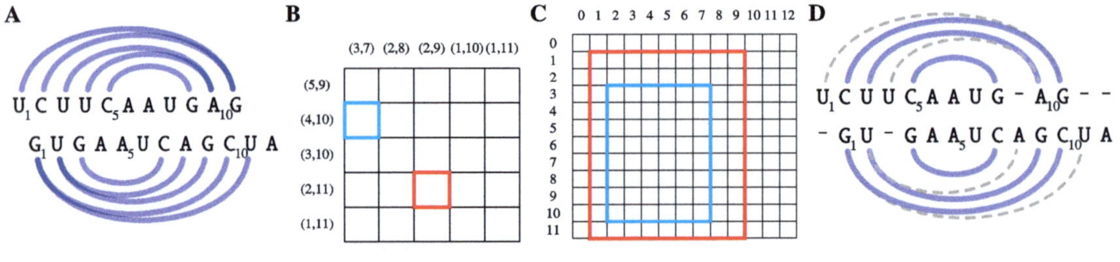

Fig. 2 LocARNA's dynamic programming pairwise alignment algorithm, illustrated for an example instance. The pairwise algorithm operates on two input sequences, each annotated by a set of "probable" candidate base pairs (**a**); their selection is illustrated in Fig. 3. The outcome is an alignment of the sequences and a prediction of matches of the candidate base pairs as shown in (**d**). The algorithm successively calculates scores of sub-alignments that match two base pairs and the enclosed sub-sequences and tabulates them in a two-dimensional matrix (**b**) (indexed by the candidate base pairs). Such scores are calculated by filling a second two-dimensional matrix over sequence positions. For example, the score for base pairs (2, 11) and (2, 9), red cell in (**b**), is calculated by successively computing entries in the red rectangle shown in (**c**). The computation of such entries resembles Needleman-Wunsch sequence alignment, extended by cases for matching base pairs. The computation of the red entries thus depends on entries for enclosed base pair candidates like the cyan entry (which in turn is derived from computing a submatrix over sequence positions). Note that LocARNA typically restricts the computed entries in the matrices further (cf. Figs. 4 and 5)

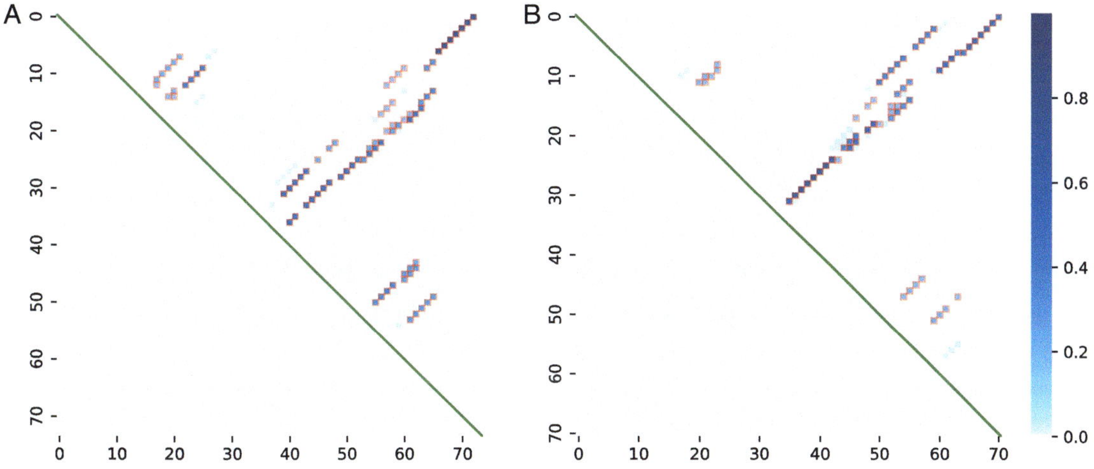

Fig. 3 Base pair probabilities of (**a**) tRNAs AE004237.1_2976-2903 (**b**) J01435.1_264-194. The heatmap colors show the probabilities of base pairs (i, j). These probabilities from single sequence folding allow effectively restricting the folding space in simultaneous alignment and folding. The red boxes mark base pairs with probabilities greater equal 10^{-2}. For this threshold, LocARNA would consider only the highlighted base pairs. This strategy significantly speeds up LocARNA while preserving accuracy

alignable positions i (of the first sequence) and j (of the second sequence). In typical dynamic programming alignment algorithms, one would technically limit this difference for all matrix entries (i, j). In consequence, such *banding methods* compute only bands (around the diagonal) of the DP matrices. Thus, already before version 2.0, LocARNA implemented two "ad-hoc" banding strategies that are illustrated in Fig. 4. The first one (Fig. 4a) restricts the computed matrix entries around the matrix diagonal (option `--max-diff`). For the case of realignment, a second strategy was suggested [24] that restricts the alignment space relative to existing alignments (Fig. 4b; option `--max-diff-aln`). It allows the fast realignment of initial alignments, whether these come from a database or are computed by less costly alignment methods. The second banding strategy especially helps to overcome problems of the first one with large insertions and deletions.

Probabilistic Banding in LocARNA 2.0 Recent LocARNA replaces the default banding strategy to allow more flexible and stronger banding that automatically adapts to the lengths and sequence similarity of the input sequences. This allows LocARNA 2.0 to run (for specific applications, even much) faster with default parameters than its predecessors. It builds on the computation of trace probabilities in sequence alignments of the input sequences. After computing the probability for each matrix cell (i, j) that occurs on the trace of the alignment, this technique applies a threshold probability to decide about the matrix cells computed by the computationally more costly SA&F algorithm. Figure 5 illustrates the effect for two pairs of example sequences.

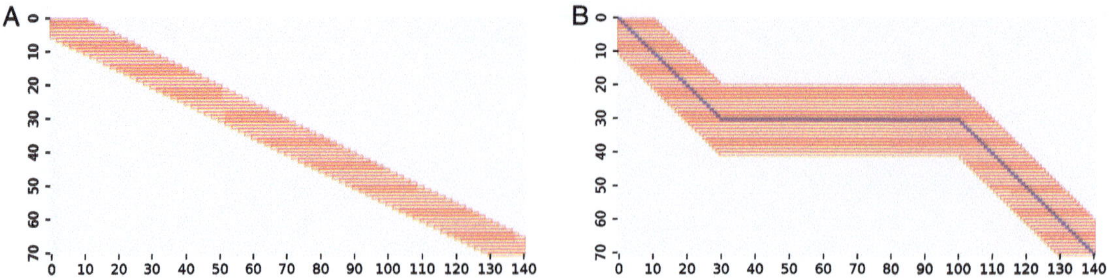

Fig. 4 Banding by a maximal position difference $\Delta = 10$. The (due to the banding strategy) computed matrix cells are highlighted in red. (**a**) in the case of unequal sequence lengths, here n=70 and m=140, the length difference can be distributed by $|i - j \cdot n/m| \leq \Delta$. Before version 2.0, LocARNA mainly used this banding method (for global alignment), which is still available with argument `--max-diff` Δ. (**b**) This approach still struggles with unequally distributed insertions. In an attempt to overcome this in the case of realignment, given an initial alignment (in the example with one large insertion; trace in blue), LocARNA can restrict the difference to this trace (with argument `---max-diff-aln` Δ). In practice, this allows LocARNA to quickly realign the given initial alignment (which could e.g., be computed by less accurate methods). Going beyond these ad-hoc or very specialized strategies, LocARNA 2.0 implements a probabilistic banding technique that can be universally applied and thus allows easier and more flexible use of strong banding (*see* Fig. 5)

Probabilistic banding was similarly used before by SA&F tools like RAF [3] and Dynalign [7]. The strategy is applicable because trace probabilities from sequence alignments can be computed quickly by a forward/backward partition function algorithm (modeling alignment probabilities in a statistical mechanics approach; i.e., assuming Boltzmann distribution of the alignments based on their sequence alignment scores). This technique works typically well because high-scoring alignments from SA&F are unlikely to be strongly improbable in the sequence alignment model.

Progressive Alignment by `mlocarna` Figure 6 depicts the progressive alignment for the running example. As pre-computation, `mlocarna` calculates the base pair probability matrices of all input sequences. For each input RNA, it stores the sequence and pair probabilities (above the threshold), in a file in the `input` subdirectory. These files are written in LocARNA's pp format and can serve as direct input to the tool `locarna`. The first row of Fig. 6 represents the state after the pre-computation. For illustration, we also show the predicted secondary structure (minimum energy structure) of each sequence.

The progressive scheme starts by constructing a phylogenetic tree ("guide tree") from all pairwise distances between the input sequences. These distances are obtained by computing (optimal) pairwise alignments of all quadratically many sequence pairs.

The guide tree (blue arrows in the figure) dictates the order of constructing the multiple alignment from the input sequences. The progressive alignment therefore first aligns the closest two sequences; then (in the example) the next two closest sequences.

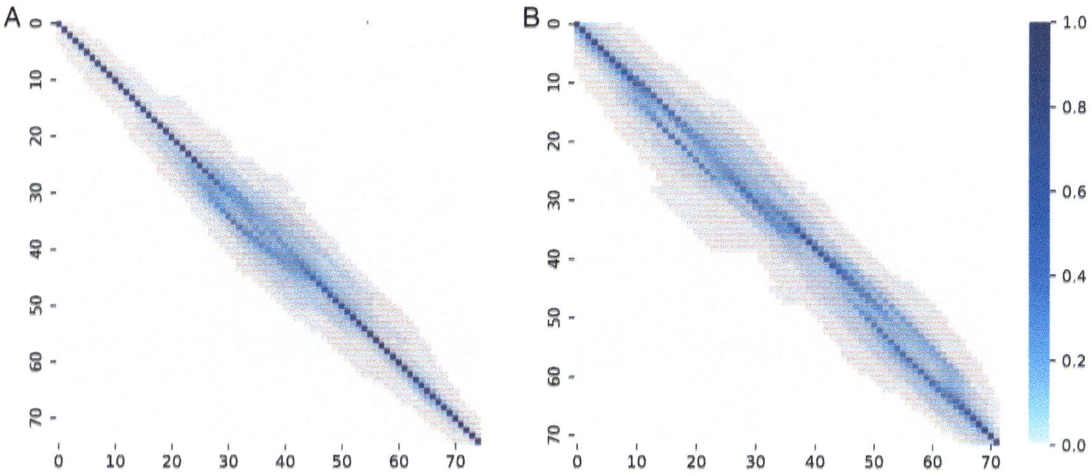

Fig. 5 Trace probabilities from pairwise sequence alignments of (**a**) tRNAs AE004237.1_2976-2903 with X03715.1_288-361i and (**b**) J01435.1_264-194 with V00654.1_12038-12108. The color indicates the probability that an alignment trace passes each matrix cell (i, j). Such probabilities are computed from forward/backward partition functions. These sequence-alignment inferred trace probabilities allow effectively restricting the alignment space in simultaneous alignment and folding. This banding method significantly speeds up LocARNA 2.0 while preserving accuracy. For trace probability threshold 10^{-5}, LocARNA's simultaneous and folding algorithm would consider only the entries in red boxes

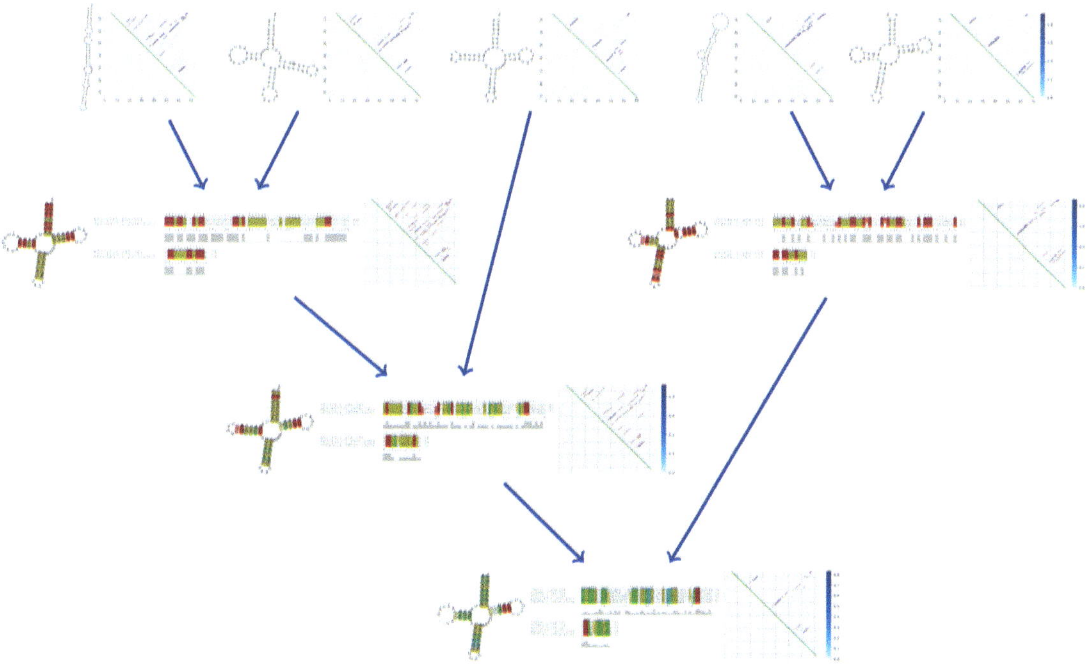

Fig. 6 Progressive alignment of the five tRNAs from the running example. The figure shows base pair probability heatmaps ("dot plots") and minimum free energy structures of the input structures in the first row. The blue arrows represent the guide tree and thus show the order of progressive alignment steps that combine two sequences, a sequence and an alignment or two alignments into a larger alignment. For each intermediate or final alignment, we show the RNAalifold consensus structure, the alignment figure drawn by RNAalifold and the 'averaged' base pair probabilities as heat map. In these heat maps, the base pairs with probability $\geq 10^{-3}$, which are (or, for the final alignment, would be) considered by `locarna` in the next progressive step, are highlighted

This procedure then comes to the point where a sequence has to be combined with an alignment—or more generally, where given two alignments, we want to derive a larger alignment comprising all their sequences. Following the general idea of progressive alignment, LocARNA derives this larger alignment by aligning the two alignments. Here, one can essentially apply the same pairwise SA&F algorithm as in the case of two input sequences. However, in extension, the sequence similarities are computed from the input alignment columns and the structure prediction is based on (in principle) average base pair probabilities—technical details are found in [23].

These progressive alignment steps are repeated in the order of the tree until an alignment of all sequences is derived. In the course of this scheme, intermediary alignments (and their averaged base pair probabilities) are derived. Especially when aligning larger sets of RNAs (or even more in clustering applications), this information can become valuable. Therefore, `mlocarna` stores this data in subdirectory `intermediates`—as ClustalW aln files and as files in pp-format, which contain the alignment and its (consensus) base pair probabilities.

3.4 Structure and Anchor Constraints

An important feature of LocARNA's alignment tools is the support of constraints that allow users to guide the alignment due to prior knowledge. There are two major types of prior knowledge that can be encoded as constraints.

- *First,* information about the structure of input RNAs. Here, one can provide structure constraints strings (for each input RNA) in the syntax of the Vienna RNA package (compare option `-C`, `--constraint` of RNAfold). These constraints are then directly applied in the prediction of the input sequences' base pair probabilities. In this way, they limit the considered structure space in the SA&F algorithm.
- *Second,* prior knowledge about homologous positions. LocARNA can be forced to align specific positions of the input sequences by specifying anchors (thus, imposing *anchor constraints*). It is noteworthy that anchor constraints could be inconsistent to each other or conflict with other heuristics. In such cases, LocARNA will not produce an alignment.

Specification of Constraints For their use in `mlocarna`, both types of constraints can be specified in extension of the standard fasta format via "constraint lines". Here is an example input for `mlocarna` of two sequences together with structure and anchor constraint annotation in the extended fasta-like format.

File example-w-constraints.fa:

```
>A
GACCCUGGGAACAUUAACUACUCUCGUUGGUGAUAAGGAACA
..((.(....xxxxxx...................))).xxx #S
..........000000.......................111 #1
..........123456.......................123 #2
>B
ACGGAGGGAAAGCAAGCCUUCUGCGACA
.(((....xxxxxx.......))).xxx #S
........000000...........111 #1
........123456...........123 #2
```

In this example, the two structure constraints (lines annotated #S), specify that certain sequence positions must be unpaired (`x`) and others must not be crossed by base pairs (corresponding brackets; the exact syntax and semantics of these constraints strings is defined by RNAalifold.) For each sequence, we specify named anchors, where the names should be read top-down over the anchor lines #i. That is, the file defines names `01,02,...,06,11,12,13` (of length two) for according positions of each sequence. Positions of equal names must be aligned to each other. Moreover, names must be specified in lexicographic order.

A simple mlocarna run on this file will by default take the constraints into account and produce an alignment that is guided by the restricted structure space and aligns all anchors:

```
A         GACCCUGGGAACAUUAACUACUCUCGUUGGUGAUAAGGAACA
B         A--CGGAGGGAAAGCA---AGCCUUCUGCG---------ACA
#A1       ..........000000.......................111
#A2       ..........123456.......................123

alifold   ..((((((((...........)))))..)).)......... (-12.85 =  -3.70 +  -9.15)
```

While the alignment clearly satisfies the anchor constraints, note that the alifold structure does not necessarily satisfy the structure constraint annotation (as seen in this case). This happens since the RNAalifold predicts the unconstrained best structure from the alignment; recall that the structure constraints affect only the input structure space.

Anchor Specifications as Bed Format Annotation Anchors can alternatively be specified in bed format. This is useful to annotate larger alignments and/or include information from sequence annotation. Using this method, the same anchor constraints as in file example-w-anchors.fa (by lines tagged #1, #2) could be specified by

File example-anchors.bed:

```
A  10   16   first_box
B  8    14   first_box
A  39   42   ACA-box
B  25   28   ACA-box
```

As suggested by the example, anchor regions receive arbitrary names and contig (or sequence) names. Again, the idea is to enforce the alignment of equally named regions. This bed file can be used in combination with a fasta input file that defines the sequences and potential structure constraints. For example, to reproduce the above example, we define

File example-wo-anchors.fa:

```
>A
GACCCUGGGAACAUUAACUACUCUCGUUGGUGAUAAGGAACA
..((.(....xxxxxx..................))).xxx #S
>B
ACGGAGGGAAAGCAAGCCUUCUGCGACA
.(((....xxxxxx.......))).xxx #S
```

and then call mlocarna as follows:

```
mlocarna --anchor-constraints example-anchors.bed  example-wo-anchors.fa
```

Realignment Another way to specify prior knowledge on the sequence alignment space is provided by LocARNA's realignment mode. In this mode (option `--realign`), `mlocarna` is called with input in the form of an alignment in ClustalW format, e.g.,

```
mlocarna --realign example-realign.aln
```

Then, `mlocarna` uses the unaligned sequences of the alignment as input (i.e., it removes all gap symbols). The real benefit of this input mode is that it allows defining anchor constraints as "consensus constraints" in the input, e.g.,

File example-realign.aln:

```
CLUSTAL W

vhuU           AGCUCACAACCGAACCCAUU------------UGGGAGGUUGUGAGCU
fdhA           CGCCACCCUGCGAACCCAAUAUAAAAUAAUACAAGGGAGCAG-GUGGCG
fruA           --CCUCGAGGGGAACCCGAA------------AGGGACCCGAGAGG--
#A1            ..*..........CCC............................5..
#S             ((((((.((((...(((.................))).)))).))))))
```

Note that the "names" in the anchor annotation string (line #A1) are arbitrary. All symbols apart from blank and dot are interpreted as defining an anchor for the respective position of each sequence. On this input, mlocarna then produces an alignment that preserves the matches in the marked alignment columns. For example, running

```
mlocarna --realign example-realign.aln
```

yields the alignment

```
vhuU           AGCUCACAACCGAACCCAUU------------UGGGAGGUUGUGAGCU
fdhA           CGCCACCCUGCGAACCCAAUAUAAAAUAAUACAAGGGAGCAG-GUGGCG
fruA           --CCUCGAGGGGAACCCGA------------AAGGGACCCGAGAGG--
#A1            ..0..........123............................4..
```

Banding in Relation to an Alignment Note that realignment moreover gives access to a special type of banding heuristic that restricts the alignment space in relation to a given alignment. The central idea is illustrated (for pairwise alignment) in Fig. 4b but is available, generalized to multiple alignment, in mlocarna.

For the purpose of limited (and therefore fast) realignment of an existing alignment, we specify a maximal distance to the alignment. To realign our example alignment in distance 1 (in addition to respecting the constraint annotation), one uses the following syntax:

```
mlocarna --realign example-realign.aln --max-diff 1 --max-diff-aln .
```

For this simple example, we obtain exactly the same alignment as above, which changed the original alignment only slightly (only in distance one). Nevertheless, limiting the realignment distance guarantees limited modification and speeds up the alignment, which becomes important to handle a large amount of larger alignments (for example, whole genome realignment in REAPR [24]).

4 Notes

Benchmarking the performance of LocARNA 2.0 in comparison to the previous release.

A benchmark on Bralibase 2.1 provides a general impression of the performance of LocARNA 2.0, in absolute terms as well as compared to the latest previous release 1.9.2.3. For this purpose, without aiming at a comprehensive benchmark, we consider only the sets k2 and k10 for pairwise and 10-way alignments, respectively. The benchmark scripts are provided with the source of the LocARNA package and provide experimental support (through a make file Makefile-benchmark) to reproduce the results, benchmark LocARNA on different benchmark sets, or measure performance with different options.

The Bralibase 2.1 benchmark [25] suggested to compute the two alignment statistics SPS and SCI. The SPS is a sum-of-pairs similarity measure to the reference structure in the benchmark set. The SCI (structure conservation index) is defined as ratio between the average minimum free energy of the single RNAs and the RNAalifold minimum free energy of the alignment. Thus, in this benchmark, it measures the alignment's suitability to guide the prediction of *some* common RNA structure. In addition, the structure prediction accuracy is measured by MCC [12] (Matthews correlation coefficient), which measures the similarity of the predicted structure to the Rfam-derived reference structure.

Table 1 summarized the results for LocARNA 2.0 and the previous release. Since LocARNA 2.0 only marginally improves these measures on average, users of older versions should not expect generally more accurate alignments from LocARNA2.0 in standard global alignment. Here, apart from the generally improved usability and higher flexibility, the main advantage is found in computational efficiency. As such, the benchmarks show dramatically improved run-times and space consumption. LocARNA 2.0 computes k2 in about 40% of the time and in 55% of the space; for k10, it takes 0.06% of the time (15-fold speed up) and 44% of the space. To make this comparison realistic for experienced users of previous versions of LocARNA, we used the old version with argument `--max-diff 60`, which activates a moderate ad-hoc banding heuristic. Note that the performance increase over old LocARNA with default parameters is even higher (for k10, roughly, a total factor of 30 in time and 3.5 in space). The time and

Table 1
Benchmark on k2 and k10 of Bralibase 2.1, comparing LocARNA 2.0 to the previously released version 1.9.2.3 Set k2 consists of 8976 pairwise alignment; k10, of 845 10-way alignments. Time is reported as total user time of running all instances in parallel on Intel(R) Core(TM) i7-10810U CPU; since this machine distributes to 12 cores, the real time is about 1/12 of the reported user time; space, as maximum resident set size of the parallel execution

	LocARNA 1.9 (`--max-diff 60`)		LocARNA 2.0 (default)	
	k2	k10	k2	k10
SPS	0.86	0.91	**0.89**	0.92
SCI	1.01	0.92	1.01	0.92
MCC	0.81	0.87	**0.82**	0.87
Time (min)	294	1143	**110**	**76**
Space (MB)	49	70	**26**	**30**

space improvements in LocARNA 2.0 can be mostly attributed to its probabilistic banding strategy.

References

1. Bernhart SH, Hofacker IL, Will S, Gruber AR, Stadler PF (2008) RNAalifold: improved consensus structure prediction for RNA alignments. BMC Bioinform 9:474
2. Dal Palù A, Möhl M, Will S (2010) A propagator for maximum weight string alignment with arbitrary pairwise dependencies. In: Principles and Practice of Constraint Programming – CP 2010. Springer, Berlin, pp 167–175
3. Do CB, Foo C-S, Batzoglou S (2008) A max-margin model for efficient simultaneous alignment and folding of RNA sequences. Bioinformatics 24(13):i68–i76
4. Do CB, Mahabhashyam MS, Brudno M, Batzoglou S (2005) ProbCons: probabilistic consistency-based multiple sequence alignment. Genome Res 15(2):330–340
5. Gardner PP, Giegerich R (2004) A comprehensive comparison of comparative RNA structure prediction approaches. BMC Bioinform 5:140
6. Gardner PP, Wilm A, Washietl S (2005) A benchmark of multiple sequence alignment programs upon structural RNAs. Nucleic Acids Res 33(8):2433–2439
7. Do CB, Mahabhashyam MS, Brudno M, Batzoglou S (2007) Efficient pairwise RNA structure prediction using probabilistic alignment constraints in Dynalign. BMC Bioinform 8: 130
8. Havgaard JH, Torarinsson E, Gorodkin J (2007) Fast pairwise structural RNA alignments by pruning of the dynamical programming matrix. PLOS Comput Biol 3(10):e193
9. Hofacker IL, Bernhart SH, Stadler PF (2004) Alignment of RNA base pairing probability matrices. Bioinformatics 20(14):2222–2227
10. Höchsmann M (2005) The tree alignment model: algorithms, implementations and applications for the analysis of RNA secondary structures. Accessed 19 Sep 2022
11. Lorenz R, Bernhart SH, Höner Zu Siederdissen C, Tafer H, Flamm C, Stadler PF, Hofacker IL (2011) ViennaRNA Package 2.0. Algorithms Mol Biol 6:26
12. Matthews BW (1975) Comparison of the predicted and observed secondary structure of t4 phage lysozyme. Biochem Biophys Acta 405: 442–451

13. McCaskill JS (1990) The equilibrium partition function and base pair binding probabilities for RNA secondary structure. Biopolymers 29(6–7):1105–1119
14. Miladi M, Raden M, Will S, Backofen R (2020) Fast and accurate structure probability estimation for simultaneous alignment and folding of RNAs with Markov chains. Algorithms Mol Biol 15(1)
15. Otto C, Möhl M, Heyne S, Amit M, Landau GM, Backofen R, Will S (2014) ExpaRNA-P: simultaneous exact pattern matching and folding of RNAs. BMC Bioinform 15(1):404
16. Otto W, Will S, Backofen R (2008) Structure local multiple alignment of RNA. Gesellschaft für Informatik e. V.
17. Sankoff D (1985) Simultaneous solution of the RNA folding, alignment and protosequence problems. SIAM J Appl Math 45(5):810–825
18. Siebert S, Backofen R (2005) MARNA: multiple alignment and consensus structure prediction of RNAs based on sequence structure comparisons. Bioinformatics 21(16):3352–3359
19. Thompson JD, Higgins DG, Gibson TJ (1994) CLUSTAL W: improving the sensitivity of progressive multiple sequence alignment through sequence weighting, position-specific gap penalties and weight matrix choice. Nucleic Acids Res 22(22):4673
20. Torarinsson E, Havgaard JH, Gorodkin J (2007) Multiple structural alignment and clustering of RNA sequences. Bioinformatics 23(8):926–932
21. Will S, Joshi T, Hofacker IL, Stadler PF, Backofen R (2012) LocARNA-P: accurate boundary prediction and improved detection of structural RNAs. RNA 18(5):900–914
22. Will S, Otto C, Miladi M, Möhl M, Backofen R (2015) SPARSE: quadratic time simultaneous alignment and folding of RNAs without sequence-based heuristics. Bioinformatics 31(15):2489–2496
23. Will S, Reiche K, Hofacker IL, Stadler PF, Backofen R (2007) Inferring non-coding RNA families and classes by means of genome-scale structure-based clustering. PLoS Comput Biol 3(4):e65
24. Will S, Yu M, Berger B (2014) Structure-based whole-genome realignment reveals many novel noncoding RNAs. Genome Res 23(6):1018–1027
25. Wilm A, Mainz I, Steger G (2006) An enhanced RNA alignment benchmark for sequence alignment programs. Algorithms Mol Biol 1(1):1–11
26. Zuker M, Sankoff D (1984) RNA secondary structures and their prediction. Bull Math Biol 46(4):591–621

Chapter 11

Evolutionary Structure Conservation and Covariance Scores

Florian Eggenhofer and Christian Höner zu Siederdissen

Abstract

Effective homology search for non-coding RNAs is frequently not possible via sequence similarity alone. Current methods leverage evolutionary information like structure conservation or covariance scores to identify homologs in organisms that are phylogenetically more distant. In this chapter, we introduce the theoretical background of evolutionary structure conservation and covariance score, and we show hands-on how current methods in the field are applied on example datasets.

Key words Multiple sequence alignment, Secondary structure, RNA families, Covariance models, Unsupervised model construction

1 Introduction

Evolution is a balance between change and conservation in natural organisms. Biologically functional elements that enable the survival of an organism are often conserved. Changes in the environment might render these elements or combinations of them dysfunctional, resulting in a loss of fitness for the species.

Fitness can be restored or even increased by the evolutionary process of adaptation. This process works through random change and selection for those changes that improve fitness. The random changes are introduced by mutational events that affect the genome of the organism, which consists of the polymer deoxyribonucleic acid, in short DNA of nucleotides.

Genomes store the DNA information with which the cell interacts with its environment. These polymers are created by transcribing genes into an RNA molecule, and some of these RNAs are processed further by translation into proteins. Mutational events that change the DNA are also further propagated to these molecules. The effects introduced by mutations can be of variable strength. The translation step is somewhat resilient to point mutations due to the degeneracy of the genetic code, where multiple

Ronny Lorenz (ed.), *RNA Folding: Methods and Protocols*, Methods in Molecular Biology, vol. 2726, https://doi.org/10.1007/978-1-0716-3519-3_11, triplets encode a single amino acid. The biochemistry of different amino acids or the exact position of the mutation affect the impact of a mutation as well. For non-coding RNAs, the RNA structure due to base pairing is oftentimes more important than the actual sequence, leading to covariance, discussed further below.

In this chapter, we will discuss statistical properties of structural non-coding RNAs and how structural features increase the power of homology search algorithms compared to completely sequence-based methods. We are particularly interested in the additional statistical information gained from covariation in base pairs. The central topic of this discussion will be the concept of RNA families, sets of related RNA sequences with a common structure, and how to construct the so-called covariance models for them. A covariance model captures both the statistical properties of the set of aligned sequences and the statistical properties underlying the secondary structure.

We then expand our discussion towards questions of quality control, as well as visualization and comparison of whole families as opposed to individual sequences and structures. The combination of these topics leads us to fully automated methods for the construction of novel families from single sequences with desirable properties such as a deep phylogeny and a comprehensive set of family members. Furthermore, we present published tools, databases, and online resources and describe their use and fields of application.

Subheadings 1.1–1.3 give a very short biological introduction. In Subheadings 1.4–1.7, we present a high-level overview of the computational theory of the algorithms we discuss here. Finally, in Subheadings 1.8–3.6, we give usage examples of programs for the construction of RNA families, quality control, and visualization.

1.1 Evolutionary Events

The genome stores the genetic information of an organism as chains of nucleotides in the form of DNA. Several of these chains are contained within a cell, forming chromosomes. Mainly bacteria, but also certain other organisms also contain smaller DNA molecules, known as plasmids. Our computational interest is in non-coding genes, that is, substrings of the chains of nucleotides that do not have a protein as a final product.

Certain evolutionary events lead to mutations in the form of point mutations in which single nucleotides are changed, as well as insertions and deletions. We only consider small-scale changes that leave a gene still present, but lead to changes in sequence, structure, or function. The accumulation of mutations can create increasingly divergent organisms of one population, which can at some point lead to speciation.

The different types of ancestry or homology relations between genes are depicted in Fig. 1. Orthologs are genes arising from a common ancestor gene that serves the same biological function in organisms of different species. In this case, the original biological

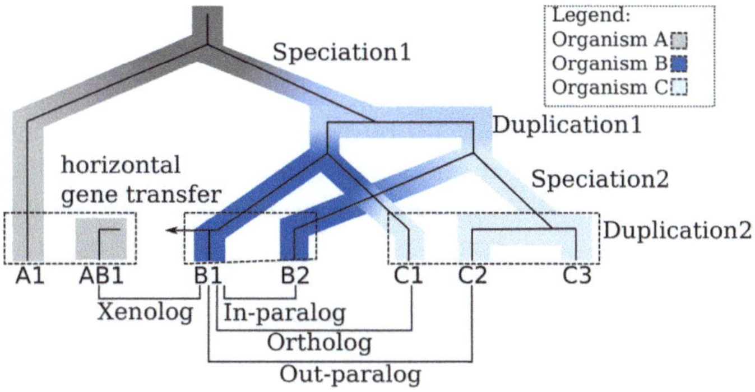

Fig. 1 Homology relations between genes derived from a common ancestor gene adopted from [7]. Paralogous genes originate from a gene duplication and have different biological functions within the same organism. Orthologous genes arise from speciation events and serve the same biological function in different species. Xenologous genes are acquired by horizontal gene transfer and serve the same biological function in different organisms. The different types of homology relations further complicate accurate annotation, especially if functionally different, but still statistically related, homologs must be correctly annotated

function is conserved, while the organisms diverge into different species. A set of orthologous (structured) non-coding RNA genes forms a *canonical RNA family*. However, it does not follow that every set of orthologous genes forms a single family in the computational sense, and neither does it follow that a single family only contains orthologous genes [14]. As we will see later, statistical properties will be an important decision factor in models for RNA families.

Paralogs are genes arising from a common ancestor gene that serve different biological functions in the same species. They are created by a gene duplication event, in which the whole gene is duplicated and will exist as two copies. Gene duplications are events of interest, since they enable one of the copies to change via mutations, while the other copy conserves the original biological function of the gene. Since different biological functions lead to different patterns of sequence and structure conservation, phylogenetically well-separated paralogs should be modeled separately.

Xenologs are genes that serve the same biological function and have been introduced in the organism by horizontal gene transfer. Horizontal gene transfer means that the gene has been acquired via foreign DNA that has been integrated into the genome of the organisms. This phenomenon is very common in prokaryotes.

1.2 Structures of RNA

The biological function of molecules depends on their spatial structure. The structure ensures that the correct chemical groups are in place to catalyze a chemical reaction or to facilitate the interaction with ligands or macromolecules. If the structure of a molecule is altered, then the chemical properties and possible interaction partners consequently change with it.

Ribonucleic acid (RNA) is a polymer that consists of nucleotide monomers, covalently linked together in a sequence of adenine (A), cytosine (C), guanine (G), and uracil (U). Each of the monomers is composed of different chemical groups and side chains. Changing the sequence of nucleotide monomers also changes the chemical groups of the resulting polymer. The chemical groups of the polymer are interacting, attracting and repulsing each other, and giving rise to the structures into which the polymer molecule can fold itself.

The structure of RNA molecules is often misunderstood as to be unique for each individual (RNA) gene. In fact RNA molecules can assume a whole ensemble of alternative structures, also depending on external factors like temperature and concentration of ions, ligands, and the time of observation. One example is riboswitches, whose mechanism is based on structural changes in response to changes in the environment [38].

The structural ensemble refers to all spatial conformations of foldings the RNA can assume in the three-dimensional space. This space is commonly referred to as tertiary structure. Structured RNAs have a convenient secondary structure representation in the form of a tree or graph, which exchanges the more exact three-dimensional representation for a computationally more convenient and efficient two-dimensional one. The change from three-dimensional structures to sequences of characters enabled dynamic programming solutions to structure prediction (e.g., with the `ViennaRNA` package [25]), or homology search, as discussed in this chapter.

The monomers of RNA form non-covalent interactions that are also known as base pairs. The monomers Adenosine and Uracil, as well as the monomers Guanine and Cytosine, form stable, non-covalent interactions with each other that are referred to as canonical base pairs. Canonical base pairs often dominate the observable binding patterns of folding due to their high interaction energy and therefore the overall structure a molecule will assume.

There are multiple software tools that predict secondary structure from sequence, e.g., `RNAfold` [25] or `UNAfold` [27]. Moreover, there are multiple experimental protocols that investigate the secondary structure of RNAs, based on chemical probing [24] or biophysical methods. The knowledge about the structure is essential to trace the evolution of RNA.

1.3 From Covariation to RNA Families

Covariation is the single most important structural feature of RNA families that can be used to improve the sensitivity and specificity of homology search tools.

Biological function is connected to a specific structure. For an evolutionary correct, structured view on RNA, it is essential to know that many alternative sequences can fold into the same structure. That means that the sequence can mutate while the structure and therefore the biological function are maintained.

Indeed, it can be observed that even over large taxonomic distance, RNAs in different species have similar structure. A famous example is the cloverleaf structure of tRNAs, required for correct interaction with the ribosome and the messenger RNA. The sequence of tRNAs in different species varies, but the structure remains conserved. This is possible due to the importance of canonical base pairs for the structure. A base pair disrupted by a mutation of one the two participating nucleotides can be restored by a compensating mutation of the pairing partner (*see* Fig. 2). Mutations that maintain a base pair are called covariant base pairs. The phenomenon is called covariance and is not to be mistaken with the measure from statistics and probability theory.

As an example, consider the framed base pair formed between Adenosine (A) and Uracil (U) in Fig. 2. A mutation of the Adenosine (A) to Guanine (G) would allow a so-called non-canonical wobble base pair (G–U) of lower stability, but a corresponding mutation of the right nucleotide to Cytosine (C) retains a canonical base pair with the ability to keep a stack of pairs stable.

These mutational patterns lead to the formation of recognizable consensus structures between homologous sequences. In Fig. 2, these are simple. Both sequences form the same hairpin structure. In an alignment of both sequences, their common structure can then be used to increase the performance of algorithms designed to find more homologous sequences that match this

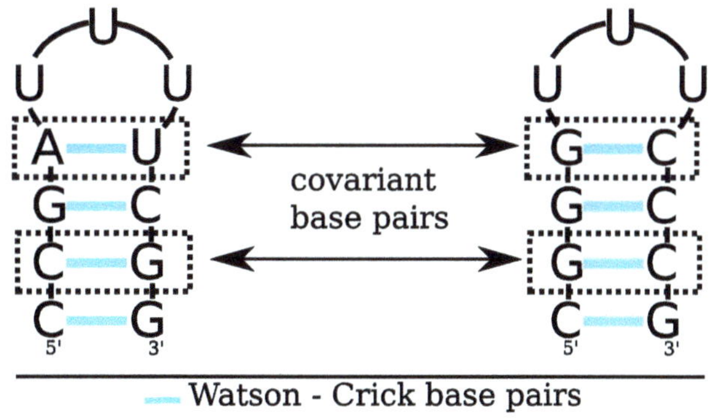

Fig. 2 Covariation is the conservation of structure relevant for the biological function of RNA by compensatory mutations. The figure, adopted from [7], depicts a toy example of the same hairpin loop and stem of two homologous sequences. The two pairs where a mutation of one nucleotide has been compensated for by a second mutation of the pairing partner have been highlighted

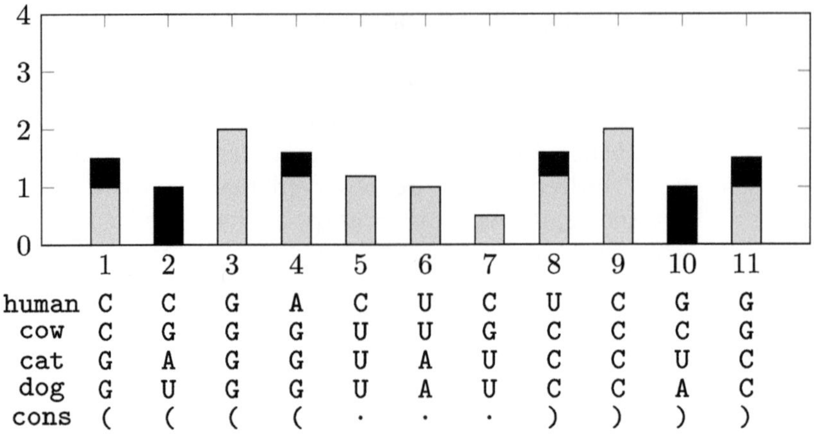

Fig. 3 Amount of information in bits for each column in a multiple alignment with structural information. The gray bars indicate the information from individual sequence-based columns. The gain of information from covariation is shown in black. For visualization purposes, half of the gain is shown in each paired column. This provides a convenient means to evaluate the per-column increase in information available from covariation. The amount of information gained is greatest for the paired columns 2 and 10. Here, the information gain from sequence alone is 0 *bits*, as each nucleotide occurs with frequency $1/4$. However, since all nucleotides pair, the total amount of information, taking structure into account, is 2 *bits* $= -\log_2(4/16)$ as each of the four canonical base pairs occurs once. Base pairs with perfectly conserved sequence information, such as columns 3 and 9, do not benefit from encoding the structural information. Columns 5, 6, and 7 are unpaired and do not profit from encoding structural information in the form of base pairs at all

pattern. The alignment in Fig. 3 depicts both sequences and their consensus structure, as well as multiple additional sequences that fit this pattern.

1.4 Alignments

Accurate alignments of homologous RNA sequences are important both for predicting the consensus secondary structure, and to construct RNA family models for homology search. However, the lower reliance on sequence conservation, and higher reliance on structure conservation in many RNA families, is a potential problem. Alignments without structure might well have low quality and not be useful for either structure predication or as a basis for homology search. Unfortunately, including structure in RNA alignment algorithms has high running time costs already for pairs of sequences.

Performing experiments to elucidate the structural is even more prohibitive due to high costs in particular for genome-scale analyses.

1.4.1 Sequence Alignment

Using sequence information alone and excluding structural information yields algorithms with lower runtime, but fails in regions where there is mainly structure similarity present. Examples for multiple sequence alignment tools are `clustalo` [39], `MAFFT` [22], `muscle` [6], and `probalign` [35]. For use with RNA, it is recommended to prioritize the use of structure-aware alignment algorithms. A switch to alignments based on sequence information alone should be reserved for very large data sets that are intractable otherwise, or for certain strategies of sequence and structure alignments discussed next.

1.4.2 Sequence and Structure Alignments

Alignment tools that are secondary structure-aware are the tools of choice for alignments of non-coding RNAs. They combine the alignment of sequences with prediction of a consensus secondary structure. These algorithms have high asymptotic and wall-clock running time complexity; therefore, different strategies were developed to obtain RNA sequence alignments with structure. Important strategies are fold-then-align, align-then-fold, and heuristics that improve the running time of the simultaneous alignment and structure prediction problem.

Examples of the fold-then-align strategy are `STRAL` [3], `MARNA` [45], while `RNAalifold` [1] and `PETfold` [37] rely on pre-computed sequence alignments.

Algorithms that align and fold simultaneously are the method of choice for non-coding RNA alignments since sequence based alignments fail for structural RNA with lower sequence similarity but conserved structure [13]. The Sankoff algorithm [36] gives an exact implementation of simultaneous alignment and folding for pairs of sequences. The exact algorithm has a prohibitive running time of $O(n^3m^3)$ with space requirements of $O(n^2m^2)$, where n and m are the lengths of the sequences to be aligned.

Due to the importance of this problem for high-quality RNA sequence alignments, and number of heuristics have been developed, e.g., `dynalign` [18], `foldalign` [40, 42], and `LocARNA` [45, 46].

Finally, a multiple alignment is performed by wrapper tools for the pairwise case as even heuristics for this problem are prohibitive with more than two inputs. The wrapper for `LocARNA` is called `mLocarna` and for `foldalign` called `FoldalignM`.

1.5 Covariance Models

The `Infernal` [30] package implements a set of algorithms that support the use of position-specific stochastic context-free grammars (SCFGs) in RNA homology search. The use of a context-free, as opposed to a linear, grammar allows the inclusion of long-range interactions in the form of base pairs. Context-free grammars also allow for the explicit modeling of base pair nesting on the one hand, and juxtaposition of base pairs on the other. Said otherwise, context-free grammars allow modeling pseudoknot-free RNA secondary structures. Here, we provide only a short introduction into the model underlying `Infernal`. Detailed theory on SCFGs can be found in Durbin et al. [4, Chapter 10], while the `Infernal` manual [31] gives an in-depth guide to using and understanding the software.

The major difference between `Infernal` [30] covariance models and `HMMER` [5] Hidden-Markov models is the inclusion of long-range interactions in the form of base pairs. The hypothesis is that structured non-coding RNAs show an additional preserved signal in the form of covariation (*see* Figs. 2 and 3) that can be exploited to gain both higher sensitivity and specificity. Covariance models for non-toy families are fairly complex. Figure 12 shows part of the guide tree of the tRNA family model. Each node of the guide is further sub-divided into individual states, which carry character emission scores, and transition edges between them (cf. Fig. 13).

In order to be able to *score* hits according to their relevance in a search, some sort of scoring model is needed. `Infernal` models are position-specific log-odds models with *bit* scores. For each consensus i, we have the relative frequency $F_i(n) \in [0, 1]$ for each character $n \in \{A, C, G, U\}$. We also need a background model $B(n) \in [0, 1]$ that gives the average frequency of a character, which is typically $B(n) = 0.25$. The position-specific score can now be calculated as $\log_2(F_i(n)/B(n))$. In case of base pairs, both columns are considered together, and the contribution is $\log_2(F_{i,j}(n, m) \ / \ (B(n)B(m)))$.

As examples, consider the multiple alignment of Fig. 3. Let us assume that all columns of the alignment are part of the consensus structure, that is, the majority of sequences (possibly weighted using a phylogenetic model) has an aligned character in each column. While both column 1 and column 2 form part of a base pair, the actual composition of characters is quite different. In column 1, there are exactly two choices with 50% probability each: `C` or `G`. As such, a sequence match in columns 1 to `C` or `G` provides $\log_2(0.5/0.25) = 1.0$ bit of information, while `A` or `U` would yield a score of $-\infty$ at this position and not be allowed to match. Column 2 on the other hand allows matching all four nucleotides with probability 25%, but a match will not provide any gain of information as $\log_2(0.25/0.25) = 0$ bit.

`Infernal` models make use of prior models that modify the column-based scoring in such a way that every nucleotide may always match, though with negative score contribution in the case of nucleotides that are encountered with frequency less than 25% in the model. In addition, models include options to ignore columns for a cost, or insert additional nucleotides "between" matching columns.

Returning to both column 2 and column 10, we can notice that while both individual columns do not provide any sequence information, taken together they do contribute to the model. This is the main difference between purely sequence-based HMMs as implemented with, for example, `HMMER` models and covariance models we are using here. Since covariance models can model the base pair between columns 2 and 10, we can now ask what the additional gain from the structural contribution is. Of the $4 \times 4 = 16$ possibilities, we observe the four canonical base pairings with a probability of 25% each, and the other 12 options not at all. This yields information beyond what purely sequence-based models provide: $\log_2(0.25/0.25^2) = 2$ bit. For visualization purposes only, we divide the additional structural information equally between both columns and as such observe 1 bit of information each for columns 1 and 10. Figure 3 gives the additional structure-based information gain in the form of black bars on top of the sequence-based information.

`Infernal` makes use of multiple steps to find homologs in a genome, given a covariance model for an RNA family. It starts with HMM-based prefilters. While these have the disadvantage of not taking into account base pairing information, they are extremely fast. If the genome has size L, and the model N consensus columns, an HMM will take $O(LN)$ steps. The prefilters are setup to provide a balanced set of candidates as high likelihood of containing actual homologs, but not so many to become useless as a speedup heuristics.

Following prefilter steps are two CM-based steps. Here, the main difference is the scoring scheme. The first step returns a single optimal parse or predicted structure π and its score in bits for each candidates sequence s passing the HMM filter: $S_{\text{CYK}} = \max_\pi \log_2(P(s, \pi|\text{CM})/B^L)$. As a last step, the *Inside* score is calculated for all candidate sequences that score high enough. *Inside* "integrates out" the actual parse π and considers the log-sum of all parses: $S_{\text{Inside}} = \log_2(P(s|\text{CM})/B^L) = \log_2(\sum_\pi(P(s, \pi|\text{CM})/B^L))$. These calculations come at a cost. Asymptotically, Aligning a sequence to a CM costs up to $O(LN^3)$, and has been estimated to be $\approx O(LN^{2.4})$ [28, 29].

The information gain from structure on top of sequence is typically in the range of 10 to 20 bits, with sequence information between 50 and 200 bits. Figure 2 in [28] gives a more detailed comparison for a large number of RNA families. As each additional bit of information halves the chances of finding a match in a purely random sequence, this gain from structure can be substantial.

1.5.1 Clans

Each RNA family should be sufficiently unique that genomic or database hits from different families with a reasonably high score do not overlap. While this is good system for quality control purposes, it also presents a challenge due to a number of reasons.

The stochastic covariance model itself is composed of a set of sequences that have to be similar enough to yield a consensus structure and either covariation for paired nucleotides, or a statistical signal for unpaired positions (cf. Fig. 3 for a toy model). As families diverge more and more, it becomes infeasible to construct a single RNA family that is sensitive enough to detect even remote homologs, but still specific enough to only detect true homologs. One option is to split such a family into a set of related families. Each such family retains both high specificity and sensitivity. However, they will with high likelihood detect sibling sequences belonging to the split set of families. This runs counter to the idea that high scoring hits should not overlap. An alternative is to annotate two previously existing families as belonging to the same clan.

The Rfam database has dealt with this problem with the introduction of *Rfam clans* [14]. Clans are sets of families that could be unified into a single family but have not with good reason. Clan members either constitute a set of families with highly diverse secondary structures, or different function but similar enough sequences. As of January 2019 with `Rfam` version 14.1, there are a total of 113 clans in the database. The largest clan, with 11 families in total, is the *SNORD29* clan.

To provide a sense of the similarity expected between members of the same clan, Table 1 gives an overview of the RNAseP clan. This clan has six members, and is of particular interest, since the member families are structurally complex. The member families are also not well separated. The first indication is that the trusted cutoff values (TC) of true homologs of each family are very close to the noise cutoff (NC) values, which indicates the highest scoring false

Table 1
Cross-comparison of RNA families within the *RNAseP* clan. On the main diagonal, we give the *lowest* score of a true homolog for each family. The off-diagonal scores are the *highest* given a sequence in one family tested against another family in the same clan. In addition, under thresholds, we give the gathering (GA) threshold, as well as trusted (TC), and noise (NC) cutoff for each model. Except for the plasmodial RNAseP, these values are virtually identical. This means that the least fitting true homolog of each model as given by the trusted cutoff (TC) can not be distinguished from the highest scoring false hit (NC) in a sequence data base. The *null3* null model has been disabled for the creation of this table. Numbers have been rounded. A value of—denotes that no hits above the reporting thresholds are found. Hence, there are no false positives that have a reasonably high score

Name	Accession	Fasta query accession numbers						Thresholds		
		9	10	11	30	373	1577	GA	TC	NC
Nuclear	9	29	–	–	41	–	58	28	28	28
Bacterial a	10	–	165	77	–	89	–	100	101	100
Bacterial b	11	–	90	206	–	22	–	97	97	97
MRP	30	23	–	–	56	36	–	50	50	50
Archaeal	373	–	90	30	23	109	–	84	84	84
Plasmodium	1577	–	–	–	–	–	86	100	156	92

hit in a large sequence database. A cross-comparison where all sequences of each family are aligned against all RNAseP models shows that individual families, say the bacterial compared to the archaeal family will likely yield overlapping hits.

In general, establishing a clan relationship between two RNA families requires extensive manual work, as additional homology information has to be obtained. Such sources of information can be alternative, more specialized databases.

1.6 Comparison of RNA Families with CMCompare

As already stated in our short introduction to RNA family clans (Subheading 1.5.1), no two RNA families should annotate the same nucleotides of a genome with a significant score, as such double annotation would produce uncertainties in an automated genome annotation pipeline. Furthermore, querying an RNA family database with a single sequence could yield two or more families as likely homologs of the sequence in question.

In principle, it is possible to prevent multiple annotation by querying all sequences from all family models against all models as a form of quality control. Performing this query will yield the same cutoffs as used in `Rfam` covariance model (CM) files. In particular, one should expect that the lowest score of an actual family sequence of the full alignment should be equal to the *gathering threshold*. The highest score of any non-family sequence should be equal to the *noise cutoff*. A family with high quality should have a gathering threshold much higher than the noise cutoff. With a large enough data base, this should provide reliable scores to determine membership or false positives.

However, apart from problems where these cutoffs are close together, and do not provide a strong signal, one also requires a large set of family sequences, instead of just the `Infernal` covariance models, and a large amount of processing power, as `Infernal` itself is a costly algorithm to run (cf. Subheading 1.5).

As such, it is desirable to have a quality measure that is independent of individual sequences and requires just two covariance models to determine whether these two models are likely to be well separated from each other. If a model is well separated from all others models, we say that this model has high discriminatory power, as there are *no* possible sequences that score high in both this model and any other.

Our `CMCompare` software [20] and web service [8] use a dynamic programming approach to calculate an optimal local alignment between pairs of RNA family models. This allows us to perform the same kind of cross-comparison calculation as shown in Table 1, but *without* having to rely on family member sequences.

The idea underlying `CMCompare` is best explained with a thought experiment. Given a model M, and a sequence s, the CYK algorithm yields the optimal alignment π, and a score based on $P(s, \pi|\mathrm{CM})$. This is indeed just the CYK step of `Infernal`. Given a fixed s, we can compare two models M_1 and M_2 by calculating $P(s, \pi_1|\mathrm{CM}_1)$ and $P(s, \pi_2|\mathrm{CM}_2)$. If we now take multiple sequences s_k, say from all seed alignment sequences for each model CM_i, we are able to recreate Table 1. In a further (and this is the thought experiment) step, we extend the set of sequences to the set of all possible sequences, i.e., $s \in \mathcal{A}^*$.

Now we can calculate the so-called *link sequence* and *link score*:

$$\mathrm{Link}(\mathrm{CM}_1, \mathrm{CM}_2) = \mathrm{argmax}_s(\min(\mathrm{CYK}(\mathrm{CM}_1, s), \mathrm{CYK}(\mathrm{CM}_2, s)|s \in \mathcal{A}^*))$$

as described in [20]. We score each sequence s with both models and take the minimum of both scores for each sequence. We then return the sequence s with the highest score in total and call it the link sequence s_{Link}, and the resulting score $\min(\mathrm{CYK}(\mathrm{CM}_1, s_{\mathrm{Link}}), \mathrm{CYK}(\mathrm{CM}_2, s_{\mathrm{Link}}))$ the link score of the two models.

If this score is high, then these two RNA family models are not well separated if they encounter the link sequence or a similar sequence in a genome, a situation that quality control procedures should prevent, if possible.

Instead of actually trying to generate either a large set of sequences, or to use large sequence databases, the `CMCompare` algorithm instead performs a dynamic programming alignment of the two models. We have that `CMCompare`$(\mathrm{CM}_1, \mathrm{CM}_2) \leq \mathrm{Link}(\mathrm{CM}_1, \mathrm{CM}_2)$ for the CYK model, but with much better asymptotic and wall-clock running time costs. `CMCompare` will typically provide a lower score, than running `Infernal` directly, as we use some heuristics to further speed up the comparison. Having a lower bound means that pairs of models that score high are very likely to be problematic in terms of the uniqueness condition for annotation, as `Infernal` will more likely find sequences in genomes that are not uniquely annotated.

In addition, the link score as calculated by `CMCompare` has another very convenient property. Since it is calculated using the same log-odds scoring model that s used by `Infernal`, link scores can be immediately compared to `Infernal` hit scores from genomes, gathering thresholds, as well as trusted and noise cutoffs, or for that matter, the scores provided in Table 1.

While there are no hard rules at which point a given `Link Score` marks two families are having too much overlap, there are two very good indicators. The first indicator is the comparison with the gathering threshold, and the noise cutoff. If the `Link Score` is much lower than either for both models, there is no big danger of overlap. The second indicator is much more task-specific. When searching for remote homologs, low bit scores and high E-values are to be expected. Putative homologs with bit scores close to or lower than the `Link Score` should be considered very carefully as the probability of false positive hits with this score is increasingly high.

The running time of CMCompare is $O(NM)$, where N and M are the number of consensus columns in the two models to be compared. Cross-comparing family member sequences using Infernal on the other hand has approximate asymptotic costs of $O(KL(N^{2.4} + M^{2.4}))$, where K is the total number of sequences used for the comparison, and L the average length of each. The exponents have been determined empirically [28, 29].

1.7 RNAlien: Unsupervised RNA Family Model Construction

Establishing a novel RNA family requires a number of non-trivial steps. If prior information is thin, say when a new RNA sequence has been isolated with putative novel function and no established homology, all steps we describe in this chapter and more have to be performed. RNAlien [9] and its web service are designed to automate this step as much as possible. Ideally, one starts RNAlien with a single sequence and obtains a set of family member sequences, their homology, a structural multiple alignment, and a covariance model without having to interfere manually in the process at all.

The major steps of this process are shown in Fig. 4. The only information known for the initial sequence, from which a family is to be built, is the organism it came from. Knowledge about the origin is helpful to reduce the initial search space, but not necessary in cases where even that information is unavailable. As no initial structure is available, a de novo structural model needs to be created. RNAlien starts with a sequence-based search using blastn [2] in the taxonomic neighborhood. The initial consensus structure is predicted using mLocARNA (cf. Subheading 1.4 and [46]). Once the initial structure is established, a first covariance model can be created.

The model can then be expanded. Locally, a user could make use of Infernal, which provides very high sensitivity and has a higher chance to yield remote homologs. The web service is more resource-constrained and expands the model with additional blastn searches. At every iteration, RNAlien tries to expand the phylogenetic depth of the model with the inclusion of more remote taxonomic subtrees. It also keeps track of several quality measures as a trade-off between sensitivity, specificity, and running time of RNAlien has to be taken into account.

Finally, once a fixed point has been reached where no further improvement of the model is possible, a final covariance model is created and returned together with all supporting information. This includes the structured alignment and taxonomy of the hits.

RNAlien has been extensively tested by reconstructing known RNA families from the Rfam 12 [32] database. Tests have been performed with a smaller set of 56 families, of which all have a known, experimentally derived structure. Both specificity and recall of the automatically reconstructed families are high. In 98% of the cases, at least half of the sequences being collected by RNAlien are identified as homologs by the original model. In 62%, even all sequences are identified as homologs [9, Fig. 2A].

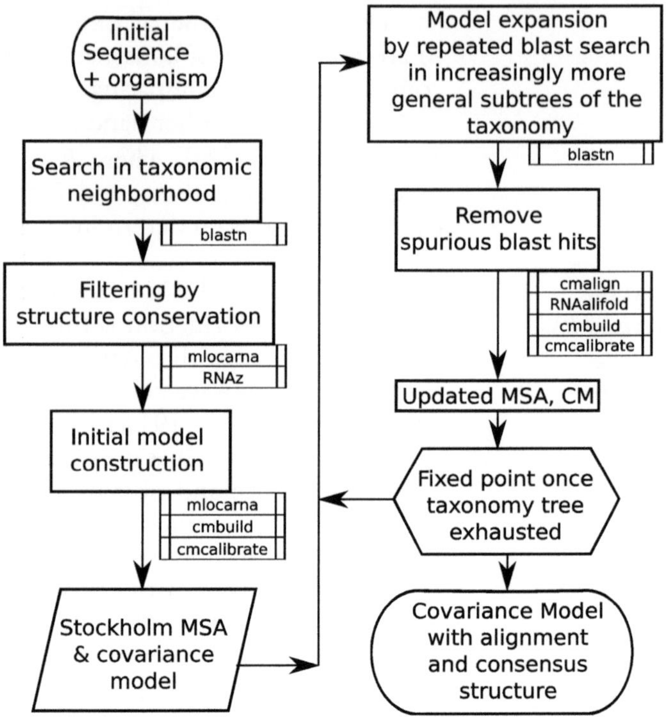

Fig. 4 The major steps done by the `RNAlien` pipeline. `RNAlien` performs an initial `BLAST`-based search, followed by the automated construction of a structural alignment. This initial alignment is then refined until a given taxonomic search space has been fully searched. Finally, a covariance model with supporting information is returned. This Figure has been taken from [9] under the terms of the Creative Commons Attribution License

The sensitivity of `RNAlien` families is slightly lower than hand-curated ones [9, Fig. 2B]. Nevertheless, in 78% of the tested families, at least half of the known homologous sequences are correctly recognized by `RNAlien` as belonging to the reconstructed family.

Many users will want to make use of the `RNAlien` web service located at http://nibiru.tbi.univie.ac.at/rnalien, which requires only the initial RNA sequence, nothing more. The deep taxonomic sequence data required for `RNAlien` make it less amenable for installation by the casual user.

While we demonstrate `RNAlien` as general purpose tool for RNA family model construction, two alternatives have become available. `GLASSgo` [26] is a specialized solution for bacterial small RNAs. In addition, a new family building pipeline by the `Rfam` team is available on request: https://twitter.com/RfamDB/status/1139103527668764672.

1.8 RNA Family Databases

RNA families are collected in the `Rfam` [14, 21, 32] database. `Rfam` provides for each family the seed and full structural alignments, and the covariance model. Seed alignments are hand-curated and should be trustworthy. They necessarily contain fewer candidates than full alignments, which can be of very large size. For the largest alignment in `Rfam` 14, the `tRNA` family, there are 954 sequences in the seed alignment but 1 429 129 sequences in the full alignment. Full alignments are constructed using `Infernal` and a covariance model created from the seed alignment.

Moreover, there are visualizations that show the taxonomy of species that have a homologous sequence in that family and a lot of additional context information. RNA families with a common ancestor RNA are grouped into clans. `RNAcentral` [41] takes a more sequence-centric approach and enables querying for sequences with multiple different identifiers used by e.g., `Refseq` [33] or `ENA` [19] and connects them to potentially known RNA families and context information. The database is also accessible via a programmatic API that allows for automatic queries.

2 Materials

We will use `bioconda` [16] to install the tools for the hands-on parts throughout the manuscript. Please first install `miniconda3` by following the instructions found here:// https://conda.io/projects/conda/en/latest/user-guide/install/index.html. The tutorials assume a Linux-based environment.

The installation of the required tools in their corresponding conda environments and the data necessary to reproduce the examples in the method section are listed below. Requirements for *Family construction with a set of homologous sequences* (*see* Subheading 3.1):

```
conda create -c conda-forge -c bioconda -n locarna locarna
wget https://zenodo.org/record/4506477/files/tRNA.fa
conda create -c conda-forge -c bioconda -n cmv cmv
conda create -c conda-forge -c bioconda -n r2r r2r
conda create -c conda-forge -c bioconda -n infernal infernal
wget ftp://ftp.ncbi.nlm.nih.gov/genomes/all/GCF/000/005/845/GCF_000005845.2
   ↪ _ASM584v2/GCF_000005845.2_ASM584v2_genomic.fna.gz
```

Requirements for *Family construction starting from a single sequence* (*see* Subheading 3.2):

```
conda create -c conda-forge -c bioconda -n rnalien rnalien
conda create -c conda-forge -c bioconda -n infernal infernal
wget https://zenodo.org/record/4506477/files/ctminput.fa
```

The major tool in this section can be used in an online or offline mode, please see refer to the note Subheading 4 to learn more.

Requirements for *Using* `CMCompare` *for RNA family quality control* (*see* Subheading 3.4):

Opposed to most other software we use in the tutorial sections, `CMCompare` is not available in `conda`. Its current home is: http://www.bioinf.uni-leipzig.de/Software/cmcompare/.

```
mkdir ~/tmp && cd ~/tmp
wget https://zenodo.org/record/4532364/files/cmcompare-legacy
chmod 755 cmcompare-legacy
wget https://zenodo.org/record/4532364/files/rnasep.tgz
cd -
conda create -c conda-forge -c bioconda -n infernal infernal
```

Requirements for *Visualization of RNA families* (*see* Subheading 3.6):

```
conda create -c conda-forge -c bioconda -n cmv cmv
conda create -c conda-forge -c bioconda -n infernal infernal
```

3 Methods

The number of known RNA families has been constantly increasing over the past years. There are different scenarios that lead to the discovery of a new family. The first scenario is that there is already an established set of sequences from different species that are known to have a common biological function and are thus homologous to each other. In the second scenario, a large set of sequences (e.g., genomes) is screened for novel homologs. Sequences with high structure and sequence similarity are candidates for inclusion in known families. In some cases, experimental testing for common biological function might even be performed. Finally, novel RNA sequences that have no known homologs can be discovered. Sources may be experiments such as RNA-seq data or northern blots. RNA family databases can be screened for inclusion, maybe as a remote homolog, in an existing family, but the possibility that the sequence is indeed member of a hitherto unknown family should be considered, and tested for, as well. In the following, all three cases will be discussed with hands-on examples.

3.1 Family Construction with a Set of Homologous Sequences

The structural alignment of a small set of tRNA sequences from the `Rfam` [14, 21] tRNA family is demonstrated in the following steps.

`mLocARNA` is installed via `bioconda` in its own environment. The environment is activated, and the alignment is performed.

```
conda activate locarna
mlocarna --stockholm tRNA.fa
```

We will visualize the resulting Stockholm alignment by using `StockholmV` from the `CMV` [10, 34] package. The Stockholm format shows the alignment together with the consensus secondary structure and enables encoding extra annotation for columns and lines (*see* Fig. 5). The `-f` parameter enables to select from `pdf`, `png`, `ps`, and `svg` as output formats, while the `-x` parameter allows to extract the consensus sequence and secondary structure of the alignment for secondary structure visualization via `R2R` [44] or `forna` [23].

```
conda activate cmv
StockholmV -s tRNA.out/results/result.stk -f pdf -x r2r
mv 1.aln.pdf tRNA.pdf
mv 1.r2r tRNA.r2r
```

The consensus sequence and structure is visualized with `R2R` (*see* Fig. 5). We can observe that the tRNA sequences fold into the cloverleaf shape (Fig. 6).

```
conda activate r2r
r2r tRNA.r2r tRNAstructure.pdf
```

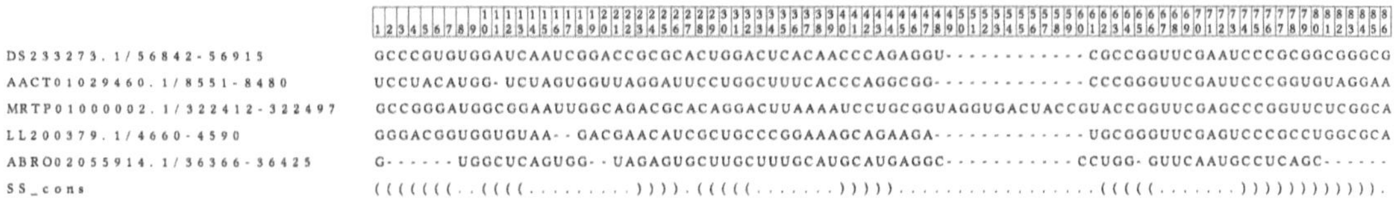

Fig. 5 The Stockholm alignment visualization by `CMV` shows a `mLocARNA` alignment of tRNA sequences with a consensus sequence encoded in dot-bracket string below and the column index above. The structural alignment yielded the cloverleaf structure typical for tRNAs

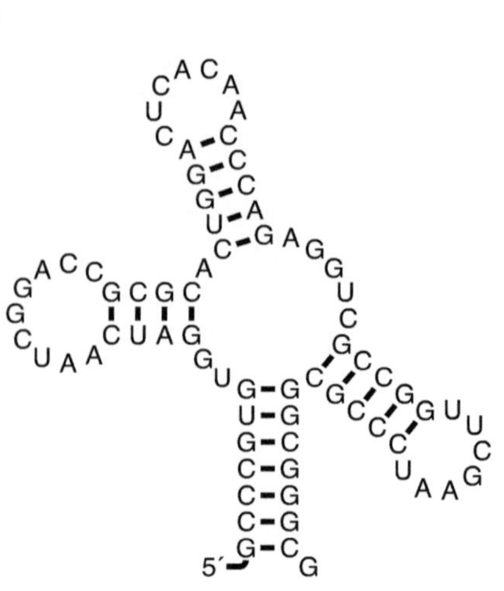

Fig. 6 The figure shows the consensus structure and sequence for a tRNA sequence alignment, extracted by `CMV` and visualized with `R2R`

In the next step, the structural alignment is used to build an `Infernal` covariance model.

```
conda activate infernal
cmbuild --refine tRNA.cm tRNA.stk
```

The covariance model needs to be calibrated to yield E-values during homology search. This step requires a lot of computation time, therefore, it is recommendable to request a number of threads equal to the number of CPU cores available (which is 5 here).

```
cmcalibrate --threads 5 tRNA.cm
```

The model can now be directly used for homology search with `Infernal cmsearch`. We will use our model to search the *Escherichia coli K-12 MG-1655* genome for tRNAs matching to our tutorial model.

```
gunzip GCF_000005845.2_ASM584v2_genomic.fna.gz
cmsearch --threads 5 tRNA.cm GCF_000005845.2_ASM584v2_genomic.fna
```

The resulting significant hits can now be inspected in detail (*see* Subheadings 3.4 and 3.3).

`RNAlien` contains the `cmsearchtobed` tool allowing to convert the search results into a bed file for genome browser visualization.

3.2 Family Construction Starting from a Single Sequence

RNA-seq experiments frequently yield novel RNAs for which no annotated homologs are known. `RNAlien` [9] uses both sequence and structure information to search for homologous sequences and to construct a structural RNA alignment and family model and can be installed with all dependencies and required tools via `bioconda`.

```
conda activate rnalien
```

The input sequence(s) to be used for RNA family model construction are passed to `RNAlien` in multi fasta format. We will use a sequence from the `Rfam` family for *Cyanobacteria transfer-messenger RNA* (with Rfam family https://rfam.xfam.org/family/RF01851) for our hands-on example.

In addition to the input Fasta file, which is supplied via the parameter `-i`, the number of cores to be used is set with `-c`, the name for the output directory with `-d`, and the path where to write the output with `-o`. For using the offline `BLAST` database, it is additionally necessary to provide the directory path with the `-b` parameter (*see* the Notes Subheading 4) on how to download and use the database. Optionally, the species from which the input RNA originates can be specified via the `-t` parameter as a NCBI taxonomy id [12]. The id can be retrieved by querying the species name on the NCBI taxonomy website (https://www.ncbi.nlm.nih.gov/taxonomy).

```
RNAlien -i ctm.fa -d ctm -c 2 -j -o /output_path/
```

`RNAlien` performs an iterative search as described in Subheading 1.7, which searches for potential homologs of the input RNA. The search process of `RNAlien` may take up to 24h and yields the following results, which we provide for users not willing to wait.

All non-redundant sequences: https://zenodo.org/record/4506477/files/ctm.fa, their structural alignment with consensus secondary structure: https://zenodo.org/record/4506477/files/ctm.stockholm, a calibrated RNA family model: https://zenodo.org/record/4506477/files/ctm.cm, and an evaluation of the constructed model.

The result files are deposited directly in the top-level directory of the model construction. We will copy and rename them for use in Subheading 3.3.

```
cp ctm/result.fa ctm.fa
cp ctm/result.stockholm ctm.stockholm
cp ctm/result.cm ctm.cm
```

The whole process of collecting and analyzing candidates as potential family members is tracked in the Log file. The Log file lists all used initial parameters and, stepwise for each iteration, the number of collected candidates as well as the aggregated evaluation output. Inspecting the Log file for our example output we, can observe that three distinct potential homologs were collected.

```
Iteration: 10
Message: Modelconstruction complete: Out of queries or taxonomic tree
    ↪ exhausted
Global search iteration: 10
Upper taxonomy id limit: not set, Collected members: 3
Reevaluation of potential members iteration: 11
```

The evaluation contains the results of RNAz, which computes the properties of a structural RNA alignment and tests if it has the properties of a functional ncRNA.

```
RNAz statistics for result alignment:
  Mean pairwise identity: 87.66
  Shannon entropy: 0.19761
  GC content: 0.46628
  Mean single sequence minimum free energy: -92.23
  Consensus minimum free energy: -82.4
  Energy contribution: -81.02
  Covariance contribution: -1.38
  Combinations pair: 1.09
  Mean z-score: -2.45
  Structure conservation index: 0.89
  Background model: dinucleotide
  Decision model: structural RNA alignment quality
  SVM decision value: 1.76
  SVM class propability: 0.990488
  Prediction: RNA
```

In this case, RNAz [15] predicts the alignment to originate from a non-coding RNA (see last line), which is an important indicator that the construction process obtained potentially homologous sequences. The evaluation also uses RNAcode [43] to determine if the alignment has coding potential and if the contained sequences are translated into protein, as in the case of dual function RNAs. Sequences found during the construction process are being queried for existing records at RNAcentral, which allows to obtain contextual information, e.g., experiments or reports about biological function.

The resulting sequences in Fasta format (https://zenodo.org/record/4506477/files/ctm.fa) are non-redundant within a certain sequence identity level. This means that if a potentially matching sequence is very similar to sequences that already have been collected and accepted into the alignment, then the potentially matching sequence will not be included. This measure is taken to prevent the inclusion and overspecificity resulting from species with high sequence multiplicity, e.g., hundreds of identical sequences from *Escherichia coli* strains. Therefore, to obtain all sequences matching to the newly created model, a homology search, as described in the next section, is required.

3.3 Using `Infernal` to Annotate Homologs of Known RNA Families

We will demonstrate how to screen the whole NCBI nucleotide (nt) database for potential homologs with the RNA family model created by `RNAlien` in the previous section. We will show how to convert the `BLAST` database that was downloaded in the previous step into a Fasta file, which is suitable for use with Infernal. Please note that the unpacked `BLAST nt` database has a size of about 260Gb. For an in-depth tutorial on `Infernal`, please consult [28] or the `Infernal` user guide [31].

```
blastdbcmd -entry all -db nt_v5 -out nt_v5.fa
```

Running `cmsearch` against the whole `nt` database takes several Hours; if the number of threads is omitted, it will use all available cpu cores.

```
conda activate infernal
cmsearch ctm.cm nt_v5.fa > ctm.cmsearch
```

The result of the search (https://zenodo.org/record/4506477/files/ctm.cmsearch) yields 116 significant hits indicated by exclamation mark.

3.4 Using `CMCompare` for RNA Family Quality Control

In this section, we use `CMCompare` to compare the RNA families of the RNAseP clan with each other. As already seen in Table 1, comparing the clan members based on a cross-comparison of seed sequences to each individual covariance model yields some peculiar results. The most important points are: (i) The score of the lowest-scoring true homolog of each family can be lower than the highest scoring sequence from a sibling family. This behavior can be seen with the *Nuclear (Rfam Id: 9)* family, where the lowest in-family score is 29, while sequences from the *MRP (30)* and *Plasmodium (1577)* family with lowest score have a score of 41, or 58 bits. (ii) The table is not symmetric. While there are *MRP (30)* family sequences that score 41 bits using the *Nuclear (9)* family model, the highest score of a *Nuclear (9)* sequence in the *MRP (30)* model is 23 bits. This non-symmetry is, of course, to be expected and depends on the actual construction and quality of each model.

For this short `CMCompare` tutorial, we will need `Infernal` [30], `CMCompare` [20], and the family models for the RNAseP clan from `Rfam` [21]. The tutorial for the `CMCompare` webserver [8] will yield similar results for its example, minus the comparison to `Infernal` scores. Using `CMCompare` locally has the benefit of not requiring any internet access, which can be important if the models to be compared should be kept private. In addition, the CMCompare output is very pipeline-friendly and can be post-processed with additional tools. Here, we use Infernal to further check how far the lower bound given by the Link Score is from Infernal scores.

To run the examples, it is best to use a temporary directory where CMCompare and the example files are being worked with.

```
cd ~/tmp
./cmcompare-legacy --help
tar -xf rnasep.tgz
```

For the next step, we compare families of interest with each other. The MRP and Plasmodium family share no significant hits within the sets of their seed sequences, as shown in Table 1. We test those two families to check if there is a high-scoring Link Sequence.

```
./cmcompare -q RF01577.cm RF00030.cm
# RF01577.cm    RF00030.cm     30.472     31.648
```

Since there is at least one Link Sequence with a Link Score of ≈ 30 bit, it is worthwhile to consider the models in more detail. This time, we do not use the -q switch to produce full output.

```
./cmcompare RF01577.cm RF00030.cm
# RF01577.cm    RF00030.cm     30.472     31.648
    ↪ CCUGCGCUUAAGGCCGAAUUUCAUAAAUGAUAUUGAUUU...
```

The sequence being returned is too long, with 151 characters, to display here in full, but it is available for download on the CMCompare page above.

We now have have a sequence that should yield at least 30 bit, but possibly even higher when scored using either of the two covariance models RF01577 or RF00030. This little script can be easily extended to produce Fasta files from a CMCompare run. The script will leave us with cmc.fa file with the Link Sequence. Alternatively, the cmc.fa can be created by hand by just copy-pasting the Link Sequence into a new Fasta file.

```
./cmcompare RF01577.cm RF00030.cm | \
  awk {printf ">cmcompare-%s-%s\n%s", \$1, \$2, \$5} | \
  tee cmc.fa
# >cmcompare-RF01577.cm-RF00030.cm
# CCUGCGCUUAAGGCCGAAUUUCAUAAAUGAUAUUGAUUU...
```

We now can use cmsearch from the Infernal package to obtain the CYK and Inside bit scores. We disable the null3 model to return the score without bias correction. We also assume (with `-Z 1`) a genome size of 1 Mb. While we should be careful with score and E-value, this will give us a quick hint if we should investigate the families more carefully, or if they are separated well enough for our use case. Note that the output shown in the listings has been cut down to just the important lines.

```
cmsearch --nonull3 -Z 1 --cyk RF00030.cm cmc.fa
# rank     E-value  score
#  (1) !   3.1e-10   39.5

cmsearch --nonull3 -Z 1 --cyk RF01577.cm cmc.fa
# rank     E-value  score
#  (1) !   3.1e-05   33.5
```

Finally, we calculate the Inside score as well.

```
cmsearch --nonull3 -Z 1 RF00030.cm cmc.fa
# rank     E-value  score
#  (1) !   1.1e-10   41.8

cmsearch --nonull3 -Z 1 RF01577.cm cmc.fa
# rank     E-value  score
#  (1) !     1e-05   40.4
```

The final `Link Score` between these two models is 40.4 bit as the minimum of 41.8 bit and 40.4 bit. Given that we know that the two `Rfam` models are part of the same clan, this score is acceptable. If, however, we had the same or higher scores between two supposedly very different RNA families, we would need to be careful. Scores between 20 and 40 bit could easily be the result of a remote homolog. Furthermore, the E-values for the scores would still be ≤ 0.1 for genomes of 10^9 nucleotides or more, since the above E-values are for a database of 10^6 nucleotides.

3.5 Using the `CMCompare web service` for RNA Family Quality Control

The CMcompare webserver http://rna.tbi.univie.ac.at/cgi-bin/cmcws/cmcws.cgi/cmcws.cgi allows to compare an input model against all models in Rfam or a set of models with each other. We will compare the tRNA model constructed with `RNAlien` in Subheading 3 with the entry type tRNA subset of `Rfam`. First, we select the mode to compare against models of `Rfam` as shown in Fig. 7.

Then the model of interest is being uploaded by selecting it with the file browser (*see* Fig. 8)

Upon successfully checking the uploaded model, the user can select against different subsets of `Rfam` labeled with entry types. Since our example model was created from tRNA sequences, we want to test against the tRNA subset (*see* Fig. 9).

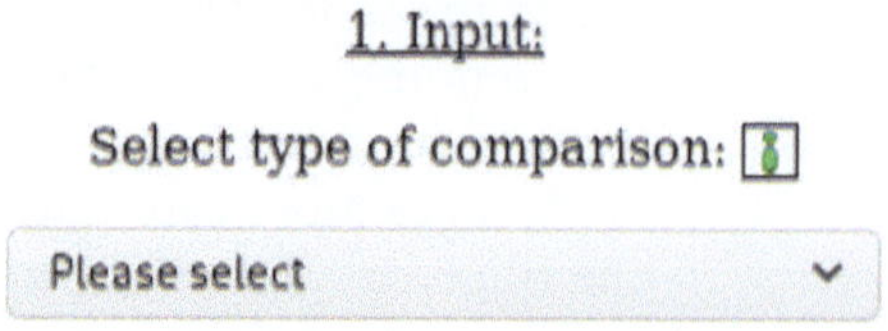

Fig. 7 CMCompare webserver offers two modes of comparison. The first mode compares the input RNA family model against all `Rfam` models, while the second mode compares a set of uploaded models with each other

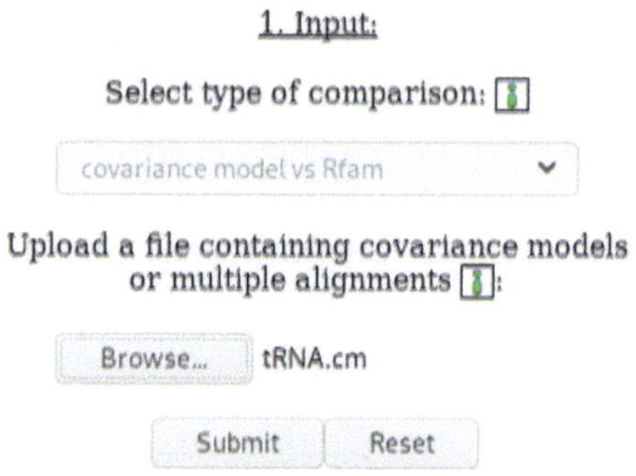

Fig. 8 The model of interest to be compared is selected via file browser

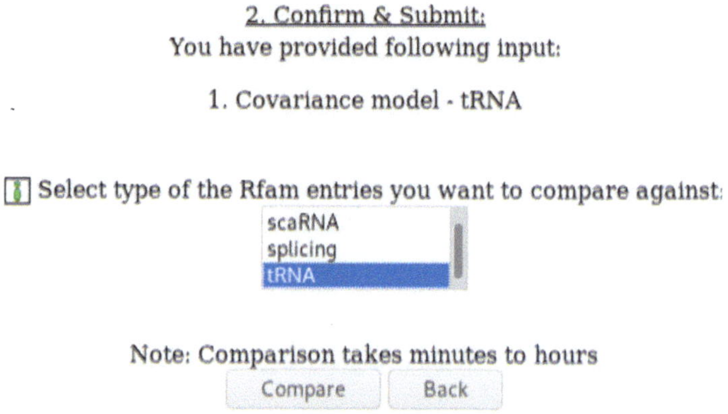

Fig. 9 Before submission, the uploaded model is validated and a subset of models to compare against can be selected

The progress of comparing the models is updated in regular intervals. The provided link of the progress page can be used to access the results at a later time (*see* Fig. 10). On completion, the page redirects to the result page.

The result page shows the top comparison results from the selected `Rfam` models sorted by `Link Score`. `CMCompare` detects the `Rfam` tRNA model as the best matching model with the highest score. A subset of the models in the list can be resubmitted to investigate all pairwise links between them (Fig. 11).

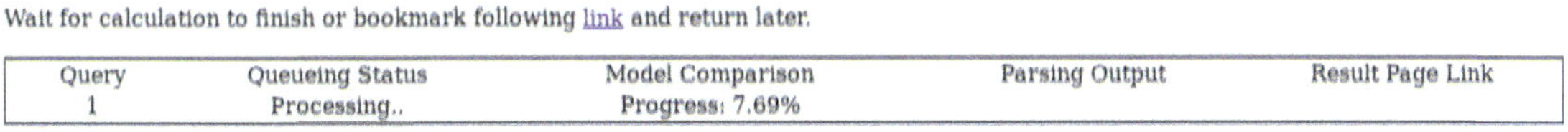

Fig. 10 The progress of the comparison is regularly updated, and a link to return to the comparison later on is provided

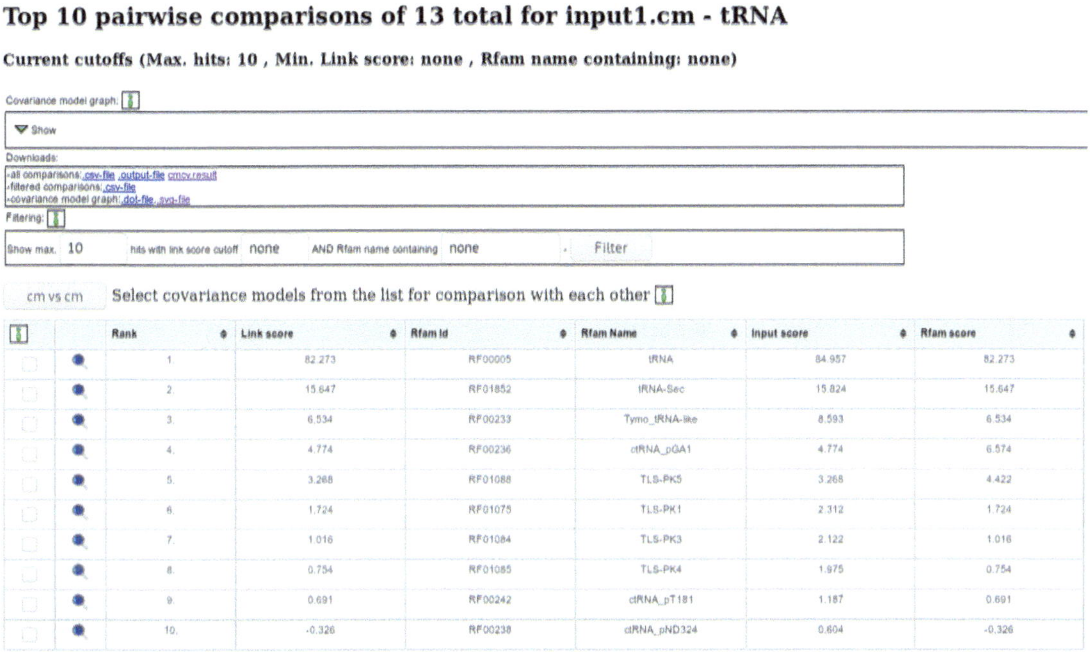

Fig. 11 The result page shows the models with the highest link scores that should be inspected first for a specificity overlap with the input model. The results can additionally be filtered and submitted for detailed comparison

Additional help and examples for the CMCompare webserver can be found on the help page: http://nibiru.tbi.univie.ac.at/cmcws/help.html.

3.6 Visualization of RNA Families

RNA family models or covariance models are encoded in a tabular format, and it can be challenging to inspect the model itself. `CMV` [10] is a tool that provides a number of options to analyze RNA family models visually using different levels of detail. Moreover, it can visualize model comparisons performed with `CMCompare` with the included `CMCV` tool.

We will visualize the tRNA model constructed in Subheading 3, by reusing the `CMV` and `Infernal` environment created there. The model is constructed first, then we create a visualization in simple detail.

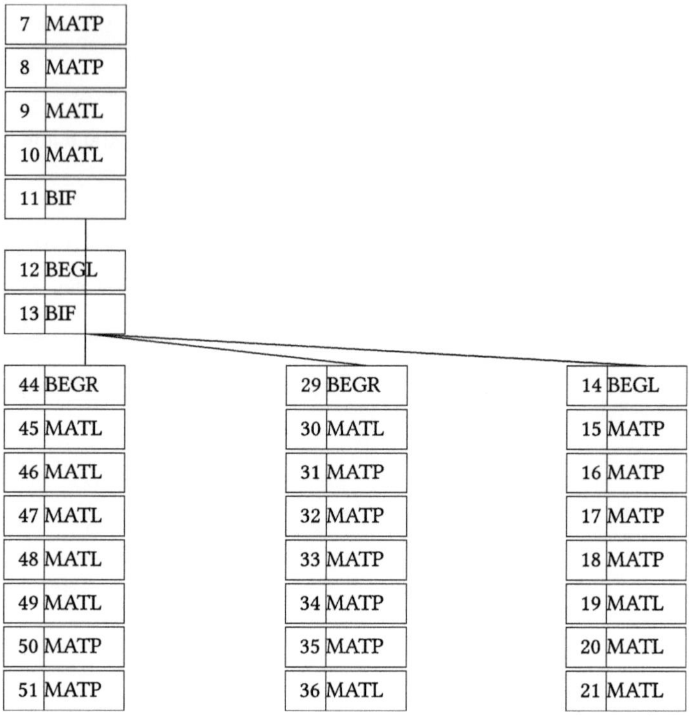

Fig. 12 An overview of part of the guide tree of a tRNA family model. Each node contains the index as well as the type of the node. This simple detail level groups the nodes according to bifurcations, indicating nested structures but leaves out (statistical) sequence information Consensus structure can be identified as paired (`MATP`) or unpaired (`MATL/R`)

```
conda activate infernal
cp tRNA.out/results/result.stk tRNA.stockholm
cmbuild tRNA.cm tRNA.stockholm
cmcalibrate tRNA.cm
conda activate cmv
CMV -m tRNA.cm -s tRNA.stockholm -f pdf -d simple -x r2r
mv tRNA.pdf tRNA_simple_detail.pdf
```

In order to have an overview of the structure of a covariance model, it makes sense to strip away statistical sequence information. This more simple level of detail is provided by a figure of the guide tree, the "spine" of the covariance model. The example of the `Rfam` tRNA model in Fig. 12 shows the acceptor stem, the two bifurcations (`BIF`), and three loop stem of the tRNA. Even though tRNAs have a single multi-branched loop with four stems, the covariance model specification allows only bifurcations. This leads to the occurrence of two bifurcations here.

For an in-depth investigation of the structure of a covariance, a detailed mode is available as well. It shows the nodes with their

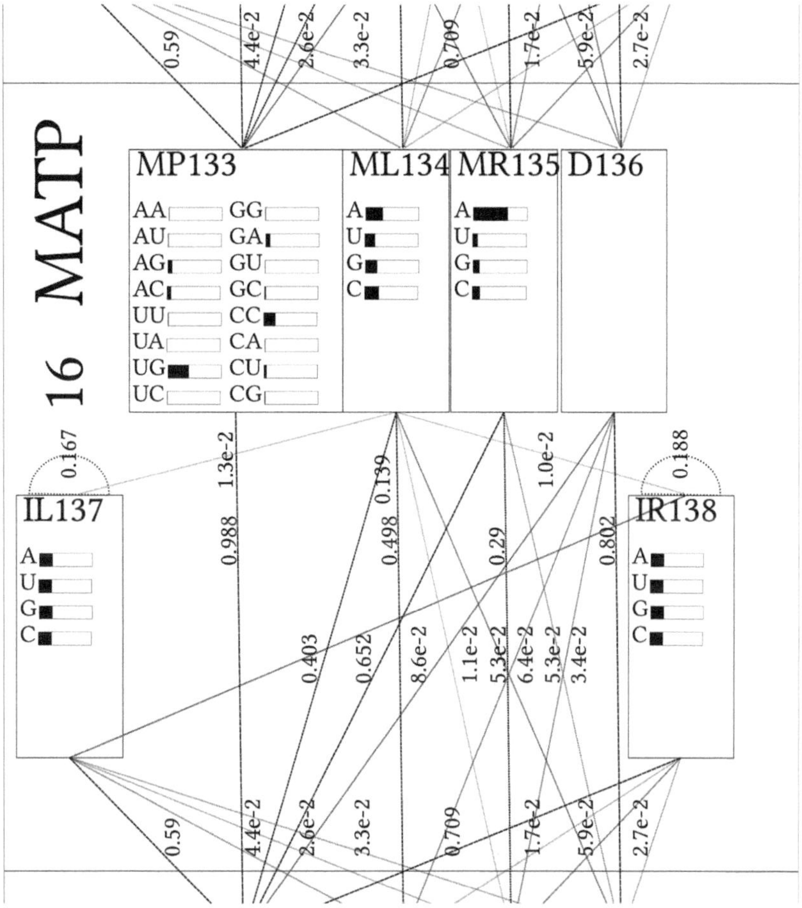

Fig. 13 Visualization with full detail shows all nodes of the model with their index and type—here the pair node 16 with internal, as well as incoming and outgoing state, transitions is shown. In addition, for each node, all states are shown. Match states are drawn in the upper half of each node, while the insertion states are drawn in the lower half. All states are displayed with emission and transition probabilities

transition probabilities between states and the emission states themselves. For each state, histograms show the character match probabilities. Figure 13) shows the pair node 16 with its (default) pair match, as well as degenerate unpaired states, which are used in case only part of the base pair can be matched to a sequence. The insertion states deal with non-consensus nucleotides. Edges give the transitions between states from this and other nodes.

```
CMV -m tRNA.cm -s tRNA.stockholm -f pdf -d detailed -x r2r
mv tRNA.pdf tRNA_detail.pdf
```

Additional parameters control the minimal probability for which transitions are displayed. Emission probabilities can alternatively be shown as scores or log probabilities. The alignment and consensus secondary structure can be visualized as demonstrated above with `StockholmV` (*see* Subheading 3). Tools for model construction and visualization can also be found in the Galaxy RNA workbench [11, 17].

4 Notes

`RNAlien` can be used in an online mode, which is recommended for trying the tool out, and an offline mode that speeds up the construction process by using a local `BLAST` [2] database. It can be downloaded as follows but please note that the `nt` (nucleotide database) has a size of about 70 gigabyte after download.

```
mkdir db_path
cd /db_path
update_blastdb.pl --blastdb_version 5 nt_v5 --decompress
```

Then the `RNAlien` command can be run as follows:

```
RNAlien -i ctm.fa -d ctm -c 2 -j -b /db_path/nt_v5 -o /output_path/
```

References

1. Bernhart SH, Hofacker IL, Will S, Gruber AR, Stadler PF (2008) RNAalifold: improved consensus structure prediction for RNA alignments. BMC Bioinform 9(1):474
2. Camacho C, Coulouris G, Avagyan V, Ma N, Papadopoulos J, Bealer K, Madden TL (2009) BLAST+: architecture and applications. BMC Bioinform 10(1):421
3. Dalli D, Wilm A, Mainz I, Steger G (2006) STRAL: progressive alignment of non-coding RNA using base pairing probability vectors in quadratic time. Bioinformatics 22(13): 1593–1599
4. Durbin R, Eddy S, Krogh A, Mitchison G (1998) Biological sequence analysis. Cambridge University Press, Cambridge
5. Eddy SR (2011) Accelerated profile HMM searches. PLoS Comput Biol 7(10):e1002195
6. Edgar RC (2004) MUSCLE: multiple sequence alignment with high accuracy and high throughput. Nucleic Acids Res 32(5): 1792–1797
7. Eggenhofer F (2016) Unsupervised construction, evaluation and visualisation of RNA family models. Wien
8. Eggenhofer F, Hofacker IL, Höner zu Siederdissen C (2013) CMCompare webserver: comparing RNA families via covariance models. Nucleic Acids Res 41(W1):W499
9. Eggenhofer F, Hofacker IL, Höner zu Siederdissen C (2016) RNAlien— unsupervised RNA family model construction. Nucleic Acids Res 44(17):8433
10. Eggenhofer F, Hofacker IL, Backofen R, Höner zu Siederdissen C (2018) CMV: visualization for RNA and protein family models and their comparisons. Bioinformatics 34(15): 2676–2678
11. Fallmann J, Videm P, Bagnacani A, Batut B, Doyle MA, Klingstrom T, Eggenhofer F, Stadler PF, Backofen R, Grüning B (2019) The RNA workbench 2.0: next generation RNA data analysis. Nucleic Acids Res 47(W1): W511–W515

12. Federhen S (2011) The NCBI taxonomy database. Nucleic Acids Res 40(D1):D136–D143
13. Gardner PP, Wilm A, Washietl S (2005) A benchmark of multiple sequence alignment programs upon structural RNAs. Nucleic Acids Res 33(8):2433–2439
14. Gardner PP, Daub J, Tate J, Moore BL, Osuch IH, Griffiths-Jones S, Finn RD, Nawrocki EP, Kolbe DL, Eddy SR, et al. (2010) Rfam: Wikipedia, clans and the "decimal" release. Nucleic Acids Res 39(suppl_1):D141–D145
15. Gruber AR, Findeiß S, Washietl S, Hofacker IL, Stadler PF (2010) RNAz 2.0: improved noncoding RNA detection. In: Biocomputing 2010. World Scientific, Singapore, pp 69–79
16. Grüning B, Dale R, Sjödin A, Chapman B, Rowe J, Tomkins-Tinch C, Valieris R, Köster J, Bioconda T (2018) Bioconda: sustainable and comprehensive software distribution for the life sciences. Nature Methods 15(7):475
17. Grüning BA, Fallmann J, Yusuf D, Will S, Erxleben A, Eggenhofer F, Houwaart T, Batut B, Videm P, Bagnacani A, Wolfien M, Lott SC, Hoogstrate Y, Hess WR, Wolkenhauer O, Hoffmann S, Akalin A, Ohler U, Stadler PF, Backofen R (2017) The RNA workbench: best practices for RNA and high-throughput sequencing bioinformatics in Galaxy. Nucleic Acids Res 45(W1):W560–W566
18. Harmanci AO, Sharma G, Mathews DH (2007) Efficient pairwise RNA structure prediction using probabilistic alignment constraints in Dynalign. BMC Bioinform 8(1):130
19. Harrison PW, Alako B, Amid C, Cerdeño-Tárraga A, Cleland I, Holt S, Hussein A, Jayathilaka S, Kay S, Keane T, et al. (2018) The European Nucleotide Archive in 2018. Nucleic Acids Res 47(D1):D84–D88
20. Höner zu Siederdissen C, Hofacker IL (2010) Discriminatory power of RNA family models. Bioinformatics 26(18):i453–i459
21. Kalvari I, Argasinska J, Quinones-Olvera N, Nawrocki EP, Rivas E, Eddy SR, Bateman A, Finn RD, Petrov AI (2018) Rfam 13.0: shifting to a genome-centric resource for non-coding RNA families. Nucleic Acids Res 46(D1), D335–D342
22. Katoh K, Rozewicki J, Yamada KD (2019) MAFFT online service: multiple sequence alignment, interactive sequence choice and visualization. Briefings Bioinform 20(4), 1160–1166
23. Kerpedjiev P, Hammer S, Hofacker IL (2015) Forna (force-directed RNA): simple and effective online RNA secondary structure diagrams. Bioinformatics 31(20), 3377–3379
24. Kubota M, Tran C, Spitale RC (2015) Progress and challenges for chemical probing of RNA structure inside living cells. Nat Chem Biol 11(12):933
25. Lorenz R, Bernhart SH, Höner zu Siederdissen C, Tafer H, Flamm C, Stadler PF, Hofacker IL (2011) ViennaRNA Package 2.0. Algorithms Mol Biol 6(1):26
26. Lott SC, Schäfer RA, Mann M, Backofen R, Hess WR, Voß B, Georg J (2018) GLASSgo—automated and reliable detection of sRNA homologs from a single input sequence. Front Genet 9:124
27. Markham NR, Zuker M (2008) UNAFold. In: Bioinformatics. Springer, Berlin, pp 3–31
28. Nawrocki EP (2014) Annotating functional RNAs in genomes using Infernal. In: RNA Sequence, Structure, and Function: Computational and Bioinformatic Methods. Springer, Berlin, pp 163–197
29. Nawrocki EP, Eddy SR (2007) Query-dependent banding (QDB) for faster RNA similarity searches. PLoS Comput Biol 3(3): e56
30. Nawrocki EP, Eddy SR (2013) Infernal 1.1: 100-fold faster RNA homology searches. Bioinformatics 29(22):2933–2935
31. Nawrocki EP, Eddy SR (2016) INFERNAL User's Guide (1.1.2)
32. Nawrocki EP, Burge SW, Bateman A, Daub J, Eberhardt RY, Eddy SR, Floden EW, Gardner PP, Jones TA, Tate J et al (2014) Rfam 12.0: updates to the RNA families database. Nucleic Acids Res 43(D1):D130–D137
33. O'Leary NA, Wright MW, Brister JR, Ciufo S, Haddad D, McVeigh R, Rajput B, Robbertse B, Smith-White B, Ako-Adjei D et al (2015) Reference sequence (RefSeq) database at NCBI: current status, taxonomic expansion, and functional annotation. Nucleic Acids Res 44(D1):D733–D745
34. Raden M, Ali SM, Alkhnbashi OS, Busch A, Costa F, Davis JA, Eggenhofer F, Gelhausen R, Georg J, Heyne S, et al. (2018) Freiburg RNA tools: a central online resource for RNA-focused research and teaching. Nucleic Acids Res 46(W1):W25–W29
35. Roshan U, Livesay DR (2006) Probalign: multiple sequence alignment using partition function posterior probabilities. Bioinformatics 22(22):2715–2721
36. Sankoff D (1985) Simultaneous solution of the RNA folding, alignment and protosequence problems. SIAM J Appl Math 45(5):810–825

37. Seemann SE, Gorodkin J, Backofen R (2008) Unifying evolutionary and thermodynamic information for RNA folding of multiple alignments. Nucleic Acids Res 36(20):6355–6362
38. Serganov A, Nudler E (2013) A decade of riboswitches. Cell 152(1–2):17–24
39. Sievers F, Wilm A, Dineen D, Gibson TJ, Karplus K, Li W, Lopez R, McWilliam H, Remmert M, Söding J, et al. (2011) Fast, scalable generation of high-quality protein multiple sequence alignments using Clustal Omega. Mol Syst Biol 7(1):539
40. Sundfeld D, Havgaard JH, de Melo AC, Gorodkin J (2015) Foldalign 2.5: multithreaded implementation for pairwise structural RNA alignment. Bioinformatics 32(8):1238–1240
41. The RNAcentral Consortium (2018) RNAcentral: a hub of information for non-coding RNA sequences. Nucleic Acids Res 47(D1):D1250–D1251
42. Torarinsson E, Havgaard JH, Gorodkin J (2007) Multiple structural alignment and clustering of RNA sequences. Bioinformatics 23(8):926–932
43. Washietl S, Findeiß S, Müller SA, Kalkhof S, Von Bergen M, Hofacker IL, Stadler PF, Goldman N (2011) RNAcode: robust discrimination of coding and noncoding regions in comparative sequence data. RNA 17(4): 578–594
44. Weinberg Z, Breaker RR (2011) R2R—software to speed the depiction of aesthetic consensus RNA secondary structures. BMC Bioinform 12(1):3
45. Will S, Reiche K, Hofacker IL, Stadler PF, Backofen R (2007) Inferring noncoding RNA families and classes by means of genome-scale structure-based clustering. PLoS Comput Biol 3(4):e65
46. Will S, Joshi T, Hofacker IL, Stadler PF, Backofen R (2012) LocARNA-P: accurate boundary prediction and improved detection of structural RNAs. RNA 18(5):900–914

Chapter 12

Developing Complex RNA Design Applications in the Infrared Framework

Hua-Ting Yao, Yann Ponty, and Sebastian Will

Abstract

Applications in biotechnology and bio-medical research call for effective strategies to design novel RNAs with very specific properties. Such advanced design tasks require support by computational tools but at the same time put high demands on their flexibility and expressivity to model the application-specific requirements. To address such demands, we present the computational framework Infrared. It supports developing advanced customized design tools, which generate RNA sequences with specific properties, often in a few lines of Python code. This text guides the reader in tutorial format through the development of complex design applications. Thanks to the declarative, compositional approach of Infrared, we can describe this development as a step-by-step extension of an elementary design task. Thus, we start with generating sequences that are compatible with a single RNA structure and go all the way to RNA design targeting complex positive and negative design objectives with respect to single or even multiple target structures. Finally, we present a "real-world" application of computational design to create an RNA device for biotechnology: we use Infrared to generate design candidates of an artificial "AND" riboswitch, which activates gene expression in the simultaneous presence of two different small metabolites. In these applications, we exploit that the system can generate, in an efficient (fixed-parameter tractable) way, multiple diverse designs that satisfy a number of constraints *and* have high quality w.r.t. to an objective (by sampling from a Boltzmann distribution).

Key words RNA design, Multi-target design, Declarative modeling, Boltzmann sampling, Fixed-parameter tractable sampling

1 Introduction

Designing molecules with novel functionality or very specific desirable properties for applications in biological fundamental research, biotechnology, and medicine is a highly complex task that typically requires interdisciplinary efforts, combining biochemical experimentation and computational design [7]. Compared to proteins, designing RNAs is particularly attractive for the construction of new biotechnological devices. On the one hand, functional RNA molecules save the detour of translation into proteins, and can

Ronny Lorenz (ed.), *RNA Folding: Methods and Protocols*, Methods in Molecular Biology, vol. 2726,
https://doi.org/10.1007/978-1-0716-3519-3_12,

therefore act more efficiently, e.g., as fast on/off-switches of gene activity. On the other hand, the design process itself can build on the well-understood combinatorics of RNA secondary structure and available computational models [16], algorithms [13, 18] and implementations [12, 15].

Despite this starting position, the computational support of RNA design remains challenging. First of all, RNA design is an optimization problem with often complex objectives with respect to multiple (secondary) structures, e.g., when the designed RNAs should switch between alternative structural states or fold via specific intermediary structures. Moreover, RNA design is computationally complex even in elementary problem variants. For example, even in a simple base pair model, one cannot efficiently design an RNA that favors a single target structure over all others—this problem is NP-hard [3].

Here we present the framework Infrared, which addresses the multiple demands of computational RNA design in several ways:

- To address the issues of computational complexity, Infrared adopts the classical decomposition of RNA design into two related sub-problems, often called **positive and negative design**. Positive RNA design aims at very specific properties (e.g., specific energy) for certain 'target" RNA structures, while negative design additionally aims to avoid similar properties for all exponentially many other, "non-target" structures. While efficient algorithms for the latter problem cannot exist [3], the system automatically derives fixed-parameter tractable algorithms to solve the sub-problem of positive design efficiently. In many cases, this can already solve negative design by searching through relatively few samples of good positive designs. To address harder negative design tasks, we demonstrate constraint generation as well as classic stochastic optimization techniques as supported by the system.
- Real-world RNA design applications typically demand for targeting thermodynamic criteria referring to multiple target structures (e.g., on and off-states of riboswitches, binding pockets of aptamers, and competing structures in specific energetic relations), potentially under side constraints like moderate GC-content or avoidance of specific sequence motifs. Thus, we designed Infrared as a library that supports programmers in developing complex and potentially novel design strategies based on a declarative framework, which supports the definition of constraints that designs must satisfy as well as features (like GC content, energy, ...) that solutions should target. This allows application programmers to harness the full power of fixed-parameter tractable sampling of designs in an easy to use system.

In this chapter, we walk the reader through the use of this library and show by examples how to develop and run design programs from IPython notebooks. Further examples are provided as part of the documentation of Infrared. (Notably, these include applications to problems outside of design, e.g., structure alignment, demonstrating the generality of this system.) Further examples are found in our published design tools based on the library for the specific applications of the generation of multi-target designs (RNARedPrint v2, as reimplementation of RNARedPrint [9]) and negative RNA design (RNAPOND [17]). These implementations illustrate the use of the library in bioinformatics tools beyond small demonstrators.

The fundamental ideas of the Infrared framework are best illustrated by showing some example code. After importing Infrared

```
import infrared as ir
import infrared.rna as rna
```

a few lines of code let us generate random (in this first example, "only" uniformly sampled) designs that can fold into the first structure of Fig. 1.

```
target = "((((((((((...))))((((....))))))))))"
model = ir.Model(len(target), 4)
model.add_constraints(rna.BPComp(i, j) for (i, j) in rna.parse(target))
sampler = ir.Sampler(model)
samples = [sampler.sample() for _ in range(10)]
```

While this code doesn't yet come close to exploiting the power of the system, it demonstrates the main structure of using Infrared's FPT sampling engine. After a short prelude and defining the input, the user sets up a constraint model. Here, one defines the number of sequence positions, i.e., `len(target)` and that there are four possible values per position (encoding the nucleotides). Moreover by the constraints `BPComp`, we require the nucleotides of base pairs to be complementary. Now, any assignment of nucleotides to sequence positions that satisfy the constraints describes a valid design. Finally, we construct a sampler for this constraint model and generate ten samples.

The real power of the Infrared framework shows only when we extend our initial simple example. First, generating uniformly sampled sequences for multi-target design, analogously to

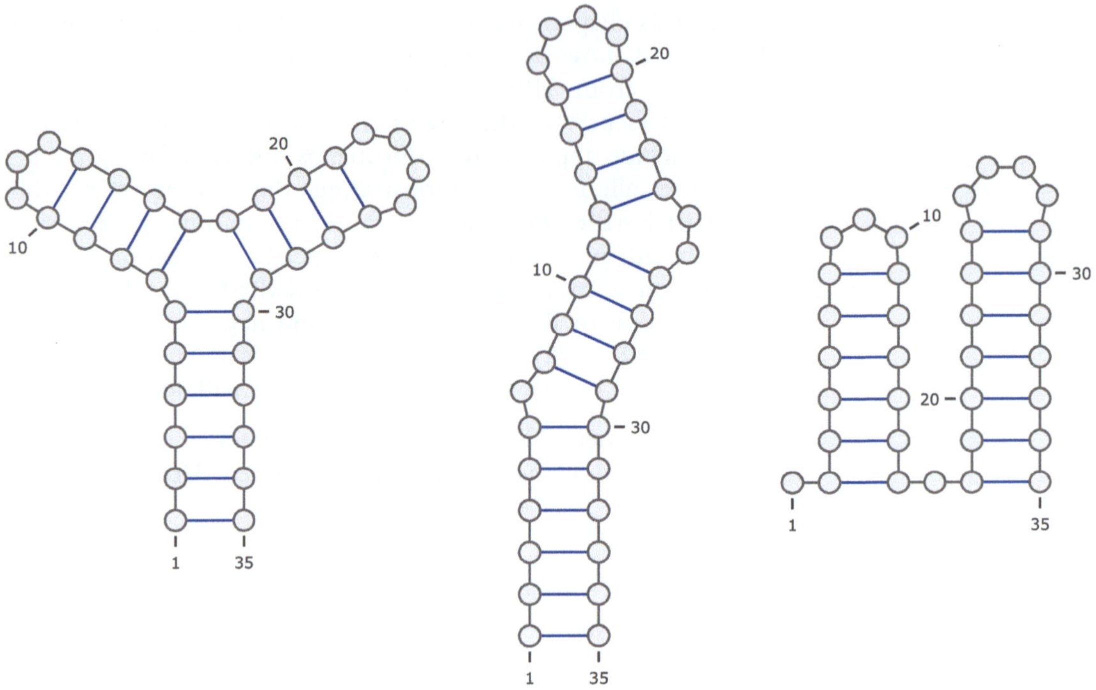

((((((((((...))))((((....))))))))))
((((((.((((((((....))))..))))))))))
.((((((...)))))).(((((((....)))))))

Fig. 1 Three RNA target structures for RNA design, which will serve us as running examples. They are shown in 2D representation (by VARNA [4]) and as dot-bracket strings. The latter represent base pairs by balanced parentheses

RNABluePrint [8], requires only a slight extension of the above model. To target the three structures of Fig. 1, we define

```
targets = ["((((((((((...))))((((....))))))))))",
           "((((((.((((((((....))))..))))))))))",
           ".((((((...)))))).(((((((....)))))))"]
```

Then, we add constraints for all these structures to our model. For this purpose, we rewrite line 3 of the previous example to loop over the targets

```
for target in targets:
    model.add_constraints(rna.BPComp(i, j) for (i, j) in
     ↪ rna.parse(target))
```

and generate samples from this model in exactly the same way as before; only now, the sequences are compatible to all three target structures.

Second, another few lines of code, let us go beyond uniform sampling. We can sample high quality designs from a Boltzmann distribution controlled by energies of the target structures after imposing (base pair) energy functions:

```
for target in targets:
    model.add_functions([rna.BPEnergy(i, j, False) for (i, j) in
    ↪  rna.parse(target)], 'energy')
```

The involved computation and algorithmic structure required by the multi-target case is handled transparently "under the hood". Internally, the problem is solved by an efficient algorithm that automatically adapts in complexity to the dependency network due to the multiple target structures.

Chapter Overview

For preparation, we describe the installation of the library and recommend additional software. In Methods Section, we systematically develop a complex multi-target design application using the Infrared framework. We start by targeting positive design objectives in increasingly complex settings. This is followed by a discussion of several approaches that integrate negative design objectives. Thus, we show-case constraint generation (in the way of RNAPOND [17]), stochastic optimization based on the library and finally, a full-fledged application to the design of an AND-riboswitch.

2 Material: Installing Infrared

We recommend installing Infrared using the package manager Conda as described below. Alternatively but less conveniently, the software can be compiled and installed from its source found on Gitlab https://gitlab.inria.fr/amibio/Infrared. In this chapter, we describe the release 1.1 of the software.

2.1 Package Manager Conda Installation

Unless Conda is already installed on your system, we recommend to install it in the form of Miniconda from https://conda.io/en/latest/miniconda.html. The page contains installation instructions for Windows, MacOs, and Linux.

2.2 Infrared Installation

To use Conda, it typically has to be activated by running the shell command

```
conda activate
```

in a terminal. All required and recommended software can be installed from the command line by

```
conda install -c conda-forge 'infrared=1.1' jupyter pip
pip install viennarna
```

The commands install the packages `infrared, viennarna` and `jupyter`; the flag `-c conda-forge` specifies the Conda channel. For the largest part of the tutorial, only the package `infrared` is required. For energy evaluation and RNA structure prediction (in negative design), we require the Vienna RNA package [12].

3 Methods

One can run all code of this tutorial from an IPython notebook using `jupyter` (or a comparable system). The tutorial notebook is available from Infrared's Gitlab repository

https://amibio.gitlabpages.inria.fr/Infrared/md_Doc_2Tutorials_2Bookchapter.html

Using `jupyter-notebook`, it is loaded by the command

```
jupyter-notebook "Bookchapter Tutorial.ipynb"
```

3.1 Elementary Use of Infrared—A Simple Design Model

Recall the first example from the introduction. We want to generate sequences compatible with a single target structure. In Infrared, this idea can be expressed in the form of a constraint/function network model.

Assume that we want to design a sequence of length $n = 35$. Then, we set up a new model with n variables $X_0, \ldots, X_{n-1}$ that each can take one of four different values (representing the nucleotides A,C,G,U by integers $0, 1, 2, 3$).

```
model = ir.Model(n,4)
```

This is Infrared's way to model the solutions of the design task, namely nucleotide sequences of length n. Next, we restrict the solution space by enforcing compatibility to a target structure (again from the running example). This is accomplished by adding a complementarity constraint for each base pair

```
target = "((((((((((...))))((((....))))))))))"
model.add_constraints(rna.BPComp(i, j) for (i, j) in rna.parse(target))
```

The constraint `BPComp(i, j)` is *valid* (or *satisfied*) if the variables X_i and X_j represent complementary nucleotides (i.e., one pair of A-U, C-G, G-C, G-U, U-A, U-G). Technically, solutions are assignments that assign to each of the variables $X_0, \ldots, X_{n-1}$ a value of its domain $\{0, 1, 2, 3\}$; moreover, they must be *valid* assignments that satisfy all the constraints. Once we have a model that defines the solution space, we can instantiate a sampler

```
sampler = ir.Sampler(model)
```

This object allows us to draw sample solutions by calls to `sample()`, e.g., we obtain 10 sampled designs by

```
samples = [sampler.sample() for _ in range(10)]
```

Recall that nucleotides are represented by numbers, such that we typically want to convert sampled assignments to sequences (over A, C, G, U); for this purpose, one typically applies Infrared's function `ass_to_seq`, like

```
sequences = [rna.ass_to_seq(sample) for sample in samples]
```

Without further specifications, such solutions will be sampled from the uniform distribution.

3.2 Imposing Additional Constraints—Sequence Constraints in IUPAC Code

Thanks to the constraint paradigm, more complex tasks can be specified compositionally. This allows us to tailor the solution space very naturally by imposing additional constraints. In design problems, it is often useful to express prior knowledge on the sequence design space in IUPAC nucleotide encoding. To provide just one (arbitrary) example, we could specify the loop regions as RYY and GNRA, as well as the left and right most base as strong (`S`) by the constraint string

```
iupac_sequence = "SNNNNNNNNNRYYNNNNNNNNGNRANNNNNNNNNS"
```

Corresponding constraints can then be added to our model by

```
for i, x in enumerate(iupac_sequence):
    model.add_constraints(ir.ValueIn(i, rna.iupacvalues(x)))
```

Here, we make use of Infrared's function `iupacvalues()`, which returns the possible nucleotides values allowed by a IUPAC symbol, and the constraint `ValueIn`, which constrains the allowed values of a variable.

3.3 Functions and Features—Control of GC Content

Infrared offers many ways to build on this initial design model and tailor it towards more sophisticated design and very specific applications. For a start, we want to generate sequences with control over their GC content.

In Infrared, we can define the GC content as a feature of solutions, which itself is specified by a set of functions. For this purpose, we extend the previous model by

```
model.add_functions([rna.GCCont(i) for i in range(n)], 'gc')
```

This defines one function `GCCont` for each variable (or nucleotide) and adds them to the model. Each single function has value of 1, if the respective variable X_i represents C or G and value of 0, otherwise. Infrared automatically derives a feature with the name `gc` that sums over all these function values. Thus, it counts the G and C nucleotides (reflecting the GC content as the *feature value*).

3.3.1 Shifting the GC Content Distribution

Infrared offers two major ways to make use of this feature: shifting the distribution and targeting specific values. In the first and simpler case, one can control the sample distribution by setting the weight of the feature before constructing the sampler and drawing samples, e.g.,

```
model.set_feature_weight(1, 'gc')
sampler = ir.Sampler(model)
samples = [sampler.sample() for _ in range(1000)]
```

Plotting sequence logos and histograms of the GC contents of these samples shows that it is no longer uniformly distributed, but the sample distribution is shifted towards higher GC content. Negative weights shift to lower GC content, while 0 keeps the distribution uniform (Fig. 2 shows the effect in the additional presence of IUPAC constraints). Technically, the samples s are generated from a Boltzmann distribution, depending on the weight α_{gc} of feature gc, i.e., their probabilities are proportional to the Boltzmann weights based on their GC contents GC(s):

$$\exp(\alpha_{gc} \cdot \mathrm{GC}(s)).$$

3.3.2 Excursion 1: A First Glimpse at the Infrared Sampling Engine

Having seen a few sampling examples, we are ready to talk about the computations behind the scenes. First observe that variables not involved in base pairs can apparently be sampled independently (and in a rather straightforward manner: either uniformly or with probability proportional to their Boltzmann factor).

It's more interesting how to sample variables X_i and X_j connected by a base pair (i, j). Since determining the variables simultaneously does not generalize well, we handle one after the other. Choosing the value of the variable X_i before X_j requires looking ahead to the possible solutions in combination with X_j. In uniform sampling, we can choose based on the number of possible solutions for each value in X_i's domain. For example, choosing G leaves two values for X_j to satisfy the complementarity constraint, whereas C leaves exactly one; consequently, G must be selected two times as often as C if we want to sample uniformly. Remarkably, sampling from a Boltzmann distribution works analogously: instead of "counting" solutions (each uniformly by 1), we need to accumulate their Boltzmann weights. Conversely, uniform sampling turns out to be a special case of Boltzmann sampling with weight zero.

To inform the choice of X_i (before choosing X_j), we marginalize the sums of Boltzmann weights $\exp(\alpha_{gc} \cdot \mathrm{GC}(s))$ over the possible values of X_j. In other words, we utilize partial partition functions.

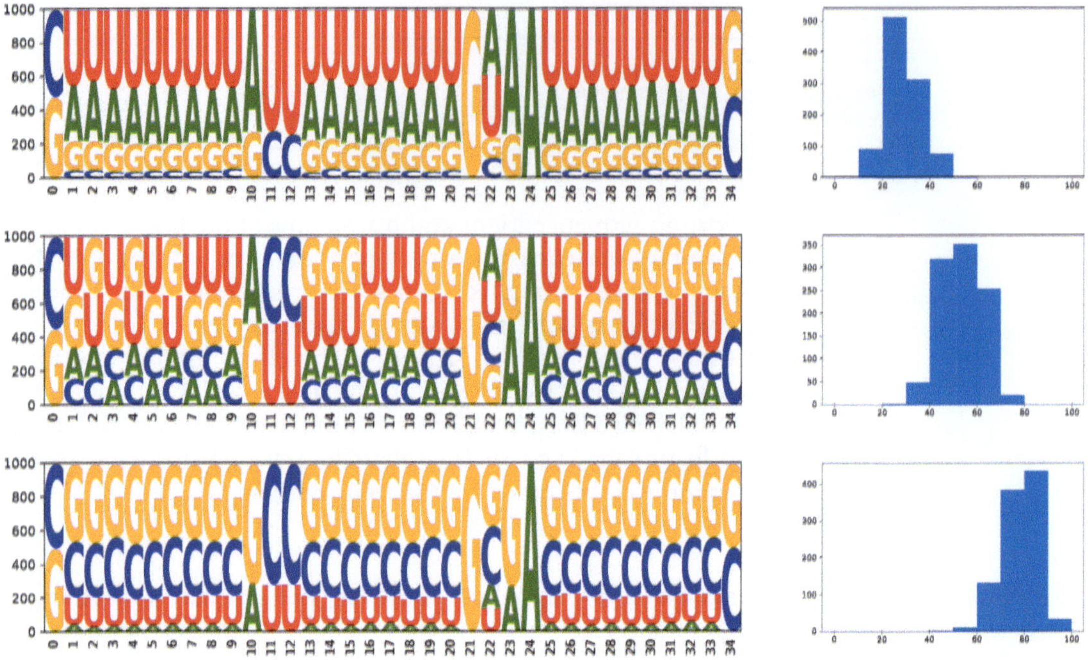

Fig. 2 Sequence logos and GC-content histograms for each time 1000 samples with respective weights -1, 0, and 1 (top, middle, bottom) of the feature `gc`. The sequences are compatible with the first example target structure and the IUPAC string `SNNNNNNNNNRYYNNNNNNNNGNRANNNNNNNNNS`

Value of X_i	Possible values of X_j	Partition function (GC content)
A	U	$\exp(\alpha_{gc} \cdot 0)$
C	G	$\exp(\alpha_{gc} \cdot 2)$
G	C, U	$\exp(\alpha_{gc} \cdot 2) + \exp(\alpha_{gc} \cdot 1)$
U	A, G	$\exp(\alpha_{gc} \cdot 0) + \exp(\alpha_{gc} \cdot 1)$

For sampling, the value of X_i is selected with probability proportional to its corresponding partition function from the table. After X_i is determined, the choice of a value for X_j becomes comparably simple.

For the previous examples, Infrared's solving mechanism indeed boils down to fixing an arbitrary order of the variables and precomputing partition functions as described for the variables in base pairs. In the presence of more complex dependencies between variables, Infrared still chooses values for the variables one-by-one in a predetermined optimized variable order. Given this order, it precomputes partial partition functions. To generate samples from the desired Boltzmann distribution, it then chooses values with partition function-derived probabilities (in constant time per variable, resulting in linear time sampling). The key to efficient sampling in Infrared is thus to make the precomputation as efficient as possible. We are going to discuss further details of the Infrared engine, when we progress to network models with more complex dependencies as they result from simultaneously targeting several RNA structures.

3.3.3 Targeting Specific GC Content

Instead of directly tweaking the weight (to shift the GC distribution as we discussed before), we can even ask Infrared to target a specific feature value. For example, we generate targeted samples with a GC content of 75% ± 1% from our model by

```
sampler = ir.Sampler(model)
sampler.set_target( 0.75 * n, 0.01 * n, 'gc' )
samples = [sampler.targeted_sample() for _ in range(1000)]
```

Note the use of `Sampler`'s method `targeted_sample()` in place of `sample()`. This method provides access to an automatic mechanism that returns only samples within the tolerance from the target. To make such a rejection strategy effective, we iteratively sample, estimate the current mean of the feature value, and then update the feature weight. Concretely, Infrared implements a form of multi-dimensional Boltzmann sampling [2] as applied in RNARedPrint [9].

3.4 Controlling Energy—Multiple Features

Analogous to the GC content, one can as well target the energy of the target structure, since (like the GC content) RNA energy can be modeled as a sum of function values. This holds even for the detailed nearest-neighbor energy model of RNAs, where energy is composed of empirically determined or trained loop energies [1, 16]. Here, we focus on the much simpler base pair energy model, which has been demonstrated to be an effective proxy for the Turner (nearest neighbor) model in design applications [9].

In this model that we originally suggested for RNARedPrint [9], every type of base pair receives a different energy. To define the feature `energy`, we impose a function for each base pair `(i, j)` in the target structure and moreover distinguish terminal and non-terminal base pairs, by `(i-1, j+1) not in bps`.

```
model.add_functions([rna.BPEnergy(i, j, (i-1, j+1) not in bps)
                     for (i, j) in bps], 'energy')
```

In the same way as for the feature `gc`, we can set the weight of the new feature `energy` and in this way shift the distribution of base pair energies. Infrared can even simultaneously control the energy and the GC content, since it samples (depending on the weights α_{gc} and α_{energy}) from the two-dimensional Boltzmann distribution, where probabilities are proportional to

$$\exp(\alpha_{gc} \cdot \mathrm{GC}(s) + \alpha_{energy} \cdot \mathrm{BPEnergy}(s)).$$

We observed [9] that, for a fixed target structure, the energies of sequences in the base pair energy model are strongly correlated to their energies in the Turner model. Thus shifting the distribution of base pair energies indirectly shifts the distribution of nearest neighbor energies.

For example, we generate sequences with high affinity to the target structure and specific GC content, extending previous code and ideas:

```
model.set_feature_weight(-2, 'energy')
sampler = ir.Sampler(model)
sampler.set_target(0.75*n, 0.01*n, 'gc')
samples = [sampler.targeted_sample() for _ in range(10)]
```

We arrived at the functionality of IncARNation [14] as an pplication of the Infrared library, requiring only a few lines of Python code. We remark that IncARNation implements the stacking energy model, which is slightly more complex than the base pair model. This model is as well available in Infrared, but was not shown to have clear advantages for sampling. To use stacking energy, it suffices to replace the functions `rna.BPEnergy` by `rna.StackEnergy`:

```
model.add_functions([rna.StackEnergy(i, j)
    for (i,j) in bps if (i+1,j-1) in bps], 'energy')
```

3.5 Targeting Turner Energies—Customized Features

This correlation between the base pair and Turner energy models can be moreover exploited in Infrared to directly target Turner energies. We will make use of the Vienna RNA package to evaluate the Turner energies using its function energy_of_struct.

```
from RNA import energy_of_struct
```

Note that the tutorial notebook provides a workaround for Windows users, since the RNA module cannot be imported on Windows.

We start by adding a custom feature that evaluates Turner energy and tell it to control the feature `energy` (this tells Infrared to internally use base pair energy as a proxy to manipulate the distribution of Turner energy):

```
model.add_feature('Energy', 'energy',
    lambda sample, target=target:
        energy_of_struct(rna.ass_to_seq(sample), target))
```

Consequently, we can set a target for the Turner energy of the target structure. In our example, we simultaneously target specific GC content:

```
sampler = ir.Sampler(model)
sampler.set_target(0.75*n, 0.05*n, 'gc')
sampler.set_target(-10, 0.5, 'Energy')
samples = [sampler.targeted_sample() for _ in range(10)]
```

This example provides a first demonstration of the targeting flexibility due to Infrared's multi-dimensional Boltzmann sampling engine, which can simultaneously target several features. We already see how this generalizes over advanced design tools like IncARNation.

3.6 Multiple Target Structures—Complex Dependencies

Due to the compositionality of constraints (and functions) in Infrared, defining multiple targets does not look any different than defining a single target structure. Thus, let us right away define a model to target the structures of Fig. 1, which were previously defined in the list `targets` with length `n`.

```
model = ir.Model(n,4)
model.add_functions([rna.GCCont(i) for i in range(n)], 'gc')
for k, target in enumerate(targets):
    bps = rna.parse(target)
    model.add_constraints(rna.BPComp(i, j) for (i, j) in bps)
    model.add_functions([rna.BPEnergy(i, j, (i-1, j+1) not in bps)
                         for (i, j) in bps], f'energy{k}')
```

Note how we just add constraints and functions for each target structure, but define different feature name (`energy0`, `energy1`, `energy2`), such that we could control them separately. As well, note that we are going to control GC content. Here, we could add further constraints like the ones for a specific IUPAC sequence.

As before (for a single target), we can go on by defining Turner energy features and specific targets. For the latter, one uses the method `set_target` of `Sampler` to specify a target value and a tolerance for a feature.

```
for k, target in enumerate(targets):
    model.add_feature(f'Energy{k}', f'energy{k}',
        lambda sample, target=target:
            energy_of_struct(rna.ass_to_seq(sample), target))
sampler = ir.Sampler(model)
sampler.set_target(0.75*n, 0.01*n, 'gc')
sampler.set_target( -15, 1, 'Energy0')
sampler.set_target( -20, 1, 'Energy1')
sampler.set_target( -20, 1, 'Energy2')
```

Finally, as expected, Infrared will indeed generate sequences that are compatible to all structures and hit the prescribed target energies and GC content.

```
samples = [sampler.targeted_sample() for _ in range(5)]
```

At this point, we arrived at reimplementing the essential functionality of RNARedPrint [9] in the Infrared framework. A Infrared-based reimplementation of this tool ("RNARedPrint 2.0") with command line interface is provided at https://gitlab.inria.fr/amibio/RNARedPrint.

3.6.1 Excursion 2: A Deeper Dive into Infrared's Sampling Engine

As we have seen, Infrared supports multiple target structures and allows us to define specific target energies for them. While the modeling syntax of Infrared even makes this appear quite natural, additional structures and targets increase the complexity of sampling. In turn, this affects the run times for generating designs with Infrared. Before we can make this relationship more precise, we need to provide some technical details of Infrared's computational engine.

Generation of Samples from a Multi-dimensional Boltzmann Distribution

Infrared implements a general solving strategy based on tree decompositions and cluster tree elimination (CTE) [10]. Such techniques are well known in the (larger) context of constraint processing [5]; more recently, we described this approach specialized to multi-target RNA design for our approach RNARedPrint [7]. The cluster tree elimination scheme yields a fixed-parameter tractable algorithm to compute (partial) partition functions, which lets us generate samples from a multi-dimensional Boltzmann distribution. In our example, this means that we can efficiently generate samples with probabilities proportional to

$$\exp(\alpha_{\mathrm{gc}} \cdot \mathrm{GC}(s) + \sum_{k=0}^{2} \alpha_{\mathrm{energy}k} \cdot \mathrm{energy}k(s)).$$

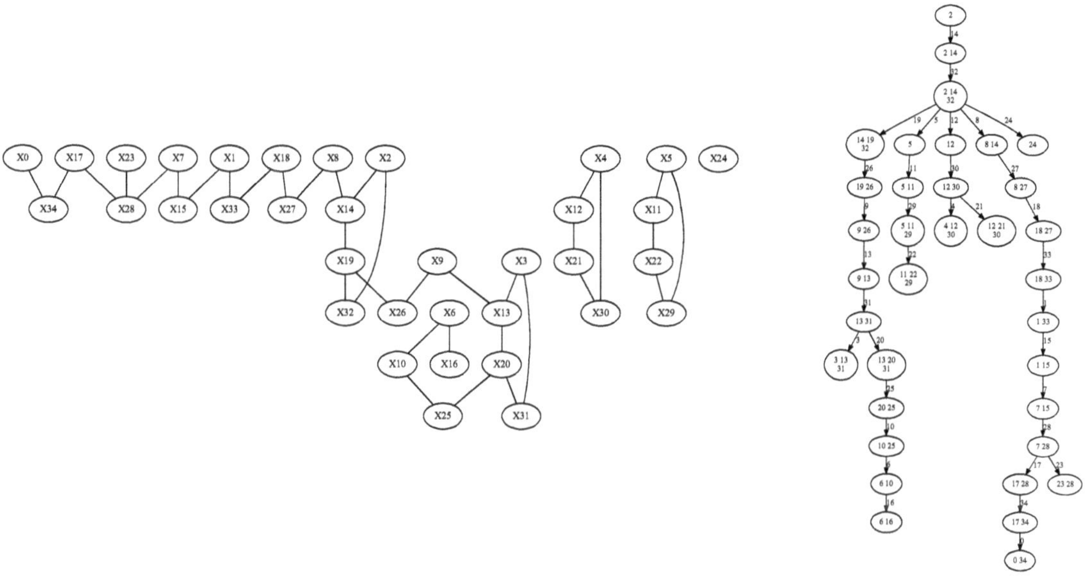

Fig. 3 (Left) Dependency graph of the multi-target design model showing the dependencies between the variables $X0, \ldots, X34$ of this model **(Right)** A tree decomposition of this graph that puts the variables into bags (only variable indices shown). The directed edges are labeled with the indices of variables introduced by the child bag. This decomposition has a tree width of two, since it's largest bags contain three (tree width plus one) variables

Due to the CTE scheme, the computation is based on a tree decomposition of the network of the dependencies induced by the constraints and functions in the model (aka *dependency graph*); *see* Fig. 3. The concept of tree decomposition allows us to recursively compute partial partition functions for the subtrees of the tree decomposition by processing the variables in its bags in bottom-up order. Moreover, this computation can be performed efficiently due to dynamic programming (which tabulates partial results that would otherwise be re-computed redundantly).

After all partition functions are computed, each sample is generated in a backtrace running from the root to the leaves. In this way, whenever a new variable is introduced in the depth-first/top-down traversal of the tree decomposition, its value can be chosen with the correct probability to generate the desired multi-dimensional Boltzmann distribution.

This solving strategy explains why Infrared can sample from one network faster than from the other. Tree-like networks of dependencies are processed quickly, while complex cyclic dependencies require tree decompositions with more variables per bag, since valid tree decompositions must satisfy certain conditions w.r.t. the dependencies in the network (which in turn guarantee the correctness of the dynamic programming evaluation).

Finally, since the computation requires to enumerate all possible sub-assignments in each bag, the computation time is exponential in the maximum number of variables per bag—this complexity is commonly described in terms of tree width, which is defined as this number minus 1.

Targeting Specific Properties

For targeting very specific properties like certain GC content and energies of the target structures, Infrared utilizes the just described sampling engine to iteratively sample from a multi-dimensional Boltzmann distribution, evaluate the generated distribution w.r.t. the target properties of the single targeted features and update their weights. By suitable updates of the weights, it is possible to shift the distribution towards the targeted feature values and increase the probability to satisfy these targets (within the given tolerance). During this entire learning procedure, Infrared returns samples inside of the tolerance range and rejects all others. Thus, Infrared implements a variant of multi-dimensional Boltzmann sampling [2].

As a consequence of this entire mechanism, the sampling efficiency in Infrared is a result of the complexity of the constraint network as well as the (in)dependence of the targeted features and the demanded tolerances. For practical applications of Infrared, it is thus generally advantageous to be aware of the properties of the solving strategy and the resulting dependencies between these factors. This is especially important, since the framework allows modeling extremely hard problems, while (as we demonstrate) it is useful for a wide range of applications in practice. Its specific properties make the system attractive for a variety of complex design applications but as well intrinsically influence its applicability.

3.7 Negative Design by Direct Sampling

Good RNA designs typically must satisfy certain constraints and show high (or specific) affinity to the target structures, but as well must avoid high affinity to all non-target structures. Objectives of the latter type are called negative design criteria (whereas the former are called positive design criteria). Intuitively, negative design seems harder than positive design, because it aims at disfavoring exponentially many structures. Indeed, as we discussed before, negative design was shown to be NP hard in relevant settings.

For our negative design examples, we make use of the Turner energy model and other functionality provided by the Vienna RNA package, which we interface via its Python module `RNA`. We import the module by

```
import RNA
```

A classic (single-target) negative design objective requires the target structure to have minimum free energy among all structures of the designed RNA, which can be tested using the Vienna RNA package:

```
def is_mfe_design(sequence, target):
    fc = RNA.fold_compound(sequence)
    return fc.eval_structure(target) == fc.mfe()[1]
```

In many cases, sampling (targeting only positive design criteria) can be sufficient to satisfy negative design criteria. For example, we can easily find designs for any of our three target structures by "generate-and-test":

```
sampler = ir.Sampler(single_target_design_model(target))
sampler.set_target(0.7 * n, 0.1 * n, 'gc')
for i in range(50):
    seq = rna.ass_to_seq(sampler.targeted_sample())
    if is_mfe_design(seq, target):
        print(f"{i} {seq}")
```

where `single_target_design_model(target)` returns a model for the given target structure with "gc" feature and a bias to good base pair energy as we have developed it before (*see* Notes). Based on this code, we find 10–20 solutions from generating 50 (biased) samples for each of our example target structures (in less than 0.1 second on a current notebook).

A similar approach can still be sufficient to optimize other negative criteria for single-target design like the frequency of the target structure (*see* Notes for function `target_frequency`). The following code regularly finds designs with $>80\%$ target frequency for our example targets in comparable run-time (roughly, twice as long).

```
sampler = ir.Sampler(single_target_design_model(target))
sampler.set_target(0.7 * n, 0.1 * n, 'gc')
best = 0
for i in range(100):
    seq = rna.ass_to_seq(sampler.targeted_sample())
    freq = target_frequency(seq, target)
    if freq > best:
        best = freq
        print(f"{i} {seq} {freq:.6f}")
```

Note that in both cases, we control the GC content of our designs. This is just one example of how the core idea of finding designs by sampling can be extended by further targets or constraints. (Adding prior knowledge as IUPAC string would be another.) Generally, this approach will be most successful for problems with a medium-sized solution space. This would also apply to many design problems, where only parts of the RNA should be redesigned, while others are kept constant.

3.8 Larger Single-Target Designs by Constraint-Generation

For larger and harder single-target design instances, we suggested a constraint-generation strategy in RNAPOND [17]. Here, we demonstrate a slightly stripped down version of this approach, which requires only a few lines of code using Infrared.

The method starts with the attempt to solve negative design by sampling (=positive design) as in the subsection before. When we fold the suggested designs, it can be observed that some base pairs, that don't occur in the target, occur more frequently than others. This motivates us to identify these most frequent disruptive base pairs and forbid them in the next round of sampling. This strategy is iterated until a design (according to the mfe criterion) is discovered. In the code presented here, we decide to terminate the design attempt, when the problem becomes inconsistent or a certain complexity of the constraint model is exceeded.

```
def cg_design_iteration():
    model = single_target_design_model(target)
    model.add_constraints(rna.NotBPComp(i, j) for (i, j) in dbps)
    sampler = ir.Sampler(model, lazy=True)
    if sampler.treewidth() > 10 or not sampler.is_consistent():
        return "Not found"
    ctr = Counter()
    found, sol = False, None
    for i in range(100):
        seq = rna.ass_to_seq(sampler.targeted_sample())
        fc = RNA.fold_compound(seq)
        mfe, mfe_e = fc.mfe()
        if fc.eval_structure(target) == mfe_e:
            sol = seq
        ctr.update(rna.parse(mfe))
    ndbps = [x[0] for x in ctr.most_common() if x[0] not in bps]
    dbps.extend(ndbps[:2])
    return sol
dbps, seq = [], None
while seq is None:
    seq = cg_design_iteration()
print(seq)
```

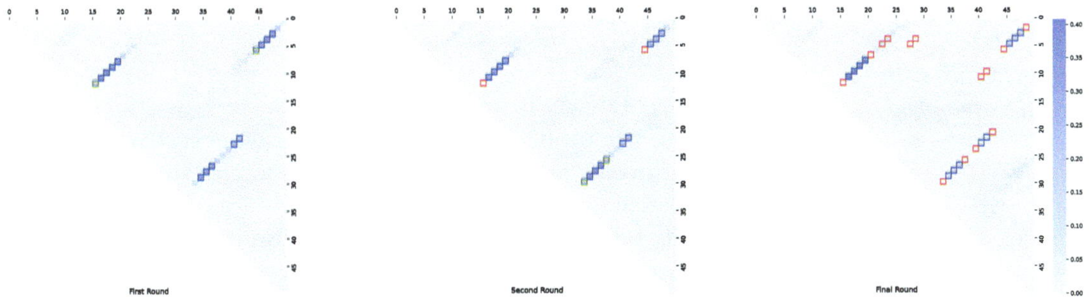

Fig. 4 Base pair frequencies and disruptive base pairs in the first, second, and final iterations of a typical run of our RNAPOND-like constraint generation design strategy targeting `"..(((..((((.....)))).((...(((.....)))...))...)))..".` In our triangular matrix visualization, each point refers to one potential base pair (i, j); base pair frequencies and the generated sample are color coded and the target base pairs are highlighted by blue squares. By adding disruptive base pairs in each iteration (red squares), the frequencies of the base pairs are shifted towards the target ones

We tested the code to find optimal "mfe" designs for instances from the Eterna100 benchmark set [11]. Note that this code does not yet match the performance of the more refined tool RNAPOND, but it is still quite effective. For example, an optimal design for instance 39 ("Adenine") is typically found after seven to nine iterations.

Figure 4 provides intuition of the effect of adding disruptive base pair constraints in our strategy. Following the code above, RNA sequences are first sampled without further limitation than the target structure. Positions corresponding to the target structure have a greater chance to form a base pair according to the MFE criteria. However, as shown in the first plot in Fig. 4, some non-target base pairs are observed frequently in the MFEs of sampled sequences. As seen in the second plot, the two most frequent non-target base pairs in the first round are forbidden (via constraint `NotBPComp`) in the second round of sampling. Then, additional constraints are imposed on two novel disruptive base pairs discovered in the following sampling round. After *seven* rounds of sampling with 14 disruptive base pairs, the successful design

`CCGACAAGGGCCAUGCGCCCGGAAACCAGUGCCUCUGAAGUCAAAGUCUG`

is found for the target structure

`..(((..((((.....)))).((...(((.....)))...))...)))...`

3.9 Negative Design by STOCHASTIC OPTIMIZATION with Partial Resampling

Infrared supports stochastic optimization strategies to optimize negative objectives for single or multi-target design. We are going to present a Metropolis-Hastings optimization strategy for optimizing the solutions from a multi-design model. It starts with a sample from the model and evaluates it by the objective function. Then, it iteratively picks a connected component of the dependency graph, constructs a model for resampling of the variables in this component, and generates a sample. Based on its evaluation by the objective function, the sample is either accepted or rejected (i.e., applying the Metropolis-Hastings criterion).

Infrared directly supports resampling of a subset of variables as a method of `ir.Sampler`. This method `resample` takes two arguments: the sub-set of variables that should be resampled and an assignment that determines all other variables. This lets us express the sketched optimization algorithm as a generic function `mc_optimize` that maximizes an objective function over the objects described by a model. The additional arguments specify the number of optimization steps and a temperature (controlling the Metropolis-Hastings criterion). Optionally, it accepts a start assignment. Note that the function requires the standard modules `random` and `math`.

```
def mc_optimize(model, objective, steps, temp, start=None):
    sampler = ir.Sampler(model)
    cur = sampler.sample() if start is None else start
    curval = objective(cur)
    best, bestval = cur, curval

    ccs = model.connected_components()
    weights = [1/len(cc) for cc in ccs]

    for i in range(steps):
        cc = random.choices(ccs,weights)[0]
        new = sampler.resample(cc, cur)
        newval = objective(new)
        if (newval >= curval
            or random.random() <= math.exp((newval-curval)/temp)):
            cur, curval = new, newval
            if curval > bestval:
                best, bestval = cur, curval

    return (best, bestval)
```

In our example, we want to minimize the multi-defect [8], which is a weighted sum of the distance of the target energies to the ensemble energy and the energy distance between the targets. Minimizing this function thus means to increase the probability of the targets and balance the target energies as much as possible. The function `multi_defect` (*see* Notes) calculates this score, given a sequence, the targets, and the weight.

Moreover, we need to describe multi-target design as an Infrared model; we already developed such a model before. The constraints of this model guarantee that the sampled and resampled sequences are compatible with the target structures. Moreover, we can control the distribution of GC content and energies due to functions and feature weights. We used this here to aim for lower energy and slightly balance the GC content.

```
n = len(targets[0])
model = ir.Model(n, 4)
model.add_functions([rna.GCCont(i) for i in range(n)], 'gc')
for target in targets:
    ss = rna.parse(target)
    model.add_constraints(rna.BPComp(i, j) for (i, j) in ss)
    model.add_functions([rna.BPEnergy(i, j, (i-1, j+1) not in ss)
        for (i, j) in ss], 'energy')
model.set_feature_weight(-0.8, 'energy')
model.set_feature_weight(-0.3, 'gc')
```

Finally, we put everything together in a call of `mc_optimize`. The lambda expression tailors the required objective function on assignments. Here, in order to minimize the multi-defect, we need to negate it.

```
best, best_val = mc_optimize(model,
    lambda x: - multi_defect(rna.ass_to_seq(x),targets,1),
    1000, 0.01)

print(rna.ass_to_seq(best), - best_val)
```

This code calculates and prints the best multi-target design and its multi-defect after 1000 iterations of the optimization procedure (at temperature 0.01). In practice, we will often restart such procedures to obtain better solutions and/or a diverse set of good solutions. In our tests, one such optimization run took a moderate run-time of a few seconds on current notebook hardware, while it can already find very good designs for the targets.

Figure 5 shows the best multi-defects in 48 runs after up to 6400 iterations. The best sequence CCCUGUGCUCCAUGGGCCCCC-GUCAGGGGACGGGG that was found in these 48 runs of optimization has a multi-defect of 1.31. For this sequence, the target structures have respective energies −16.9, −16.5, and −16.9 kcal/mol; two of the targets have minimum free energy. This small experiment yields insights into the effectiveness and convergence of the optimization procedure. For applications, it seems to suggest an optimization strategy combining restarts and moderately long runs.

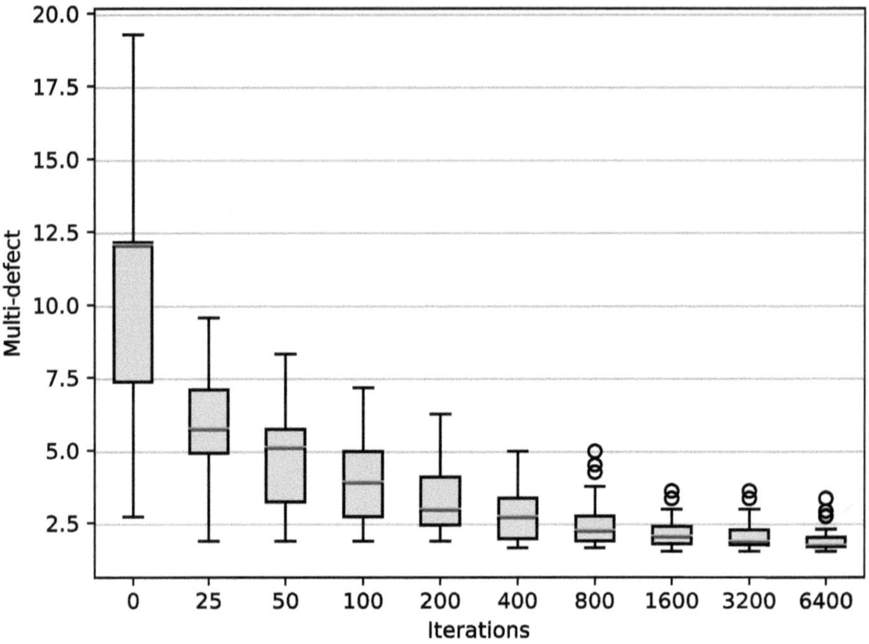

Fig. 5 Distributions of best multi-defects from 48 runs of `multi_design_optimization` without stochastic optimization (**step 1**) and after up to 6400 optimization steps at temperature 0.01. In the box plots, the boxes extend from quartile to quartile; the medians are shown in blue; and the whiskers reach 1.5 times beyond the quartile. Remaining data points are shown as circles

Remarkably, there are even solutions where all three targets have minimum free energy. For the sequence `CCCCUUGCCUCAAGGGCCCUCUUCAGAGGAAGGGG`, which was discovered by the same strategy, all three target structures have a free energy of −15.40 kcal/mol (at a multi-defect of 1.21). Even the close-to-optimal structures are all similar to the targets as can be seen from the output of `RNAsubopt` of the Vienna RNA package, which enumerates all structures within 1 kcal/mol of the minimum free energy.

```
$ RNAsubopt -s <<<CCCCUUGCCUCAAGGGCCCUCUUCAGAGGAAGGGG
CCCCUUGCCUCAAGGGCCCUCUUCAGAGGAAGGGG -15.40   1.00
((((((((((...))))((((....)))))))))) -15.40
((((((.((((((((....))))..)))))))))) -15.40
.((((((...)))))).(((((((....))))))) -15.40
((((((.(((((((......)))..)))))))))) -15.10
((((((.((((.(((....)))...)))))))))) -14.80
((((((.((((.(((.....)))..)))))))))) -14.80
((((((.((((..(((....)))..)))))))))) -14.80
((((((.((((.((((...))))..)))))))))) -14.60
(((((((((.....)))((((....)))))))))) -14.50
((((((.((((.((......))...)))))))))) -14.50
.((((((...))))))(((.((((....))))))) -14.50
```

3.10 A Real World Example: Design of a Tandem-Riboswitch

Our final example is inspired by collaborative work [6] with the goal to develop tandem constructs of two riboswitches for applications in bio-technology. In such studies, computational design is essential to generate candidate design that can be further scrutinized in lab experiments. We showcase the use of Infrared in this setting.

The single riboswitches control transcription by forming a terminator in their ground state, but they can as well bind a metabolite in an alternative aptamer structure. Upon binding, the terminator is destabilized and transcription is turned on (Fig. 6). By combining riboswitches that react to two different metabolites (here Theophylline and Tetracycline) in tandem, we aim at the design of an AND-riboswitch, which ramps up gene expression (only) in the presence of both metabolites (Fig. 7)

We use Infrared to suggest designs that connect two known aptamer constructs by a spacer region. Such design candidates could then be evaluated by biochemistry experts and tested in wet lab experiments (as e.g., done in [6]).

In preparation, we define the four structures of Fig. 6 as dot-bracket strings. Moreover, we derive sequence strings from experimentally verified functional riboswitches, but replace some nucleotides by "N" to leave additional freedom in the design (*see* definition of `seqTheo`, `aptTheo`, `termTheo`, `seqTet`, `aptTet`, `termTet` in Notes). These allow to compose the structural targets and sequence strings for the tandem construct, including a 30nt spacer (*see* Notes).

As goals for the computational design, we optimize the stability of the terminator structures (in absolute terms as well as in comparison to the structure ensemble), while keeping certain

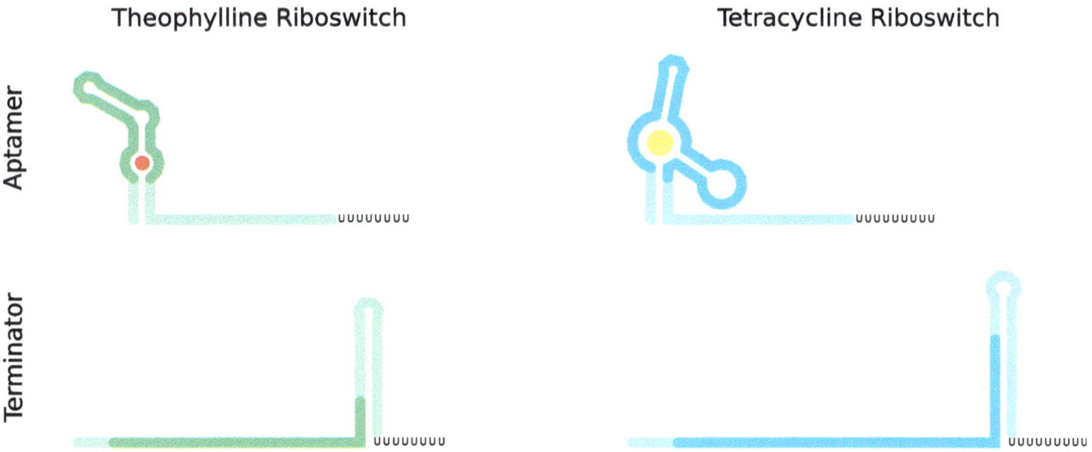

Fig. 6 States of a Theophylline (Theo) and a Tetracycline (Tet) riboswitch. For both riboswitches, we show the configuration that can bind the respective aptamer (top) and the competing terminator configuration, which terminates transcription (bottom). The regions with dark color represent functional sequences. On the other hand, the light ones are the sequences that are free for design

Fig. 7 States of an AND-riboswitch in the absence of Theo and Tet (top), in the presence of Theo and the absence of Tet (middle), and, in the presence of Theo and Tet (bottom). In our design example, we put Theophylline and Tetracycline riboswitches in tandem (Theo-Tet), connected by a freely designable 30nt space region (gray)

probabilities for the aptamer structure. Moreover, we want to avoid that the spacer region forms any stable structures or interferes with the structures of the riboswitch components. How to express such objectives as a function `rstd_objective` is described in detail in Notes Section. We make use of free energy differences between the free RNA structure ensemble (free ensemble energy) and constrained ensembles, which are constraints to form either aptamer or terminator structure, or keep the space unpaired.

Since we face a combination of positive and negative design goals under additional constraints, we are going to use stochastic optimization with resampling as we developed before for negative design. Specifically, we are going to use the built-in function `ir.mc_optimize`. Like the generic optimization function that we introduced before, it requires the objective function and a model describing tandem riboswitch designs. We develop this model in Notes Section.

Finally, we run the optimization (after adapting the objective function from `rstd_objective`) for a number of steps at and some temperature:

```
objective = lambda x: -rstd_objective(rna.ass_to_seq(x))
best, best_val = ir.mc_optimize(model, objective,
                                steps = 500, temp = 0.03)
print(rna.ass_to_seq(best), -best_val)
```

To find better and/or several design candidates, the optimization procedure can be repeated (or even run in parallel). For example, we generate designs like

```
CCGUGAUACCAGCAUCGUCUUGAUGCCCUUGGCAGCGCGGCCCGACAUCGGGCCGUGUUGUUUUUUUU ⌋
↪   CCCACACACAAUAAAGAUAAACCUACCCGGCCGUUAAAACAUACCAGAGAAAUCUGGAGAGGUGAA ⌋
↪   GAAUACGACCACCUAGCGGCUACCACCGUUAGGUGGUCGUUUUUUUUU
```

in several minutes—after 12 parallel runs. For this sequence, both terminators are very stable (energy differences of 0.14 and 0.0 kcal/mol), the stability of the aptamer structures is close to the target values (differences below 0.03 kcal/mol), and the designed spacer region behaves structurally almost neutral (low unfolding energy of 2.11 kcal/mol).

As demonstrated, Infrared appears technically well equipped to express various design constraints and target complex design objectives. Especially due to its declarative Python interface, it moreover supports to apply flexible optimization strategies that make use of its capable design sampling engine. It turns out that this allows finding good solutions to practical design tasks using only "a few lines" of code (which can still be discussed and printed in a book chapter). This allows practitioners to focus on the remaining, "true" challenges of computationally supporting the design of functional RNA molecules, as they undoubtedly exist in the realistic modeling of the design problem and the evaluation of promising design candidates that could be scrutinized in subsequent lab experiments.

4 Notes

4.1 A Model for Single Target Design

The function `single_target_design_model` accepts a target structure (as dot bracket string) and returns an Infrared model that describes designs for this target. Designs must satisfy base pair complementarity and are evaluated by GC content and base pair energy. Finally, the energy feature weight is set to prefer low energy.

```python
def single_target_design_model(target):
    n, bps = len(target), rna.parse(target)
    model = ir.Model(n, 4)
    model.add_constraints(rna.BPComp(i, j) for (i, j) in bps)
    model.add_functions([rna.GCCont(i) for i in range(n)], 'gc')
    model.add_functions([rna.BPEnergy(i, j, (i-1, j+1) not in bps)
        for (i, j) in bps], 'energy')
    model.set_feature_weight(-1.5, 'energy')
    return model
```

4.2 Target Frequency in Ensemble

The frequency of the target structure in the ensemble of the designed sequence is a good example of a negative design criterion.

```python
def target_frequency(sequence, target):
    fc = RNA.fold_compound(sequence)
    fc.pf()
    return fc.pr_structure(target)
```

4.3 Multi-Defect

The function `multi_defect` computes the multi-defect of a sequence for a list of target structures (in dot bracket format) with the help of the Vienna RNA library. The parameter `xi` controls the weight given to the difference of target energies over the difference of target energies to the ensemble energy.

```python
def multi_defect(sequence, targets, xi=1):
    k = len(targets)
    fc = RNA.fold_compound(sequence)
    ee = fc.pf()[1]
    eos = [fc.eval_structure(target) for target in targets]
    diff_ee = sum(1/k * (eos[i] - ee) for i in range(k))
    diff_targets = sum(2/(k*(k-1)) * abs(eos[i]-eos[j])
        for i in range(k) for j in range(k) if i<j)
    return diff_ee + xi * diff_targets
```

4.4 Design of a Tandem-Riboswitch

Defining the design task starts with a definition of the two components of the tandem construct.

```
seqTheo  = "NNNNGAUACCAGCAUCGUCUUGAUGCCCUUGGCAGCNNNNNNNNNNNNNN"\
           "NNNNNNNNNNUUUUUUUU"
aptTheo  = "(((((...((((((((.....)))))...)))...))))).........."\
           ".................."
termTheo = "...............................(((((((((((((....))"\
           ")))))))))))......."


seqTet   = "NNNNNAAAACAUACCAGAGAAAUCUGGAGAGGUGAAGAAUACGACCACCU"\
           "ANNNNNNNNNNNNNNNNNNNNNNNNUUUUUUUUU"
termTet  = "........................................(((((((((("\
           "(((((......)))))))))))))))........"
aptTet   = "((((((.......(((((....)))))...((((...........)))))"\
           ")))))............................."
```

These are concatenated while adding a spacer region, which are used to derive the model constraints.

```
spacerLen = 30
aptamers    = aptTheo  + "."*spacerLen + aptTet
terminators = termTheo + "."*spacerLen + termTet
sequence    = seqTheo  + "N"*spacerLen + seqTet
```

To express the objective function, we define relevant structural variants of the entire tandem construct.

```
n = len(aptTheo) + spacerLen + len(aptTet)
variants = dict(
    empty = '.'*n,
    aptTheo = aptTheo + '.'*(n-len(aptTheo)),
    aptTet = '.'*(n-len(aptTet)) + aptTet,
    termTheo = termTheo + '.'*(n-len(aptTheo)),
    termTet = '.'*(n-len(aptTet)) + termTet,
    spacer = '.'*len(aptTheo) + 'x'*spacerLen + '.'*len(aptTet)
)
```

The riboswitch tandem design objective function consists of three terms:

- The stability of the two terminator structures, measured as free energy difference between the energies of the unconstrained ensemble (empty constraint) and the ensembles that are constrained by the respective terminator structure. For each terminator, this difference has a natural interpretation as the required (additional) energy to form this terminator in the equilibrium state. Minimizing this energy targets maximal stability of the terminators.

- The difference between certain stability targets and analogous energy differences for the aptamer structures (7.0 kcal/mol for the Theo RS; 10.0 kcal/mol for the Tet RS). This targets specific unfolding energies for the aptamers. Minimizing the difference targets specific aptamer stabilities.
- The required energy to break any structures involving the spacer region. Minimizing this energy avoids interference of the spacer with other structure as well as stable structure within the spacer region.

```
def constrained_efe(sequence,c):
    fc = RNA.fold_compound(sequence)
    fc.hc_add_from_db(c)
    return fc.pf()[1]

def rstd_objective(sequence):
    efe = {k:constrained_efe(sequence,variants[k])
           for k in variants}

    term_stability = efe['termTheo'] + efe['termTet'] \
        - 2*efe['empty']
    apt_target = abs(efe['aptTheo']-efe['empty']-7) \
        + abs(efe['aptTet']-efe['empty']-10)
    spacer_unfolding = efe['spacer']-efe['empty']

    return term_stability + apt_target + spacer_unfolding
```

The model sets up the constraints due to the sequence string and the base pairs in the aptamer and terminator structures. Moreover, it sets weights for the features of base pair energies and GC content, which can guide the generation of start structures and the resampling in stochastic optimization.

```
rstd_targets = [aptamers, terminators]
n = len(rstd_targets[0])
model = ir.Model(n, 4)

for i, x in enumerate(sequence):
    model.add_constraints(ir.ValueIn(i, rna.iupacvalues(x)))

model.add_functions([rna.GCCont(i) for i in range(n)], 'gc')

for k,target in enumerate(rstd_targets):
    ss = rna.parse(target)
    model.add_constraints(rna.BPComp(i, j) for (i, j) in ss)
    model.add_functions([rna.BPEnergy(i, j, (i-1, j+1) not in ss)
        for (i, j) in ss], f'energy{k}')
model.set_feature_weight(-0.6, 'energy0')
model.set_feature_weight(-1, 'energy1')
model.set_feature_weight(-0.3, 'gc')
```

References

1. Andronescu M, Condon A, Hoos HH, Mathews DH, Murphy KP (2007) Efficient parameter estimation for RNA secondary structure prediction. Bioinformatics 23(13):i19–i28
2. Bodini O, Ponty Y (2010) Multi-dimensional Boltzmann sampling of languages. In: Discrete Mathematics and Theoretical Computer Science, DMTCS Proceedings vol. AM, 21st International Meeting on Probabilistic, Combinatorial, and Asymptotic Methods in the Analysis of Algorithms (AofA'10), pp 49–64
3. Bonnet É, Rzażewski P, Sikora F (2020) Designing RNA secondary structures is hard. J Comput Biol 27(3):302–316
4. Darty K, Denise A, Ponty Y (2009) VARNA: interactive drawing and editing of the RNA secondary structure. Bioinformatics 25(15): 1974–1975
5. Dechter R (2006) Tractable structures for constraint satisfaction problems. In: Handbook of Constraint Programming Elsevier, Amsterdam, pp 209–244
6. Domin G, Findeiß S, Wachsmuth M, Will S, Stadler PF, Mörl M (2017) Applicability of a computational design approach for synthetic riboswitches. Nucleic Acids Res 45(7):4108
7. Hammer S, Günzel C, Mörl M, Findeiß S (2019) Evolving methods for rational de novo design of functional RNA molecules. Methods 161:54 – 63. Development and engineering of artificial RNAs
8. Hammer S, Tschiatschek B, Flamm C, Hofacker IL, Findeiß S (2017) RNAblueprint: flexible multiple target nucleic acid sequence design. Bioinformatics (Oxford, England) 33: 2850–2858
9. Hammer S, Wang W, Will S, Ponty Y (2019) Fixed-parameter tractable sampling for RNA design with multiple target structures. BMC Bioinform 20(1):209
10. Kask K, Dechter R, Larrosa J, Dechter A (2005) Unifying tree decompositions for reasoning in graphical models. Artif Intell 166(1–2):165–193
11. Lee J, Kladwang W, Lee M, Cantu D, Azizyan M, Kim H, Limpaecher A, Gaikwad S, Yoon S, Treuille A, Das R, EteRNA Participants (2014) RNA design rules from a massive open laboratory. Proc Natl Acad Sci U S A 111:2122–2127
12. Lorenz R, Bernhart SH, Höner Zu Siederdissen C, Tafer H, Flamm C, Stadler PF, Hofacker IL (2011) ViennaRNA package 2.0. Algorithms Mol Biol 6:26
13. McCaskill JS (1990) The equilibrium partition function and base pair binding probabilities for RNA secondary structure. Biopolymers: Original Res Biomol 29(6–7):1105–1119
14. Reinharz V, Ponty Y, Waldispühl J (2013) A weighted sampling algorithm for the design of RNA sequences with targeted secondary structure and nucleotide distribution. Bioinformatics 29(13):i308–i315
15. Reuter JS, Mathews DH (2010) RNAstructure: software for RNA secondary structure prediction and analysis. BMC Bioinform 11: 129
16. Turner DH, Mathews DH (2010) NNDB: the nearest neighbor parameter database for predicting stability of nucleic acid secondary structure. Nucleic Acids Res 38(Database issue): D280–D282
17. Yao H-T, Waldispühl J, Ponty Y, Will S (2021) Taming disruptive base pairs to reconcile positive and negative structural design of RNA. In: Research in Computational Molecular Biology—25nd Annual International Conference, RECOMB 2021
18. Zuker M, Sankoff D (1984) RNA secondary structures and their prediction. Bull Math Biol 46(4):591–621

Chapter 13

A Guide to Computational Cotranscriptional Folding Featuring the SRP RNA

Stefan Badelt and Ronny Lorenz

Abstract

Although RNA molecules are synthesized via transcription, little is known about the general impact of cotranscriptional folding in vivo. We present different computational approaches for the simulation of changing structure ensembles during transcription, including interpretations with respect to experimental data from literature. Specifically, we analyze different mutations of the *E. coli* SRP RNA, which has been studied comparatively well in previous literature, yet the details of which specific metastable structures form as well as when they form are still under debate. Here, we combine thermodynamic and kinetic, deterministic, and stochastic models with automated and visual inspection of those systems to derive the most likely scenario of which substructures form at which point during transcription. The simulations do not only provide explanations for present experimental observations but also suggest previously unnoticed conformations that may be verified through future experimental studies.

Key words RNA secondary structure, Cotranscriptional folding, Folding kinetics, SRP

1 Introduction

During RNA transcription, a polymerase adds single nucleotides to the 3' end of the molecule. The rate at which nucleotides are attached varies from 20 to 200 nt s^{-1} (nucleotides per second), but the processes are not necessarily continuous: pausing during transcription can last for multiple seconds [1].

In comparison, the fastest formations of helices can be on the order of microseconds, so-called branch migration processes happen on the order of milliseconds, and large structural rearrangements may take seconds, hours, or much longer. As a consequence, the 5' end of RNA molecules can form secondary structures temporary during transcription, and those structures may either remain stable for some time or fold into another (potentially thermodynamically optimal) secondary structure.

Presumably, cotranscriptional folding can be used as a mechanism to encode multiple functions into a single molecule or simply

Ronny Lorenz (ed.), *RNA Folding: Methods and Protocols*, Methods in Molecular Biology, vol. 2726,
https://doi.org/10.1007/978-1-0716-3519-3_13,

to encode a delay before an RNA molecule switches into its active conformation. However, to which extent nature harnesses the potential of cotranscriptional folding is still largely unknown. It is possible that many RNA molecules must form exactly one thermodynamically stable structure and that sequences (as well as interactions with other cellular components) evolve to minimize the effects of cotranscriptional folding traps.

In the last decades, the potential importance of cotranscriptional folding has been confirmed experimentally, e.g., the work reviewed by Schärfen and Neugebauer [2], but it is still remarkably difficult to pin down the relevant intermediate structures that a given molecule exhibits during transcription.

This chapter demonstrates how RNA folding software can be used to investigate the effects of cotranscriptional folding and assist with the interpretation of experimental data. We describe a computational analysis of cotranscriptional folding at different levels of detail using existing RNA folding software. Specifically, we compare minimum free energy (MFE) folding, stochastic sampling from the equilibrium distribution, stochastic sampling from the cotranscriptional distribution, and a deterministic model of cotranscriptional folding.

2 Materials

2.1 Hardware

To follow our tutorial, command line programs have to be executed on a computer providing a `Unix` terminal. We recommend using a `Linux` or `Mac` operating system, but also the `Windows subsystem for Linux (WSL)` is possible (*see* **Note 1**).

2.2 Software

2.2.1 Scientific Software

The required software is free for educational and scientific use, and the source code can be obtained from the following resources:

1. `ViennaRNA Package` [3]
   http://www.github.com/ViennaRNA/ViennaRNA
2. `Kinfold` [4]
   http://www.tbi.univie.ac.at/RNA/Kinfold
3. `DrTransformer` [5]
   http://www.github.com/ViennaRNA/drtransformer
4. `DrForna` [6]
   http://viennarna.github.io/drforna/

For a smooth installation experience (especially regarding `ViennaRNA Package Python` bindings), we recommend using `bioconda` [7], rather than compilation from source. We show step-by-step instructions in Subheading 3.2. The API reference manual for the `ViennaRNA Package` is available at https://www.tbi.univie.ac.at/RNA/ViennaRNA/doc/html/index.html. It provides a more technical description of individual functions, especially regarding the ViennaRNA Package Python bindings used throughout this tutorial.

2.2.2 Dependencies

The following scripting languages and packages are used in this tutorial:

1. Python>=3.8 – https://www.python.org/
   (a) pandas for data processing [8]
   (b) matplotlib for plotting data [9]
2. R – https://www.r-project.org/ [10]
   (a) optparse for command line argument parsing
   (b) ggplot2 for plotting the data [11]
   (c) viridis for color-blind-friendly palettes [12]
   (d) reshape2 and scales for various internal data formatting [13]
3. git (optional) – https://git-scm.com/
4. conda (optional) – https://conda.io/

2.3 Data and Scripts

The files provided for following this tutorial are:

1. Four different SRP sequences [14, 15]
2. Cotranscriptional SHAPE reactivity data files [15, 16]
3. Scripts for data processing, file format conversion, analysis, and plotting

We provide all required files and scripts in a git repository at https://github.com/ViennaRNA/DrTutorial.

3 Methods

3.1 Notation

The instructions for execution of command line programs or scripts in a Unix terminal are wrapped in boxes and always preceded with ">," a greater sign that is *not* part of the program call:

```
> echo "Hello world"
```

Python code is presented in boxes with line numbers, e.g.:

```
print("Hello world")
```

3.2 Installation

Instead of compiling and installing the required software in Subheading 2.2 from source, the ready-made binary packages can be directly installed via conda. After conda is installed and activated, e.g., see https://conda.io/, the following channels must be set – in this order – to obtain the latest versions of the software:

```
> conda config --add channels defaults
> conda config --add channels bioconda
> conda config --add channels conda-forge
> conda config --set channel_priority strict
```

The ViennaRNA Package (including Kinfold) and DrTransformer are installed using the following commands:

```
> conda install viennarna
> conda install drtransformer
```

3.3 Setup of the Data and Script Repository

All data files and scripts used throughout this chapter are provided in a git repository available at https://github.com/ViennaRNA/DrTutorial. We recommend using git to get a local copy of these files, i.e., by cloning the repository:

```
> git clone https://github.com/ViennaRNA/DrTutorial.git
```

Using git has the main advantage that users can contribute to future versions of this tutorial. Alternatively, the directory can be downloaded and extracted from the corresponding ZIP archive file. Both ways result in a directory DrTutorial/ which contains:

- A folder sequences/ with the SRP input sequences in FASTA format
- A folder SHAPE/ with the cotranscriptional SHAPE data
- A folder scripts/ with all required Python and R scripts
- A folder drconverters/ with wrappers for various (kinetic) RNA folding methods that convert their output into a common file format
- A few additional files with further meta information, such as README.md for an overview and description of the repository contents and pyproject.toml for installation

The scripts can be installed from within the `DrTutorial` directory via `pip`:

```
> pip install .
```

See **Note 2** if installation is not desired or impossible.

3.4 An Overview of Computational Methods

All of the prediction methods below use the same energy model, the `ViennaRNA` implementation of the nearest neighbor parameters [3, 17]. This enables us to focus on kinetic cotranscriptional effects, rather than subtle differences in energy models for structure prediction.

3.4.1 Minimum Free Energy (MFE) Folding

The MFE can be calculated using `RNAfold` from the `ViennaRNA Package`:

```
> RNAfold
```

The MFE is of particular interest as it is the thermodynamically most likely structure in the energy landscape. Beware that multiple MFE structures may exist for a specific molecule (*see* also **Note 3**). The `RNAfold` program also implements the computation of base-pair probabilities p_{ij} utilizing McCaskills algorithm [18]. Specifically, the command line option `RNAfold -p` stores the base-pairing probabilities in the form of PostScript dot-plot files (*see* **Note 6**).

3.4.2 Sampling the Equilibrium Distribution

For stochastic sampling from the thermodynamic equilibrium distribution, we use `RNAsubopt` from the `ViennaRNA Package`:

```
> RNAsubopt --stochBT_en 10000
```

The default mode of `RNAsubopt` is to enumerate all suboptimal structures within a given energy range above the MFE [19]. To obtain a stochastic (Boltzmann) sample from the full equilibrium distribution [20], the option `--stochBT_en 10000` samples 10,000 structures according to their equilibrium probability and reports their structure, energy, and equilibrium probability. *See* **Note 4** for alternative options to obtain stochastic samples.

3.4.3 Sampling the Cotranscriptional Distribution

For stochastic sampling from dynamically changing cotranscriptional distributions, we use the Gillespie-type simulator for nucleic acids `Kinfold`, which assumes that any transition on the secondary structure level must proceed via sequences of "elementary moves," i.e., either the insertion or the deletion of single base pairs [4]:

```
> Kinfold --cut 999 --met --fpt --logML --noShift \
         --glen 1 --grow 4000 --time 468000
```

The options (see `Kinfold --help`) are chosen to ensure compatibility with `RNAfold` predictions and to return a detailed single cotranscriptional folding trajectory using the Metropolis rate model:

$$k_{xy} = \begin{cases} k_0 e^{\frac{-(E_y - E_x)}{kT}} & \text{if } E_x < E_y \\ k_0 & \text{otherwise} \end{cases} \quad (1)$$

where k_0 is a rate parameter which is scaled by the free energy differences between starting structure x and target structure y. While we discuss appropriate default values for k_0 below, `Kinfold` only uses free energy differences to calculate rates (i.e., $k_0 = 1$) and thus expects users to specify the time for nucleotide extensions in internal simulation time units. Here, a nucleotide is attached every 4000 internal time units, and the simulation is set to stop after $117 \cdot 4000$ time units, because the molecule is 117 nt long.

For the analysis in this book chapter, we provide the script `DrKinfold` (from the `drconverters` repository, *see* Subheading 3.3) that automatically aggregates data from multiple `Kinfold` trajectories and saves only selected time points:

```
> DrKinfold -p 20 -n 10 --k0 1e6 --t-ext 0.04 --t-end 0.04
```

The options `-p 20 -n 10` are chosen to aggregate data from 20 independent `Kinfold` processes, each simulating 10 trajectories. Generally, the options shown above for `Kinfold` are the default options used by `DrKinfold`. However, the options adjusting the rate model use the parameter `--k0`, which translates internal simulation time units into seconds (*see* Eq. 1), and two parameters `--t-ext` and `--t-end` to set the simulation times during transcription and at the end of transcription in s/nt. For example, the `DrKinfold` call increased the internal simulation time units per nucleotide by a factor 10, as $10^6 \cdot 0.04 = 40{,}000$, corresponding to the `Kinfold` option `--grow 40000`.

3.4.4 Heuristic Cotranscriptional Folding

Some of the results shown below take many `Kinfold` computation hours. For a faster alternative, we use `DrTransformer`, which returns a heuristic, deterministic cotranscriptional structure ensemble [5]:

```
> DrTransformer --k0 1e6 --t-ext 0.04 --t-end 0.04
```

The rate model-related command line options of `DrTransformer` are the same as previously introduced for `DrKinfold`, although the rate model is an Arrhenius-type model:

$$k_{xy} = k_0 e^{\frac{-\beta_{xy}}{kT}} \quad (2)$$

where β_{xy} is the minimal energy barrier found among elementary paths between starting structure x and target structure y. (In the `Kinfold` landscape, where all transitions are elementary steps, the two models are equivalent; for more details, *see* [5].) Throughout the chapter, we will also use the options `--o-prune` and `--name`, where the former determines how many structures are discarded from the cotranscriptional structure ensemble after the simulation at each nucleotide and the latter sets the name of `DrTransformer` output files.

3.5 A Data Set of Four SRP RNA Variants

A drastic example on how difficult it is to determine cotranscriptional folding paths experimentally is the well-studied 4.5S RNA from *E. coli* which is better known as SRP RNA [21–23]. Due to the importance of its biological function (guiding and rearranging ribosomes to translocation sites together with SRP proteins), it is known for over 20 years that there are competing metastable cotranscriptional intermediate structures. However, how these structures look like is still under debate [14–16, 24], and small sequence-level variations can have a strong effect on cotranscriptional dynamics [14, 15].

The functional conformation (**FC**) is an extended helical structure interspersed with a few internal loops [25]. It is composed of the hairpin-enclosing helix **H2** and four stem motifs **S1**, **S2**, **S3**, and **S4** (*see* Fig. 1a). Wong et al. [24] report that a hairpin structure at the 5' end – **H1a** (*see* Fig. 1c) – is present during two proposed pause sites (U85/U87) but then rearranges into the extended helix.

Recent cotranscriptional SHAPE-Seq experiments [15] probe the secondary structures at each transcription step to provide additional details on the dynamics: After an initial helix **H1** (*see* Fig. 1b) has formed, a large part of the molecule forms only small isolated transient helices, until about 109 nt are transcribed. Then, a central helix motif with large interior loops (H2-S1-S2) forms which is also consistent with the final configuration, and a transient helix **H3** forms which encloses the complementary region to the sequence trapped in H1 (*see* Fig. 1f). Interestingly, the authors conclude that the rearrangement between the H1-H2-S1-S2-H3 structure and the final extended H2-S1-S2-S3-S4 (FC) conformation takes place in just a single transcription step from 109 to 110 nt and suggest a branch migration process via a pseudoknotted intermediate to permit such a fast rearrangement.

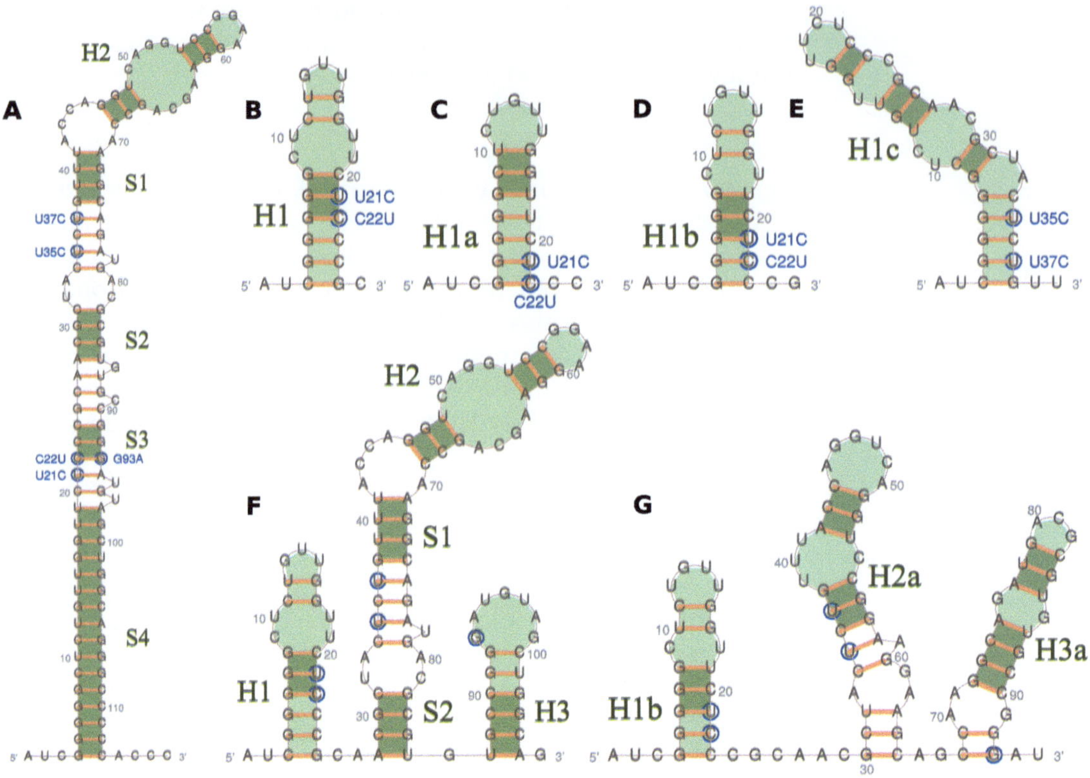

Fig. 1 Potential helices formed during SRP transcription. All structures show annotations for the mutations in SRPt (U21C), SRPr (U21C/C22U/G93A), and SRPf (U35C/U37C) indicated by blue circles around the corresponding base. Motif names refer to adjacent regions with green shade; the dark green shade indicates base-pair constraints that define a specific motif in Subheading 3.10. Helix parts not highlighted by dark green color refer to additional compatible base pairs. (**a**) The functional conformation (FC) of the SRP RNA. It is a rodlike conformation, consisting of motifs H2, S1, S2, S3, and S4. The mutation U21C destabilizes this conformation; C22U/G93A replace a more stable CG pair with a less stable UA pair. (**b**) The transient helix H1 modeled after experimental data by Fukuda et al. [14] and Yu et al. [15] is stabilized and destabilized by mutations U21C and C22U, respectively. (**c**) The transient helix H1a described and experimentally validated by Wong et al. [24] is stabilized and destabilized by mutations U21C and C22U, respectively. (**d**) The transient helix H1b serving as an alternative conformation to helix H1 and H1a during the rearrangements between H1 and H1a. (**e**) The transient helix H1c modeled after experimental data by Fukuda et al. [14] is stabilized by the mutations U35C and U37C. (**f**) An intermediate structure suggested by Fukuda et al. [14] and Yu et al. [15] which is composed of H1, H2, S1, S2, and H3. Note that H2, S1, and S2 are also part of the final functional conformation, while H1 and H3 compete with the formation of S3 and S4. (**g**) A structure composed of transient helices reported in this tutorial: H1b, H2a, and H3a. H2a is destabilized by U35C and stabilized by U37C. H3a has also been reported as eH3 (early H3) in Yu et al. [15]

Yu et al. [15] also provide additional experiments using point mutations. A sequence with the mutation **U21C** both increases the stability of H1 and destabilizes the functional conformation FC (*see* Fig. 1). As hypothesized, the U21C mutation prevents the transition into FC by the end of transcription, even though FC is still the thermodynamically favored conformation. A second mutant,

U21C/C22U/G93A, restores the original stability of H1 while maintaining Watson-Crick complementarity in FC to rescue the cotranscriptional transition into the FC structure.

Recent cotranscriptional optical tweezer experiments [14] identify transcription steps where sudden or continuous folding events take place. The authors independently reported the formation of transient H3 during the transcription of the native SRP sequence, but they also report a transient formation of **H1c** (*see* Fig. 1e) which competes with H1 before H2 forms. Interestingly, the authors also investigated the U21C mutant, which had the same overall effect as described before, and a second mutant **U35C/U37C** that should stabilize the H1c helix formed early in the transcription process. Cell viability plate assays show that the U21C mutation reduces surviving colony counts, while the U35C/U37C mutations appear indistinguishable.

In this tutorial, we investigate to which extent these experimental findings can be reproduced and predicted in silico but also present new results on the cotranscriptional folding process which – we think – are compatible with the data from experiments. For example, our results neither support the pseudoknot hypothesis from [15] nor do they confirm the H1c structure reported by [14]. Instead, our analysis reveals new insights that involve the intermediate motifs **H1a**, **H1b**, **H2a**, and **H3a** (*see* Fig. 1g). We start by defining the following four (partial) SRP RNA sequences taken from Fukuda et al. [14] and Yu et al. [15]:

```
>SRPn # "native" native E.coli SRP sequence
AUCGGGGGCUCUGUUGGUUCUCCCGCAACGCUACUCUGUUUACCAGGUCAGGUCCGGAAG
GAAGCAGCCAAGGCAGAUGACGCGUGUGCCGGGAUGUAGCUGGCAGGGCCCCCACCC
```

```
>SRPt # "trapped" mutant E.coli SRP sequence: U21C
AUCGGGGGCUCUGUUGGUUCCCCCGCAACGCUACUCUGUUUACCAGGUCAGGUCCGGAAG
GAAGCAGCCAAGGCAGAUGACGCGUGUGCCGGGAUGUAGCUGGCAGGGCCCCCACCC
```

```
>SRPr # "rescue" mutant E.coli SRP sequence: U21C, C22U, G93A
AUCGGGGGCUCUGUUGGUUCCUCCGCAACGCUACUCUGUUUACCAGGUCAGGUCCGGAAG
GAAGCAGCCAAGGCAGAUGACGCGUGUGCCGGAAUGUAGCUGGCAGGGCCCCCACCC
```

```
>SRPf # "foldingpath" mutant E.coli SRP sequence: U35C, U37C
AUCGGGGGCUCUGUUGGUUCUCCCGCAACGCUACCCCGUUUACCAGGUCAGGUCCGGAAG
GAAGCAGCCAAGGCAGAUGACGCGUGUGCCGGGAUGUAGCUGGCAGGGCCCCCACCC
```

As described in Subheading 3.3, these sequences are also provided in the `sequences/` directory of the `DrTutorial` git repository. They are given the names `SRPn.fa`, `SRPt.fa`, `SRPr.fa`, and `SRPf.fa`, respectively.

3.6 Cotranscriptional Free Energy Distributions

As a first step, we aim to identify transcription steps at which the cotranscriptional ensemble is "trapped" and thus differs from the equilibrium distribution. We consider only the free energies of structures for this analysis, as differences in free energy distributions are a simple measure that is always the result of a difference in the underlying structure ensemble. Free energies are also directly related to the probability of observing a structure at equilibrium. Specifically, the probability of any structure s is proportional to the Boltzmann factor of its energy $E(s)$, i.e.:

$$p(s) \propto exp(-E(s)/RT) \quad (3)$$

with gas constant R and temperature T. This implies that finding thermodynamically unlikely structures is difficult using stochastic sampling.

We compare the results of the computational methods `RNAfold`, `RNAsubopt`, `Kinfold`, and `DrTransformer` (Subheading 3.4) for each molecule: SRPn, SRPt, SRPr, and SRPf (Subheading 3.5). The scripts described below generate a table with *nine* columns, which correspond to the transcript length, the prediction method, the name of the input sequence, the lower and upper 25% quartiles, the median, the mean, the minimum, and the maximum values of free energy distributions. The table is stored as a single file in comma-separated value (csv) format:

```
length,method,name,Q25,Q75,Qmedian,Qmean,Qmin,Qmax
...
```

Finally, we plot properties of free energy distributions observed via thermodynamic and cotranscriptional folding methods (*see* Fig. 2). A similar figure – also for the SRP RNA – can be found in Yu et al. [15].

3.6.1 Cotranscriptional Distributions

To calculate the free energy distributions during cotranscriptional folding, the program `DrKinfold` is called for each SRP variant:

```
> DK_OPT="-p 400 -n 25 --k0 1e6 --t-ext 0.04 --t-end 0.04"
> DrKinfold ${DK_OPT} --name SRPn_kf < SRPn.fa
> DrKinfold ${DK_OPT} --name SRPt_kf < SRPt.fa
> DrKinfold ${DK_OPT} --name SRPr_kf < SRPr.fa
> DrKinfold ${DK_OPT} --name SRPf_kf < SRPf.fa
```

Here, distributions are calculated from $400 \cdot 25 = 10000$ trajectories to get a broad distribution of min/max values. (We recommend using 100 simulations for testing, as this is sufficient to reproduce most of the important results.) The options `--k0 1e6`, `--t-ext 0.04`, and `--t-end 0.04` adjust the kinetic model.

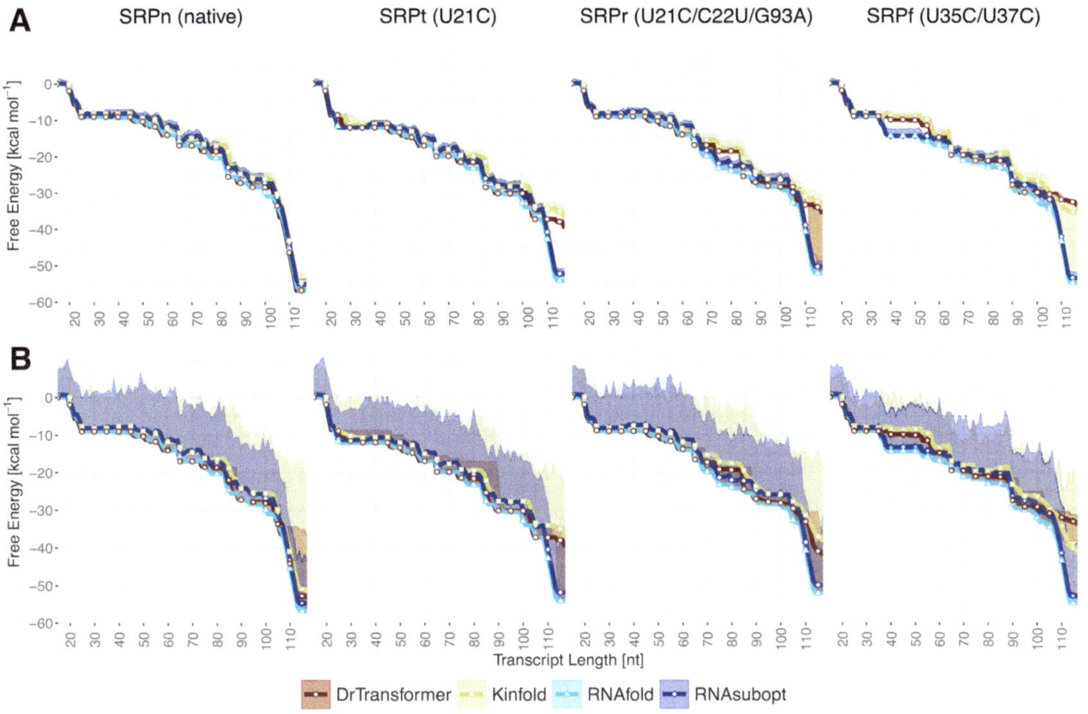

Fig. 2 Free energy distributions of thermodynamic and kinetic folding predictions. The four columns compare SRP-WT, SRP-U21C, SRP-U21C-C22U-G93A, and SRP-U35C-U37C. All plots show the MFE (`RNAfold`), 10, 000 stochastic samples from the equilibrium distribution (`RNAsubopt`), 10, 000 stochastic cotranscriptional trajectories (`Kinfold`) and the energy distribution from a heuristic method (`DrTransformer`). (**a**) Thick solid lines denote the median free energy of the prediction, while semi-opaque bands indicate the first and third quartile, respectively. (**b**) Thick solid lines denote the mean free energy of the prediction, while semi-opaque bands indicate the minimum and maximum values, respectively

Also the program `DrTransformer` is called for each SRP variant:

```
> DT_OPT="--o-prune 1e-4 --k0 1e6 --t-ext 0.04 --t-end 0.04"
> DrTransformer ${DT_OPT} --name SRPn_dt < SRPn.fa
> DrTransformer ${DT_OPT} --name SRPt_dt < SRPt.fa
> DrTransformer ${DT_OPT} --name SRPr_dt < SRPr.fa
> DrTransformer ${DT_OPT} --name SRPf_dt < SRPf.fa
```

Here, `--o-prune 1e-4` is used as a cutoff to discard low occupied secondary structures after each simulation and the remaining options adjust the kinetic model.

Both `DrKinfold` and `DrTransformer` produce `--name.drf` files, which contain a table with the output of simulations. The file suffix `.drf` denotes that these files are suitable input for the visualization software `DrForna`, which we will use in Subheading 3.8. We provide a parser script `drf_parser.py` to extract the distributions of free energies at the end of the simulation for each nucleotide, i.e., before the next nucleotide is appended during transcription. The following commands perform these computations:

```
> drf_parser.py -n SRPn -m Kinfold -o energies.csv energy SRPn_kf.drf
> drf_parser.py -n SRPt -m Kinfold -a energies.csv energy SRPt_kf.drf
> drf_parser.py -n SRPr -m Kinfold -a energies.csv energy SRPr_kf.drf
> drf_parser.py -n SRPf -m Kinfold -a energies.csv energy SRPf_kf.drf
> drf_parser.py -n SRPn -m DrTrafo -a energies.csv energy SRPn_dt.drf
> drf_parser.py -n SRPt -m DrTrafo -a energies.csv energy SRPt_dt.drf
> drf_parser.py -n SRPr -m DrTrafo -a energies.csv energy SRPr_dt.drf
> drf_parser.py -n SRPf -m DrTrafo -a energies.csv energy SRPf_dt.drf
```

where for each of the four input drf files the script is run in *energy distribution mode* (keyword: `energy`) and the options (`-n -m`) set the name and the method fields for the csv file. Note the option `-o energies.csv` to write a new output file which also includes the header information and the option `-a energies.csv` to append the output to an existing file without the header.

3.6.2 Thermodynamic Distributions

To compare the cotranscriptional folding effects with a thermodynamic approach, the equilibrium ensembles have to be calculated for each intermediate length of the growing, nascent RNAs. The procedure is simple:

1. Read the full-length input sequences.
2. Loop over subsequences for all transcript lengths.
3. Predict the MFE for the subsequences.
4. Perform stochastic sampling for the subsequences.

Instead of calling the command line executables `RNAfold` and `RNAsubopt`, we use the `ViennaRNA Package` Python interface (via the pythonimport RNA Python statement) to write a script that implements the full procedure stated above:

```
import RNA
sequence = ("AUCGGGGGCUCUGUUGGUUCUCCCGCAACGCUACUCUGUUUACCAGGUCAGGUCCGGAA"
            "GGAAGCAGCCAAGGCAGAUGACGCGUGUGCCGGGAUGUAGCUGGCAGGGCCCCCACCC")
num_samples = 10
# Initialize the "model details" object to enable unique multibranch loop
# decomposition as required for stochastic backtracking.
md = RNA.md()
md.uniq_ML = 1
# Initialize the "fold compound" object using the full-length sequence.
fc = RNA.fold_compound(sequence, md)
# Compute the MFE structure and the MFE for the full length sequence to
# rescale Boltzmann factors and avoid numeric instabilities.
(ss, mfe) = fc.mfe()
fc.exp_params_rescale(mfe)
# Compute partition function for full sequence to initialize all matrices
# for backtracking of subsequences.
fc.pf()
for i in range(1, len(sequence) + 1):
    seq = sequence[0:i]
    # Compute MFE for subsequence and print results.
    (ss, mfe) = RNA.fold(seq)
    print(f'{mfe:6.2f}')
    # Compute a fixed number of stochastic samples and print their energies.
    energies = []
    for s in fc.pbacktrack5(num_samples, i):
        energies.append(RNA.eval_structure_simple(seq, s))
    print(f'{[f"{e:6.2f}" for e in energies]}')
```

Note that the above code is a simplified version of the code implemented in the script `thermo_predict.py` that is provided in our git repository. The latter provides various additional options and prints results in the correct output format. To compute the thermodynamic energy profiles with our `RNAfold/RNAsubopt` wrapper script, use the following commands:

```
> thermo_predict.py -a energies.csv energy --mfe --samples 10000 SRPn.fa
> thermo_predict.py -a energies.csv energy --mfe --samples 10000 SRPt.fa
> thermo_predict.py -a energies.csv energy --mfe --samples 10000 SRPr.fa
> thermo_predict.py -a energies.csv energy --mfe --samples 10000 SRPf.fa
```

For each of the four SRP input sequences, the script appends (`-a`) data to the file `energies.csv`. The keyword `energy` runs the script in *energy distribution mode* to compute both the MFE (`--mfe`) and the observed distributions from 10,000 stochastic samples (`--samples 10000`).

3.6.3 Interpreting the Results

Having all energy distributions stored in the file `energies.csv` allows us to investigate the differences between thermodynamic and kinetic folding behavior. We plot the data using an `R` script provided in our git repository:

```
> plot_energy_bands.R -i energies.csv -d -s 15
```

Here, we use the `-i energies.csv` option to specify the input file and the options `-s 15` and `-d` to start the plot with transcripts of length 15. Figure 2 shows two plots for each SRP variant: the top row depicting the median energies together with the range between the 25 and 75% quartiles $Q_{[25,75]}$ (Fig. 2a) and the bottom row showing the mean energies and the corresponding minimum and maximum values (Fig. 2b).

Generally, we compare `RNAfold` with `DrTransformer` predictions and `RNAsubopt` with `Kinfold` samples. `DrTransformer` only considers local minima during a simulation, and the MFE returned by `RNAfold` is always the most dominant minimum in the equilibrated ensemble. In contrast, `RNAsubopt` and `Kinfold` both sample structures from the full secondary structure ensemble, which means the mean and median of those samples must be higher than for other methods.

For the sequence "native SRPn," the results of cotranscriptional simulations are mostly comparable with thermodynamic approaches. The medians and the $Q_{[25,75]}$ range (Fig. 2a) are similar for `RNAfold` and `DrTransformer` results, as well as for `RNAsubopt` and `Kinfold` results. Notably, around 80 and 100 nt, the `DrTransformer` simulations suggest that the MFE structure does not dominate the ensemble. This can also be seen when comparing `RNAsubopt` samples with `Kinfold` samples, especially when looking at the mean energies and the min/max range (Fig. 2b). Also, we observe that there is a point around 110 nt length, where the majority of structures is still in an energy range close to the MFE, but cotranscriptional simulations also start observing structures in a much higher energy range, compared to the thermodynamic approaches.

The mutated sequence “trapped SRPt” (mutation U21C) has an altered MFE profile during transcription: the mutation lowers the MFE after about 25 nt have been transcribed, and this effect persists until roughly 105 nt have been transcribed, then the effect is reversed, and the MFE is higher than that of the native SRP molecule. Interestingly, trapped SRPt shows the same differences between thermodynamic and cotranscriptional approaches around 80 and 100 nt that we already observed for the native SRPn molecule, indicating that the mutation has no effect on those cotranscriptional folding events. After 105 nt, the cotranscriptional simulations clearly diverge from thermodynamic predictions, as the ensemble is trapped about 15 $\text{kcal} \cdot \text{mol}^{-1}$ above the MFE, and this effect persists until the full molecule is transcribed. However, even though most samples are trapped, both `Kinfold` and `DrTransformer` observe structures in the range of the MFE.

The mutated sequence “rescue SRPr” (mutations U21C/C22U/G93A) restores the native SRPn MFE profile until about 65 nt have been transcribed. Then, cotranscriptional simulations stay in an energy regime about 5 $\text{kcal} \cdot \text{mol}^{-1}$ above the thermodynamic predictions roughly until nucleotide 85 is transcribed. Eventually, we observe a similar effect as with SRPt: after 105 nt, the majority of cotranscriptional folds is trapped about 15 $\text{kcal} \cdot \text{mol}^{-1}$ above the MFE. In contrast to SRPt, the ensemble distribution at the end of transcription is much more diverse as the MFE is now within $Q_{[25,75]}$ of both cotranscriptional methods. This indicates that some molecules fold into the MFE structure.

The mutated sequence “folding path SRPf” (mutations U35C/U37C) shows strong cotranscriptional folding effects already after 35 nt have been transcribed. The sequence mutations have lowered the MFE and equilibrium distribution between 35 and 65 nt, but the cotranscriptional ensembles do not follow this trend. Comparisons to other molecules are difficult, because neither the thermodynamic nor the cotranscriptional ensembles follow the same patterns as before. Perhaps surprisingly, between 80 and 90 nt, we observe an increase of cotranscriptional ensemble free energies. This may point to a degenerate landscape with many competing structures of similar free energy. Between 90 and 100 nt `DrTransformer` suggests that most of the ensemble is equilibrated, while `Kinfold` ensembles are trapped at a higher energy range. This effect is inverted at 105 nt, when both `Kinfold` and

`DrTransformer` ensembles are trapped in high energy conformations, but `Kinfold` has a much broader distribution that also contains the MFE configuration.

3.7 Cotranscriptional Accessibility Profiles

The previous analysis successfully identified cotranscriptional traps in the folding process, but we have no indication on what structures are actually involved. This next analysis is a first step toward visualizing the different structures that molecules adopt during transcription. Specifically, we plot the probability of being (un)-paired for each nucleotide. This measure can be plotted in one dimension (as opposed to base-pair probabilities which require a two-dimensional representation (*see* **Note 5**)), and it is also useful to relate structure prediction with RNA structure probing methods such as cotranscriptional SHAPE [26], which we will address later in Subheading 3.9. Visually observing how (un)pairedness changes during transcription can reveal structural rearrangements and thus implicitly shows the transitions between dominant secondary structure motifs. The result of this analysis is shown in Fig. 3.

Instead of calculating the probability that nucleotide i forms a base pair with any j, we calculate the opposite: that i does not form any base pair. In literature, the latter is often called **accessibility**, indicating that the nucleotides are available for interactions with the environment. Formally, we can compute the accessibility q_i of nucleotide i as:

$$q_i = 1 - \sum_{j \neq i} p_{ij}. \tag{4}$$

where p_{ij} is the equilibrium probability of a base pair (i, j) between two compatible nucleotides i and j. The equilibrium probability p_{ij} can be computed from the set of structures $S_{ij} = \{s \mid (i, j) \in s\}$ that contain (i, j) and the partition function $Z = \sum_s exp(-E(s)/RT)$, as:

$$p_{ij} = \frac{1}{Z} \sum_{s \in S_{ij}} exp(-E(s)/RT). \tag{5}$$

For kinetic folding simulations, we calculate the relative frequencies f_{ij} of observing base-pair (i, j) among all structures present at a specific time point

$$f_{ij} = \sum_{s \in S_{ij}} O(s). \tag{6}$$

where $O(s)$ is the "occupancy" of a structure – the probability of observing a structure in a kinetic simulation. The occupancy can be derived from a stochastic simulation as $O(s) = \frac{\#s}{n}$ where $\#s$ is the number of times structure s is observed in n samples at a specific time point, or it can be calculated numerically in the deterministic models.

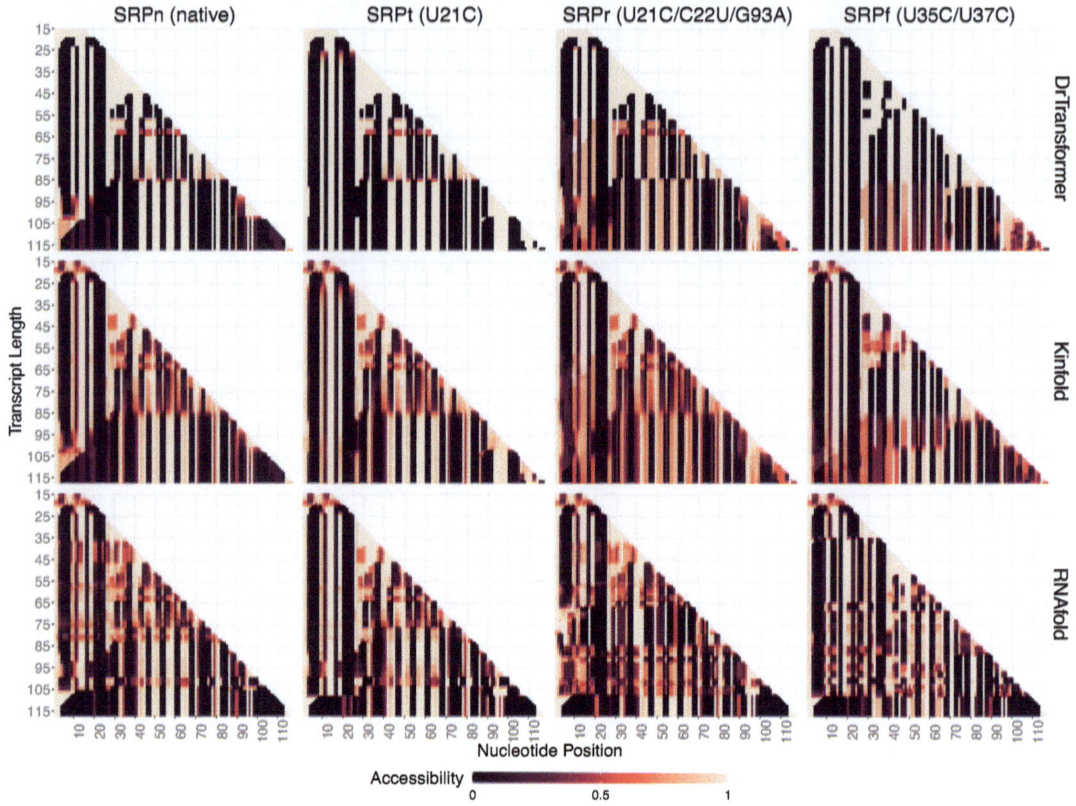

Fig. 3 Cotranscriptional accessibility profiles. The change of (un)pairedness is shown as a function of transcript length for four different SRP sequences and three different computational models

3.7.1 Cotranscriptional Accessibility Profiles

We use the *.drf output files from `Kinfold` and `DrTransformer` as produced in Subheading 3.6.1 to extract the accessibility of each nucleotide during the cotranscriptional folding process. The script `drf_parser.py` in *accessibility calculation mode* (keyword: `accessibility`) derives accessibilities from structure ensembles at the end of the simulation for each nucleotide, i.e., before the next nucleotide is transcribed:

```
> drf_parser.py -n SRPn -m Kinfold -o access.csv accessibility SRPn_kf.drf
> drf_parser.py -n SRPt -m Kinfold -a access.csv accessibility SRPt_kf.drf
> drf_parser.py -n SRPr -m Kinfold -a access.csv accessibility SRPr_kf.drf
> drf_parser.py -n SRPf -m Kinfold -a access.csv accessibility SRPf_kf.drf
> drf_parser.py -n SRPn -m DrTrafo -a access.csv accessibility SRPn_dt.drf
> drf_parser.py -n SRPt -m DrTrafo -a access.csv accessibility SRPt_dt.drf
> drf_parser.py -n SRPr -m DrTrafo -a access.csv accessibility SRPr_dt.drf
> drf_parser.py -n SRPf -m DrTrafo -a access.csv accessibility SRPf_dt.drf
```

Note the similarity to respective command line calls discussed in Subheading 3.6.1. Here, the relevant data is stored in the comma-separated value file `access.csv`.

3.7.2 Thermodynamic Accessibility Profiles

We compare the cotranscriptional with thermodynamic accessibility profiles, where the latter have to be calculated for each intermediate length of the growing, nascent RNAs. As before, the `thermo_predict.py` script uses the `ViennaRNA Package Python` API to perform all necessary computations. For each transcript length l, we compute the base-pairing probabilities p_{ij}^l to derive q_i^l according to Eq. 4. The code block required for this task is as follows:

```
# loop over all nascent transcripts
for l in range(1, len(sequence) + 1):
    # create fold_compound for subsequence
    fc  = RNA.fold_compound(sequence[0:l])
    # compute MFE
    (ss, mfe) = fc.mfe()
    # rescale Boltzmann factors
    fc.exp_params_rescale(mfe)
    # compute partition function and base pair probabilities
    fc.pf()
    # retrieve base pair probabilities
    bpp = fc.bpp()
    # initialize list for accessibilities
    q = [ 0 for i in range(0, l + 1) ]
    # sum-up probabilities to be paired
    for i in range(1, l + 1):
        for j in range(i, l + 1):
            q[i] += bpp[i][j]
            q[j] += bpp[i][j]
    # turn probabilities to be paired into actual accessibilities
    for i in range(1, l + 1):
        q[i] = 1 - q[i]
```

The script `thermo_predict.py` in *accessibility calculation mode* (keyword: `accessibility`) uses this code to calculate thermodynamic accessibility profiles:

```
> thermo_predict.py -a access.csv accessibility SRPn.fa
> thermo_predict.py -a access.csv accessibility SRPt.fa
> thermo_predict.py -a access.csv accessibility SRPr.fa
> thermo_predict.py -a access.csv accessibility SRPf.fa
```

As before, the above commands append the resulting accessibilities q_i^l for all four SRP RNA sequences to the already existing `access.csv` file.

3.7.3 Plotting and Interpreting the Results

We use the data from the generated file `access.csv` together with the script `plot_accessibility.R` from the git repository to observe the changes in accessibility as the RNA molecule is transcribed (*see* Fig. 3):

```
> plot_accessibility.R -i access.csv --start 15 -o access.pdf
```

DrTransformer predicts less variance in accessibility than Kinfold and RNAfold across all heatmaps of Fig. 3. This is expected, as the (un)pairedness for RNAfold and Kinfold is computed from the entire structure ensemble and a large amount of folding trajectories, respectively. In contrast, DrTransformer considers a relatively small number of local minimum structures during a simulation. While the information of what is paired with what is not is explicitly shown, symmetric patterns of pairedness (dark areas) typically correspond to regions paired with each other. Note that it can be helpful to compare the results here with the results from Fig. 2 as the information about energy distributions and structure ensembles together gives a much more complete impression of the underlying folding events than each method by itself.

For the native SRPn RNA sequence, DrTransformer, Kinfold, and RNAfold provide similar results. Following DrTransformer results, the first stable H1 helix forms immediately after the first 25 nt have been transcribed. A second, temporary helical structure forms between 45 and 55 nt; at 58 nt, that structure transitions into a new intermediate which builds up until approximately 84 nt have been transcribed. Finally, at 85 nt, the H2 hairpin of the final functional conformation forms and extends gradually to eventually displace H1. Note that there are competing structures, e.g., at 105 nt, H1 is not dominant, but the shades of red indicate that there is still a fraction of H1 paired in the ensemble. From the results in Fig. 2, we expect differences between cotranscriptional and thermodynamic predictions at 80 and 100 nt. Indeed, those are regions where RNAfold shows different patterns of pairedness. At 80 nt, the H2 helix is already dominant in the thermodynamic ensemble, while cotranscriptional simulations are still trapped in the H2a intermediate structure. At transcript lengths of about 100 nt, RNAfold suggests a sharp temporary transition to an ensemble with an alternative accessibility profile, but this temporary ensemble is not relevant during cotranscriptional simulations. Interestingly, RNAfold also hints toward an alternative structure between 36 and 45 nt which competes with H1, but does not play a role in Kinfold and DrTransformer simulations.

For the trapped SRPt RNA sequence, RNAfold predictions are similar to those of the native SRPn, but one can see that the mutation stabilizes the H1 helix, as less alternative structures are visible. The DrTransformer and Kinfold simulations, however, show that the RNA is cotranscriptionally trapped in structures that contain the H1 helix. While this does not effect the formation of the H2 seed (which happens at the same time as in SRPn), H1 never gets removed, and instead, the final structure contains H1, H2, and a new H3 conformation, which forms after 105 nt have been transcribed. (A faint indication of H3 forming can also be seen in the cotranscriptional simulations of SRPn.)

The accessibilities for the rescue mutant SRPr show that for this RNA many more structural alternatives exist during transcription. `DrTransformer` simulations are similar to SRPn and SRPt sequences up to about 90 nt, except for the faint red background indicating a larger variety of structures in the ensemble. After 90 nt, `DrTransformer` shows a mix of the SRPn and SRPt folding paths, suggesting that H1 is being replaced by some structures that (presumably) form the final functional conformation (H2-S1-S2-S3-S4). Like for SRPn and SRPt, the `Kinfold` data is a noisy version of the `DrTransformer` results. The `RNAfold` data, however, looks different from cotranscriptional simulations. Between lengths 65 and 85 nt equilibrium conformations do not contain the H1 helix, which seems to explain the large difference of free energy distributions observed in Fig. 2. After 85 nt, we see more similarities with cotranscriptional data, and – in combination with results from Fig. 2 – conclude that the cotranscriptional ensemble contains dominant structures from the equilibrium ensemble.

The last column in Fig. 3 shows the accessibilities of the folding path mutant SRPf. We see differences in all three prediction methods `DrTransformer`, `Kinfold`, and `RNAfold`. As originally proposed by Fukuda et al. [14], the mutations are meant to stabilize H1c, a helix competing with H1 even before H2 is forming (H2 forms typically around 65 nt). This is clearly visible in the `RNAfold` predictions at about 38 nt length (cf. SRPn), which suggest that this new helix is dominant until about 110 nt. The mutations also favor the formation of the initial H2 hairpin earlier than in other molecules. However, these results are drastically different from cotranscriptional simulations. `Kinfold` and `DrTransformer` agree that H1 is dominant throughout transcription and that the mutations stabilize a different intermediate structure between 38 and 45 nt length which is compatible with H1. After about 58 nt, the second intermediate structure (H2a) is stabilized. After 85 nt, `DrTransformer` and `Kinfold` predict opposing folding events. According to `DrTransformer`, H2a is dominant until the end of transcription, although we see a faint background that indicates H2 seed formation. In contrast, `Kinfold` suggests a transition into the functional structure (H2-S1-S2-S3-S4) at around 105 nt. Thus, `Kinfold` results suggest that a part of the ensemble reaches the functional structure at transcription end, while `DrTransformer` predicts the ensemble to be trapped in a nonfunctional form.

3.8 Visual Inspection of the Structures

Figures 2 and 3 provide a high-level overview on how ensembles change during transcription. We can use `DrForna` to see whether those changes correspond to structures that have been previously described in literature (*see* Fig. 1) or to novel conformations. `DrForna` is a Javascript application that displays the secondary structures of cotranscriptionally changing ensembles. While we

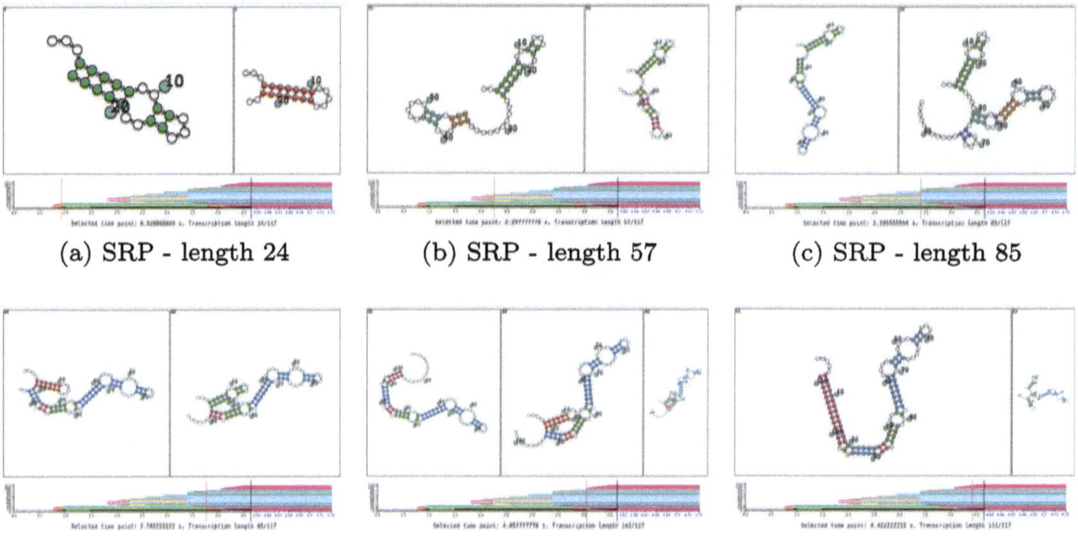

(a) SRP - length 24 (b) SRP - length 57 (c) SRP - length 85

(d) SRP - length 95 (e) SRP - length 102 (f) SRP - length 111

Fig. 4 Visual inspection of selected structure transitions. (**a**) SRP - length 24. (**b**) SRP - length 57. (**c**) SRP - length 85. (**d**) SRP - length 95. (**e**) SRP - length 102. (**f**) SRP - length 111

encourage users to visit the `DrForna` project[1] and load the respective *.drf files generated in this tutorial, we show a few selected transcription steps for the native SRPn and `DrTransformer` results in Fig. 4.

Figure 4a shows SRPn at 24 nt: H1a transitions into H1. Figure 4b shows SRPn at 57 nt: the first (unnamed) intermediate structure transitions into H2a. Figure 4c shows SRPn at 85 nt: H2a transitions into H2-S1-S2. Figure 4d shows SRPn at 95 nt: While the S3 stem is forming, there is another competition between H1 and H1a (cf. Fig. 4a). Figure 4e shows SRPn at 102 nt: The competition from the previous panel continues, but now a third structure is dominating the ensemble where S4 is (partially) formed, and neither H1 nor H1a are present. Figure 4f shows SRPn at 111 nt: The functional conformation H2-S1-S2-S3-S4 is in competition with the H1-H2-S1-S2-H3 structure that was observed by Yu et al. [15].

A key moment revealed by visual analysis is the competition between H1 and H1a, which is not only relevant during the early formation of H1, but also during the formation of the S3 stem of the final functional structure. To learn about the transition landscape between H1 and H1a in more details, we show a screenshot of `Kinfold` results visualized with `DrForna` in Fig. 5. It shows a short-lived intermediate conformation H1b, which may be relevant during the transition from H1 to H1a. This rearrangement

[1] https://viennarna.github.io/drforna/

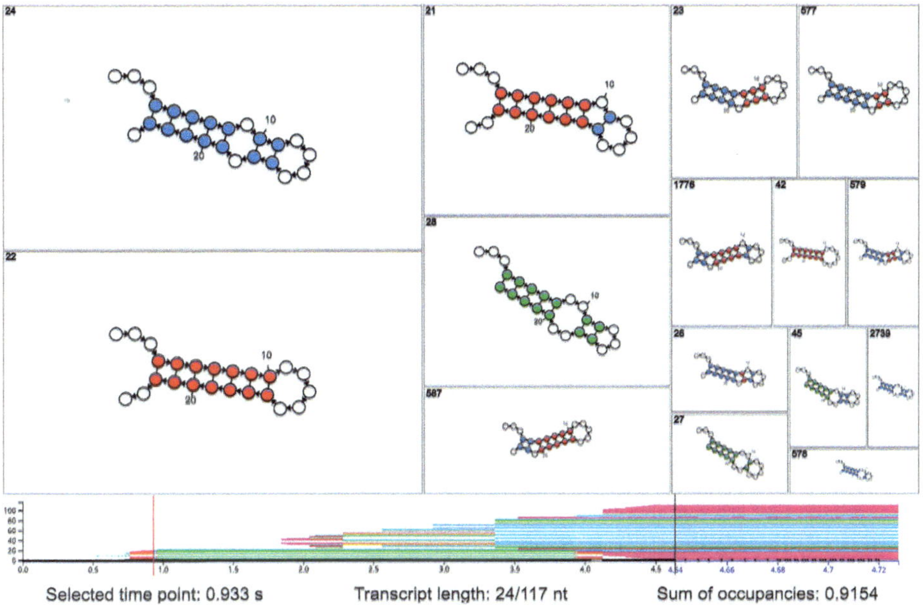

Fig. 5 A `DrForna` snapshot to learn about the transition landscape from H1a (red) to H1 (green) in detail. The analysis reveals a new transient hairpin H1b (blue) which can be generated from both H1a and H1 via a moving bulge loop

provides an alternative hypothesis on how the final H2-S1-S2-S3-S4 conformation can form in such a short time, without postulating a much more complex reconfiguration via a pseudoknotted intermediate (cf. Yu et al. [15]).

3.9 A Comparison with Experimental Data

As the analysis provided in this contribution is inspired from experimental work on the SRP RNA [14–16, 24], we provide a high-level comparison between experimental and simulated data on SRP RNA folding. To that end, we use SHAPE-Seq data provided by Watters et al. [16] and Yu et al. [15] and then modify the cotranscriptional model parameters such that they are closer to the experimental SHAPE-Seq setup. The results are compared visually in Fig. 6. A more detailed analysis of experimental data in combination with cotranscriptional predictions is beyond the scope of this contribution, but we hope that this approach may inspire further work in this direction.

3.9.1 Adjusting the Predictions to the Experiment

SHAPE measures the accessibility of nucleotides by incorporating a reactive agent into flexible regions of the backbone, e.g., paired bases are much less flexible and thus yield low reactivities. It is well known that reactivity profiles from high-throughput experimental structure probing methods in general do not correlate directly with the probability of being unpaired in the secondary structure model [27–30]. In part, this observation can be attributed to the fact that SHAPE probing also reveals tertiary interactions instead of just

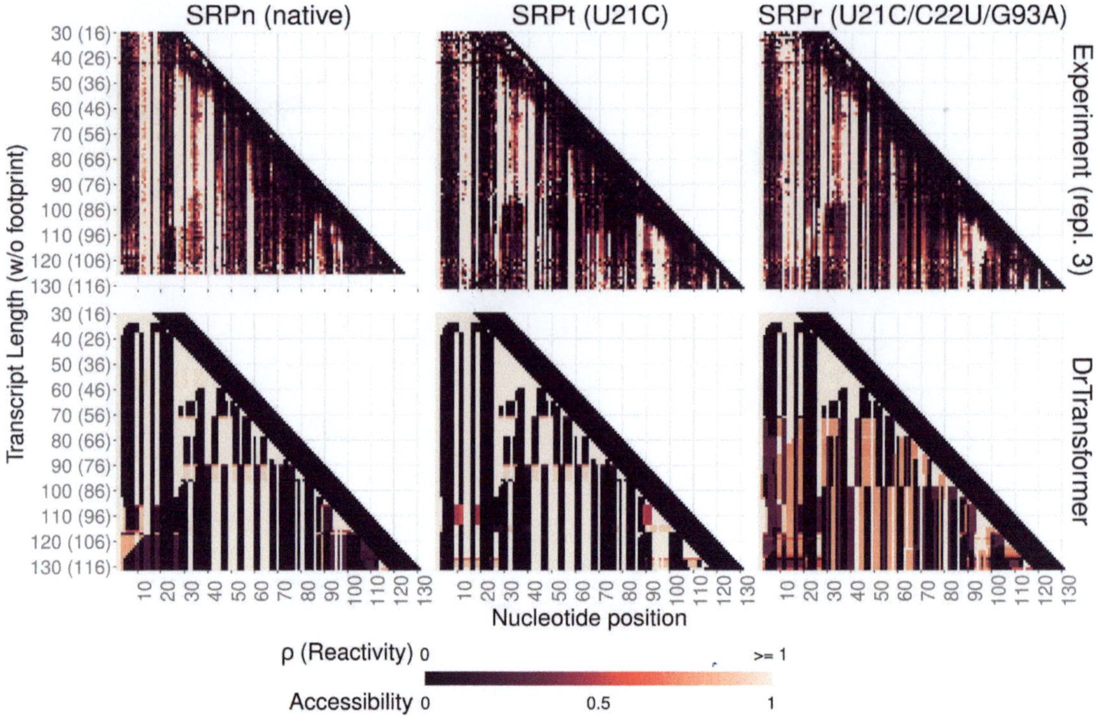

Fig. 6 Comparison against cotranscriptional SHAPE experiments. Shown in the first row of heatmaps are experimentally derived SHAPE reactivities (replicate 3) for the native SRP sequence SRPn [16] and the two mutants SRPt and SRPr [15]. The second row depicts accessibilities q_i^l as computed by cotranscriptional folding simulations with `DrTransformer`. Since the experimental protocol allows the nascent RNA sequences to refold for 30 s before they are probed, the folding simulations have been adopted accordingly. Although there is no linear relationship between SHAPE reactivities ρ and accessibilities q_i^l, both the experiment and predictions are in generally good agreement with respect to the major helices that are formed. The rightmost diagonal lines with low reactivities (resp. accessibilities) denote the RNA polymerase footprint

secondary structure, but also other known or yet unknown biases influence the reactivities, for example, potential sequence dependence of the reagent, unusually flexible bases at the end of helices or inflexible nucleotides forming noncanonical base pairs within seemingly unpaired regions, as well as biases in the relative positioning toward the 3' vs 5' end [31–33].

In the experimental setup of Watters et al. [16] and Yu et al. [15], the SHAPE reagent is added approximately 30 s after transcription stopped at the different transcript lengths. The probing reagent benzoyl cyanide (BzCN) itself can be considered very fast, as the reaction happens on the order of 250 ms (an order of magnitude slower than transcription but negligibly fast compared to the 30 s delay) [34]. Thus, we start separate `DrTransformer` simulations for each transcript length and set the simulation time for the last nucleotide to 30 s. The simulated SHAPE profile is then composed from the end of simulations for each transcript length:

```
> DP_OPT="-n SRPn -m DrTrafo"
> DrTransformer --stop 1 --t-end 30 --name SRPn_t-01 < SRPn.fa
> drf_parser.py ${DP_OPT} -o drtrafo.csv accessibility --lshape 117 SRPn_t-01.drf
> for t in {02..117}; do
>   DrTransformer --stop $t --t-end 30 --name SRPn_t-${t} < SRPn.fa
>   drf_parser.py ${DP_OPT} -a drtrafo.csv accessibility --lshape 117 SRPn_t-${t}.drf
> done
> DP_OPT="-n SRPt -m DrTrafo"
> for t in {01..117}; do
>   DrTransformer --stop $t --t-end 30 --name SRPt_t-${t} < SRPt.fa
>   drf_parser.py ${DP_OPT} -a drtrafo.csv accessibility --lshape 117 SRPt_t-${t}.drf
> done
> DP_OPT="-n SRPr -m DrTrafo"
> for t in {01..117}; do
>   DrTransformer --stop $t --t-end 30 --name SRPr_t-${t} < SRPr.fa
>   drf_parser.py ${DP_OPT} -a drtrafo.csv accessibility --lshape 117 SRPr_t-${t}.drf
> done
```

3.9.2 Preparation of Cotranscriptional SHAPE Data

Cotranscriptional SHAPE reactivity data for SRPn, SRPt, and SRPr is available at the RNA mapping database (https://rmdb.stanford.edu/) as entries SRPECLI_BZCN, SRPU21C_BZCN, and SRP21CR_BZCN, respectively [15, 16]. For convenience, we provide a copy of this data in the SHAPE/ directory of our git repository. We further use the script convert_rdat.py to convert the RDAT-formatted files into the same CSV format as that from Subheading 3.7:

```
> convert_rdat.py -i SRPECLI_BZCN_0003.rdat -o shape.csv -s SRPn -l 131
> convert_rdat.py -i SRPU21C_BZCN_0003.rdat -a shape.csv -s SRPt -l 131
> convert_rdat.py -i SRP21CR_BZCN_0003.rdat -a shape.csv -s SRPr -l 131
```

Note that here we use the experimental replicate 3, but we encourage the reader to repeat this analysis with other provided files or with the equilibrium refolding SHAPE experiments available for SRPn and SRPt. In contrast to data for the mutant SRP sequences, the SRPn data does not cover the full-length transcript when including the polymerase footprint. To compare SHAPE results that contain a different number of transcripts, we provide the command line option -l 131 that sets the expected transcript length to 131 nt (117 nt SRP plus 14 nt polymerase footprint) for each input file. Finally, all extracted data can be stored in the file shape.csv, where any missing data is indicated by NA values.

3.9.3 Plotting and Interpreting the Results

The following commands utilize the plot_accessibility.R script to plot both an accessibility and a reactivity profile for the cotranscriptional SHAPE data and the corresponding DrTransformer simulations:

```
> plot_accessibility.R -i shape.csv -o shape.pdf --SHAPE --offset=-14
> plot_accessibility.R -i drtrafo.csv -o drtrafo.pdf --offset=14 -s 16
```

In the first call of the plotting script, we provide the `--SHAPE` option to indicate that the input data is composed of SHAPE reactivities. It is reasonable to assume that SHAPE reactivities $\rho \geq 1$ originate mostly from unpaired positions. Our script by default already crops the reactivity data accordingly. Note, however, that in the publications of Watters et al. [16] and Yu et al. [15] a threshold of $\rho \geq 4$ has been applied (*see* also **Note** 7). The `--offset` option adjusts the labels on the y-axis of the plot. SHAPE data taken from Watters et al. [16] and Yu et al. [15] includes a 14 nt polymerase footprint which does not participate in cotranscriptional folding, hence the argument of -14. Simulated data on the other hand does not include a polymerase footprint, so we apply a positive offset of 14 nt to only adjust the y-axis labels for the sake of better comparison against the SHAPE data. The option `-s 16` starts the visualization at a transcript length of 16 nt. This option is not required for plotting the SHAPE data since no data for shorter sequences is available. Both resulting plots will be stored in the files `shape.pdf` and `drtrafo.pdf`, respectively.

The visual comparison of SHAPE-Seq reactivities against simulated accessibilities in Fig. 6 shows that experimental reactivities are generally lower. As expected, the last 14 nt of the SHAPE-Seq data are mostly inaccessible due to the polymerase footprint. Surprisingly, also, the presumably unstructured regions close to the $3'$ end of the nascent transcript show consistent patterns of (in)-accessibility (cf. "white triangle" in `DrTransformer` data between 40 and 60 nt transcript length). Also, the experimental reactivity patterns that correspond to H1 formation already start before H1 formation is even possible, i.e., when the nascent RNA is only 16 nt long. These effects may be consequences of partial detachment of the polymerase from the transcripts prior probing, i.e., revealing the otherwise inaccessible footprint.

Helix H1 – between positions 3 and 25 – can be seen throughout all experiments as parallel lines with low reactivity, resp. accessibility. In many cases, this includes the two additional base pairs enclosed at the tip of H1 (*see* also Fig. 1b). At about 115 nt transcript length, both SRPn and SRPr show lower reactivities within the unpaired regions of H1 hairpin, presumably because many of the probed molecules form alternative structures instead, e.g., the functional H2-S1-S2-S3-S4 conformation.

The change in reactivities at about 80–85 nt transcript length in SRPt and SRPr can be related to the predictions with `DrTransformer`, where H2 is formed as a precursor toward the functional structure (*see* also Fig. 1f). Unfortunately, the final refold to the rodlike functional form of the SRP structure (H2-S1-S2-S3-S4) is not clearly visible in the SHAPE data. The accessibility patterns in the H1 region change for all three sequences when roughly 110 nucleotides are available for folding, but it is not visible by the eye which structures are present in the ensemble. A more detailed analysis is beyond scope of this contribution but necessary to minimize the danger of cherry-picking suitably patterns from reactivity profiles. For example, reactivity profiles can also be interpreted by tracking relative changes of reactivity – the so-called upswing/downswing events [35].

3.10 Tracing Cotranscriptional Motif Formation

The final analysis method in this tutorial assumes that we know a set of particular motifs to search for, e.g., revealed by visual inspection as in Fig. 4 or also via experimental data as in Fig. 6. The formation of selected motifs in the ensemble can then be traced over time in the output of `DrKinfold` or `DrTransformer`; the results are shown in Fig. 7.

Generally, we prioritized motifs that are interesting in all of the different SRP variants for visualization in Fig. 7. Specifically, we trace motifs enclosing hairpins, H1a, H1, H2a, H2, and H3, and the stems required for the rodlike conformation, S1, S2, S3, and S4 (*see* Fig. 1). We also show the functional conformation (FC), which we define as the union of motifs H2, S1, S2, S3, and S4. A summary of investigated motifs in dot-bracket notation can be found in the file `sequences/SRP.mot`. Note that this file contains additional motifs not shown here, such as H1b, H1c, H3a, PK, and PK2.

Among the motifs *not* shown in Fig. 7 is H1c, a motif suggested by Fukuda et al. [14] to compete with H1, as well as the motifs PK and PK2 to find evidence for a pseudoknot interaction suggested by Yu et al. [15]. Both motifs PK and PK2 identify structures with six simultaneously unpaired bases. If those bases are unpaired at the same time, that would allow for a three-base-pair pseudoknotted toehold interaction. However, the results are not shown, as we found no compelling evidence for this scenario.

3.10.1 Generating Motif Plots

We repeat the `DrTransformer` and `DrKinfold` analysis from Subheading 3.6.1 with the option `--t-end 60` to follow the formation of motifs for a minute after transcription. The calls are shown below:

```
> DT_OPT="--o-prune 1e-4 --k0 1e6 --t-ext 0.04 --t-end 60"
> DrTransformer ${DT_OPT} --name SRPn_dt < SRPn.fa
> DrTransformer ${DT_OPT} --name SRPt_dt < SRPt.fa
> DrTransformer ${DT_OPT} --name SRPr_dt < SRPr.fa
> DrTransformer ${DT_OPT} --name SRPf_dt < SRPf.fa
> DK_OPT="--k0 1e6 --t-ext 0.04 --t-end 60"
> DrTransformer ${DK_OPT} --name SRPn_dk < SRPn.fa
> DrTransformer ${DK_OPT} --name SRPt_dk < SRPt.fa
> DrTransformer ${DK_OPT} --name SRPr_dk < SRPr.fa
> DrTransformer ${DK_OPT} --name SRPf_dk < SRPf.fa
```

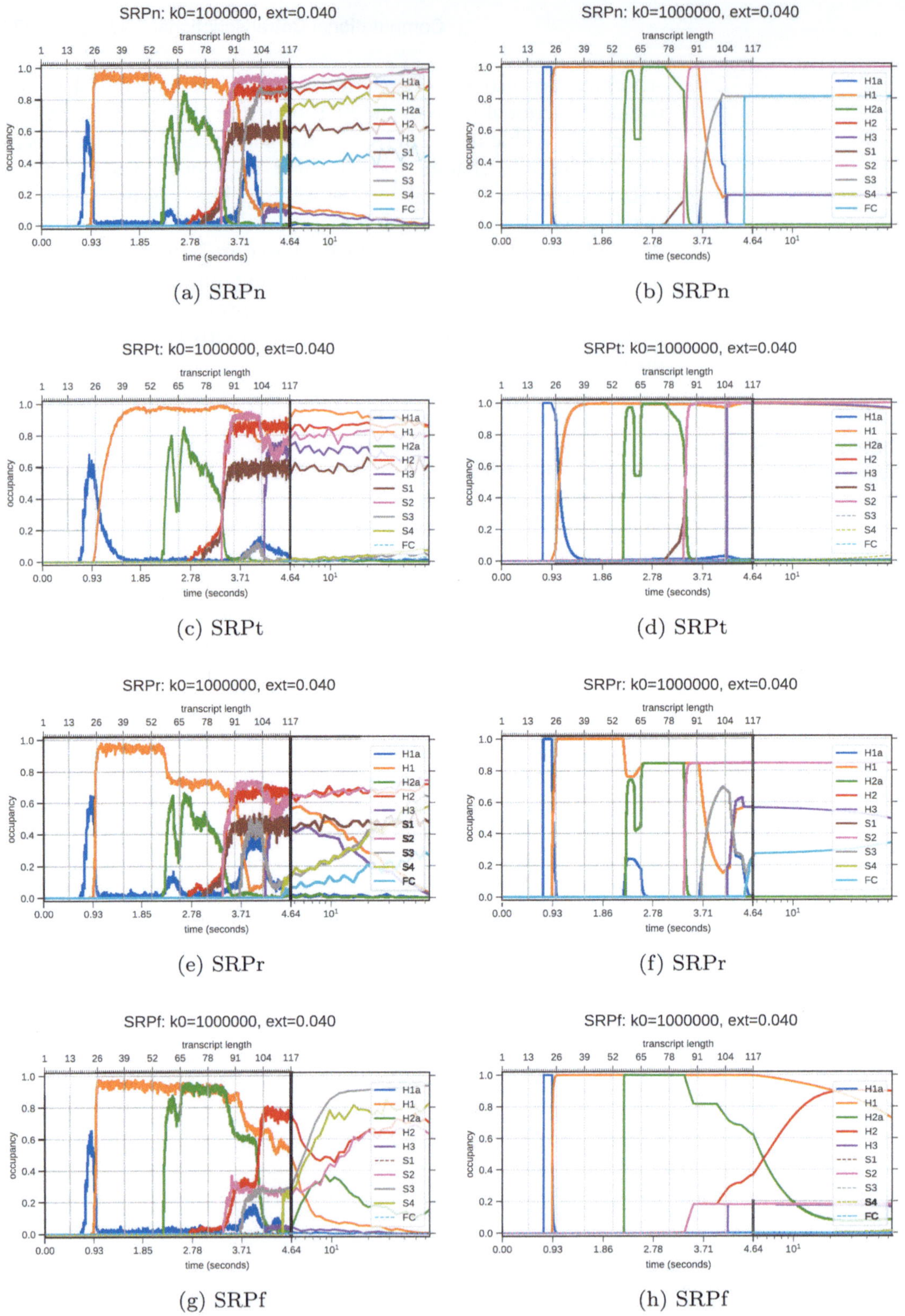

(a) SRPn (b) SRPn

(c) SRPt (d) SRPt

(e) SRPr (f) SRPr

(g) SRPf (h) SRPf

Fig. 7 Occupancy of target motifs over time in `DrKinfold` and `DrTransformer` output. The vertical black line marks the transcription of the last nucleotide and also the transition from a linear timescale to a logarithmic timescale. The end point marks 60 seconds after the last nucleotide has been transcribed. All plots compare motifs described in Fig. 1. Here, FC is defined as the union of motifs H2, S1, S2, S3, and S4. (**a**) SRPn. (**b**) SRPn. (**c**) SRPt. (**d**) SRPt. (**e**) SRPr. (**f**) SRPr. (**g**) SRPf. (**h**) SRPf

The plots to trace motifs can be generated using the DrPlotter program that is installed together with DrTransformer. We need to specify the file that contains motif specifications, as well as a list of motifs that defines which (and in which order) motifs should be plotted:

```
> DP_OPT="--motiffile SRP.mot --motifs H1a H1 H2a H2 H3a H3 S1 S2 S3 S4 FC"
> DrPlotter ${DP_OPT} --name SRPn_dt_mot < SRPn_dt.drf
> DrPlotter ${DP_OPT} --name SRPt_dt_mot < SRPt_dt.drf
> DrPlotter ${DP_OPT} --name SRPr_dt_mot < SRPr_dt.drf
> DrPlotter ${DP_OPT} --name SRPf_dt_mot < SRPf_dt.drf
> DrPlotter ${DP_OPT} --name SRPn_dk_mot < SRPn_dk.drf
> DrPlotter ${DP_OPT} --name SRPt_dk_mot < SRPt_dk.drf
> DrPlotter ${DP_OPT} --name SRPr_dk_mot < SRPr_dk.drf
> DrPlotter ${DP_OPT} --name SRPf_dk_mot < SRPf_dk.drf
```

3.10.2 Motif Plot Interpretation

The plots in Fig. 7 compare DrKinfold and DrTransformer predictions. As usual, DrKinfold is the more detailed model, but DrTransformer output can also be insightful, as it ignores base-pair-level variations.

For the native SRP molecule, we see that H1a and H1 are in competition. H1a forms first as it is the first cotranscriptionally available motif, but it is quickly replaced by H1. However, around 95–100 nt length when the S3 stem forms, H1a becomes the dominant structure again. Further visual inspection (with DrForna, not shown) reveals a causal relationship, as the H1a hairpin is compatible with S3 formation, and the newly transcribed part of the RNA can invade the stem of motif H1a to gradually extend S3 – a relatively fast process also known as three-way branch migration.

Furthermore, H2a precedes the formation of H2, and less than 1% of structures form H3 at any point in time. The motifs H2, S2, S3, and S4 are approaching more than 80% in the simulated ensemble after transcription; only, S1 is less occupied at 60%. The union of all constraints, FC, is at about 40% occupancy after transcription. The known functional structure is less dominant, as it is unlikely that all enforced base pairs are closed at the same time. In the DrTransformer output, we see that essentially all structures form S1 and S2, but only 80% form S3 and S4, as the remaining 20% of the ensemble is trapped with H1 and H3. Accordingly, FC reaches 80% (the line overlaps with formation of S4).

Not shown in Fig. 7 is the motif H1b, which is only present during the transitions from H1a to H1 and from H1 to H1a. H1c is never formed, and there is no evidence for a pseudoknotted reconfiguration, as the bases that would initiate the interaction (PK and PK2) are almost never accessible at the same time.

The trapped SRPt molecule forms mostly H2-S1-S2 in combination with H1 and H3 and cannot rearrange into the functional

FC form. The rescue mutant SRPr shows an interesting competition between H1 and H1a, but eventually all helices of FC do form at the end of transcription in 10–20% of structures. The H1/H3 structure is still at about 50% at the end of transcription, with occupancy rapidly declining within 60 seconds.

The SRPf sequence forms H2, S2, S3, and S4 at the end of transcription, but the mutations inhibit the formation of S1. Thus, it appears as if FC did not form, even though most of its structure is still intact. Interestingly, `Kinfold` predicts a second round of H2a formation after transcription, when H1 is dissolved. Note that `DrTransformer` predictions are not simply less noisy predictions compared to `Kinfold`. Here, H1 remains stable until the end of transcription, and only the S2 stem forms at 20%. All other motifs for FC are not observed on the investigated timescale. The formation of H3a is not shown, but it is predicted by both `Kinfold` (60%) and `DrTransformer` (80%) at around 90 nt length.

3.11 Concluding Remarks

We have shown how to use computational methods to find structural motifs that are present at specific points in time during cotranscriptional folding. Of course, the underlying thermodynamic nearest neighbor model makes certain simplifying assumptions about which structures are allowed and what their free energy is. Moreover, errors in free energies are exponentially amplified when calculating transition rates between structures (*see* Eq. 1). However, we showed that computational models can be useful to help with the interpretation of experimental data. We present results with $k_0 = 10^6\ s^{-1}$ in this tutorial, even though related literature used $k_0 = 10^5\ s^{-1}$ for `Kinfold`-related kinetic models [36, 37]. The main reason for choosing $k_0 = 10^6\ s^{-1}$ is the correspondence of simulations with experimental data, as we were not able to observe the same nice fits with $k_0 = 10^5\ s^{-1}$. However, this parameter also makes sense, as it is closer to the fastest reaction rates known for RNA folding.

The software discussed here is only a part of generally available cotranscriptional folding software. Among other cotranscriptional folding simulators are `Kinefold` [38], `RNAkinetics` [39], `Kinwalker` [40], `BarMap` [41], and `CoStochFold` [42]. The `drconverters` repository provides a script to convert results from `Kinefold` to the `.drf` file format. `Kinefold` is especially interesting as it allows for pseudoknotted conformations but, unfortunately, uses outdated nearest neighbor parameters.

We modeled the cotranscriptional process as continuous process that attaches nucleotides at equally spaced time intervals. However, nucleotide attachment is a stochastic process, and there are known pause site motifs in the SRP sequence. In principle, such pause sites could have been considered for cotranscriptional simulations (e.g., see `DrTransformer` command line parameters). For the SRP pause sites suggested in literature, we have not found a large influence in structure formation. This is because the pause sites are at nucleotides where the ensemble is close to the equilibrium distribution, such that there are no notable impacts on downstream folding events.

While we only compared against cotranscriptional SHAPE data here, it is relatively straightforward to plot the length of the exterior loop and compare those results with the optical tweezer results from [14]. However, an interpretation of those results seemed even more difficult without direct correspondence with experimenters and thus was beyond scope for this tutorial. We hope that this tutorial provides a foundation to investigate many more related systems and questions.

4 Notes

1. A PC with `Windows` 10 or above may work in combination with the Windows subsystem for Linux (WSL). Native Windows support without WSL is not available for the tasks presented within this chapter, since the required `ViennaRNA Package` Python bindings are at this time not available for Windows.
2. Instead of installing the scripts globally, they can be installed on a per-user base by providing `pip` with the flag `--user`:

```
> pip install --user .
```

If installation of scripts is not desirable, then all executions must be adapted to include location of scripts on disk. For instance, the command

```
> thermo_predict.py
```

becomes

```
> /path/to/DrTutorial/scripts/thermo_predict.py
```

where `/path/to/DrTutorial` is the absolute path to the DrTutorial directory. Alternatively, users can set the `PATH` environment variable such that it includes the absolute path to the `scripts` directory, e.g.:

```
DRTUTORIAL_DATA=/path/to/DrTutorial
export PATH="${DRTUTORIAL_DATA}/scripts:${PATH}"
```

To make the environment variable permanently available throughout different `Unix` terminal sessions, you may store the above lines in the appropriate config file of your terminal emulator, for instance, the `bash` terminal emulator uses the `~/.bashrc` file.

3. Beware of the fact that most RNA secondary structure prediction programs like `RNAfold` only backtrack *a single* MFE structure. However, some RNA sequences can form multiple distinct secondary structures with minimum free energy. To enumerate *all* MFE structures with `RNAsubopt`, use:

```
> RNAsubopt -e 0
```

where the `-e 0` option tells `RNAsubopt` to print all structures within a 0 $kcal \cdot mol^{-1}$ energy range above the MFE.

4. By default, the `-p/--stochBT` option of `RNAsubopt` yields a list of secondary structures s from the equilibrium distribution, randomly drawn according to their probability $p(s)$. The free energy of these structures is not part of the output due to implementation details. However, the `RNAsubopt` program comes with a second, alternative, option `--stochBT_en` that can be used as a substitute. As a result, a program call like

```
> RNAsubopt --stochBT_en 1
```

yields a single secondary structure followed by its free energy $E(s)$ in units of $\text{kcal} \cdot \text{mol}^{-1}$ and the corresponding equilibrium probability $p(s) = \frac{1}{Z} exp(-E(s)/RT)$ with temperature T, gas constant R, and partition function $Z = \sum_s exp(-E(s)/RT)$.

5. Since any nucleotide i may pair with any other nucleotide j, the number of possible base pairs (i, j) grows quadratic in the length of the sequence. Consequently, a visualization of the pairing probabilities during transcription or the change thereof would require a $four-$ dimensional plot: two dimensions to address the actual base pair, one for the length of the nascent transcript, and one for the actual pairing probability.

6. Base-pair probabilities p_{ij} can be readily computed using a command line call like

```
RNAfold -p SRPn.fa
```

This will create a *dot-plot* PostScript output file `SRPn_dp.ps` that provides a visualization of the individual pairing probabilities and stores them in a machine readable form for further processing. For each base pair (i, j) and corresponding probability p_{ij}, the PostScript file contains lines of the form

```
i j p ubox
```

where i and j are integer numbers denoting the two positions that pair (one-based) and p is a floating point number showing the square root ($\sqrt{p_{ij}}$) of the corresponding pairing probability p_{ij}.

7. For the cotranscriptional SHAPE reactivity profiles shown in Fig. 6, we applied a threshold of $\rho \leq 1$. This cropping of the data together with the limits of the color scale from 0 to 1 may conceal the full picture of potential conformations probed by the SHAPE reagent. To change the cutoff value, the `plot_accessibility.R` script takes an optional argument `--SHAPErange r` where r is the largest value of ρ that should be considered. For instance, to replicate the data shown in Watters et al. [16] and Yu et al. [15], one may use a range of $0 \leq \rho \leq 4$ by adding the parameter `--SHAPErange 4` upon creating the plot.

References

1. Pan T, Sosnick T (2006) RNA folding during transcription. Annu Rev Biophys Biomol Struct 35:161–175
2. Schärfen L, Neugebauer KM (2021) Transcription regulation through nascent RNA folding. J Mol Biol 433(14):166975
3. Lorenz R, Bernhart SH, Zu Siederdissen CH, Tafer H, Flamm C, Stadler PF, Hofacker IL (2011) ViennaRNA Package 2.0. Algorithms Mol Biol 6(1):26
4. Flamm C, Fontana W, Hofacker IL, Schuster P (2000) RNA folding at elementary step resolution. RNA 6(3):325–338
5. Badelt S, Lorenz R, Hofacker IL (2023) DrTransformer: heuristic cotranscriptional RNA folding using the nearest neighbor energy model. Bioinformatics 39(1), 01. ISSN 1367–4811. https://doi.org/10.1093/bioinformatics/btad034.
6. Tănasie AR, Kerpedjiev P, Hammer S, Badelt S (2023) DrForna: visualization of cotranscriptional folding. Bioinformatics 39(9). https://doi.org/10.1093/bioinformatics/btad555
7. Grüning B, Dale R, Sjödin A, Chapman BA, Rowe J, Tomkins-Tinch CH, Valieris R, Köster J (2018) Bioconda: sustainable and comprehensive software distribution for the life sciences. Nat Methods 15(7):475–476
8. The Pandas Development Team (2020). pandas-dev/pandas: Pandas, February 2020. https://doi.org/10.5281/zenodo.3509134.
9. Hunter JD (2007) Matplotlib: a 2d graphics environment. Comput Sci Eng 9(3):90–95. https://doi.org/10.1109/MCSE.2007.55
10. R Core Team (2022) R: A Language and Environment for Statistical Computing. R Foundation for Statistical Computing, Vienna. https://www.R-project.org/
11. Wickham H (2016) ggplot2: Elegant Graphics for Data Analysis. Springer, New York. ISBN 978-3-319-24277-4. https://ggplot2.tidyverse.org
12. Garnier, Simon, Ross, Noam, Rudis, Robert, Camargo, Antônio Pedro, Sciaini, Marco, Scherer, Cédric (2023) viridis(Lite) - Colorblind-Friendly Color Maps for R. https://sjmgarnier.github.io/viridis/. viridis package version 0.6.3
13. Wickham H (2007) Reshaping data with the reshape package. J Stat Softw 21(12):1–20. http://www.jstatsoft.org/v21/i12/
14. Fukuda S, Yan S, Komi Y, Sun M, Gabizon R, Bustamante C (2020) The biogenesis of SRP RNA is modulated by an RNA folding intermediate attained during transcription. Mol Cell 77(2):241–250
15. Yu AM, Gasper PM, Cheng L, Lai LB, Kaur S, Gopalan V, Chen AA, Lucks JB (2021) Computationally reconstructing cotranscriptional RNA folding from experimental data reveals rearrangement of non-native folding intermediates. Mol Cell 81(4):870–883
16. Watters KE, Strobel EJ, Yu AM, Lis JT, Lucks JB (2016) Cotranscriptional folding of a riboswitch at nucleotide resolution. Nat Struct Mol Biol 23(12):1124
17. Turner DH, Mathews DH (2010) NNDB: the nearest neighbor parameter database for predicting stability of nucleic acid secondary structure. Nucl Acids Res 38(suppl_1):D280–D282
18. McCaskill JS (1990) The equilibrium partition function and base pair binding probabilities for RNA secondary structure. Biopolymers 29(6–7):1105–1119. https://doi.org/10.1002/bip.360290621

19. Wuchty S, Fontana W, Hofacker IL, Schuster P (1999) Complete suboptimal folding of RNA and the stability of secondary structures. Biopolymers 49(2):145–165. ISSN 0006-3525
20. Ding Y, Lawrence CE (2003) A statistical sampling algorithm for RNA secondary structure prediction. Nucl Acids Res 31(24): 7280–7301. https://doi.org/10.1093/nar/gkg938
21. Keenan RJ, Freymann DM, Stroud RM, Walter P (2001) The signal recognition particle. Ann Rev Biochem 70(1):755–775
22. Doudna JA, Batey RT (2004) Structural insights into the signal recognition particle. Ann Rev Biochem 73(1):539–557
23. Rosenblad MA, Larsen N, Samuelsson T, Zwieb C (2009) Kinship in the SRP RNA family. RNA Biol 6(5):508–516
24. Wong TN, Sosnick TR, Pan T (2007) Folding of noncoding RNAs during transcription facilitated by pausing-induced nonnative structures. Proc Natl Acad Sci 104(46):17995–18000
25. Batey RT, Rambo RP, Lucast L, Rha B, Doudna JA (2000) Crystal structure of the ribonucleoprotein core of the signal recognition particle. Science 287(5456):1232–1239
26. Watters KE, Angela MY, Strobel EJ, Settle AH, Lucks JB (2016) Characterizing RNA structures in vitro and in vivo with selective 2-'-hydroxyl acylation analyzed by primer extension sequencing (SHAPE-Seq). Methods 103:34–48
27. Merino EJ, Wilkinson KA, Coughlan JL, Weeks KM (2005) RNA structure analysis at single nucleotide resolution by selective 2-'-hydroxyl acylation and primer extension (SHAPE). J Amer Chem Soc 127(12): 4223–4231
28. Süködsd Z, Swenson MS, Kjems J, Heitsch CE (2013) Evaluating the accuracy of SHAPE-directed RNA secondary structure predictions. Nucl Acids Res 41(5):2807–2816
29. Siegfried NA, Busan S, Rice GM, Nelson JAE, Weeks KM (2014) RNA motif discovery by SHAPE and mutational profiling (SHAPE-MaP). Nat Methods 11(9):959–965
30. Saaidi A, Allouche D, Regnier M, Sargueil B, Ponty Y (2020) IPANEMAP: integrative probing analysis of nucleic acids empowered by multiple accessibility profiles. Nucl Acids Res 48(15):8276–8289
31. Aviran S, Lucks JB, Pachter L (2011) RNA structure characterization from chemical mapping experiments. In: 2011 49th Annual Allerton Conference on Communication, Control, and Computing (Allerton), pp 1743–1750. IEEE, Piscataway. https://doi.org/10.1109/Allerton.2011.6120379
32. Weeks KM, Mauger DM (2011) Exploring RNA structural codes with SHAPE chemistry. Accounts Chem Res 44(12):1280–1291
33. Hurst T, Chen S-J (2021) Sieving RNA 3D structures with SHAPE and evaluating mechanisms driving sequence-dependent reactivity bias. J Phys Chem B 125(4):1156–1166
34. Mortimer SA, Weeks KM (2009) Time-resolved RNA SHAPE chemistry: quantitative RNA structure analysis in one-second snapshots and at single-nucleotide resolution. Nat Protoc 4(10):1413–1421
35. Xue AY, Yu AM, Lucks JB, Bagheri N (2019) DUETT quantitatively identifies known and novel events in nascent RNA structural dynamics from chemical probing data. Bioinformatics 35(24):5103–5112
36. Helmling C, Wacker A, Wolfinger MT, Hofacker IL, Hengesbach M, Fürtig B, Schwalbe H (2017) NMR structural profiling of transcriptional intermediates reveals riboswitch regulation by metastable RNA conformations. J Amer Chem Soc 139(7):2647–2656
37. Badelt S, Hammer S, Flamm C, Hofacker IL (2015) Thermodynamic and kinetic folding of riboswitches. In: Methods in Enzymology, vol 553, pp 193–213. Elsevier, Amsterdam. http://dx.doi.org/10.1016/bs.mie.2014.10.060
38. Xayaphoummine A, Bucher T, Isambert H (2005) Kinefold web server for RNA/DNA folding path and structure prediction including pseudoknots and knots. Nucl Acids Res 33 (suppl 2):W605–W610
39. Danilova LV, Pervouchine DD, Favorov AV, Mironov AA (2006) RNAKinetics: a web server that models secondary structure kinetics of an elongating RNA. J Bioinformat Comput Biol 4(02):589–596
40. Geis M, Flamm C, Wolfinger MT, Tanzer A, Hofacker IL, Middendorf M, Mandl C, Stadler PF, Thurner C (2008) Folding kinetics of large RNAs. J Mol Biol 379(1):160–173
41. Hofacker IL, Flamm C, Heine C, Wolfinger MT, Scheuermann G, Stadler PF (2010) BarMap: RNA folding on dynamic energy landscapes. RNA 16:1308–1316. https://doi.org/10.1261/rna.2093310
42. Thanh VH, Korpela D, Orponen P (2021) Cotranscriptional kinetic folding of RNA secondary structures including pseudoknots. J Comput Biol 28(9):892–908

Chapter 14

Assessing the Quality of Cotranscriptional Folding Simulations

Felix Kühnl, Peter F. Stadler, and Sven Findeiß

Abstract

Structural changes in RNAs are an important contributor to controlling gene expression not only at the posttranscriptional stage but also during transcription. A subclass of riboswitches and RNA thermometers located in the 5' region of the primary transcript regulates the downstream functional unit – usually an ORF – through premature termination of transcription. Not only such elements occur naturally, but they are also attractive devices in synthetic biology. The possibility to design such riboswitches or RNA thermometers is thus of considerable practical interest. Since these functional RNA elements act already during transcription, it is important to model and understand the dynamics of folding and, in particular, the formation of intermediate structures concurrently with transcription. Cotranscriptional folding simulations are therefore an important step to verify the functionality of design constructs before conducting expensive and labor-intensive wet lab experiments. For RNAs, full-fledged molecular dynamics simulations are far beyond practical reach because of both the size of the molecules and the timescales of interest. Even at the simplified level of secondary structures, further approximations are necessary. The BarMap approach is based on representing the secondary structure landscape for each individual transcription step by a coarse-grained representation that only retains a small set of low-energy local minima and the energy barriers between them. The folding dynamics between two transcriptional elongation steps is modeled as a Markov process on this representation. Maps between pairs of consecutive coarse-grained landscapes make it possible to follow the folding process as it changes in response to transcription elongation. In its original implementation, the BarMap software provides a general framework to investigate RNA folding dynamics on temporally changing landscapes. It is, however, difficult to use in particular for specific scenarios such as cotranscriptional folding. To overcome this limitation, we developed the user-friendly BarMap-QA pipeline described in detail in this contribution. It is illustrated here by an elaborate example that emphasizes the careful monitoring of several quality measures. Using an iterative workflow, a reliable and complete kinetics simulation of a synthetic, transcription-regulating riboswitch is obtained using minimal computational resources. All programs and scripts used in this contribution are free software and available for download as a source distribution for Linux® or as a platform-independent Docker® image including support for Apple macOS® and Microsoft Windows®.

Key words RNA, Cotranscriptional folding, Kinetics simulation, Secondary structure, BarMap, Quality assurance, Docker, BarMap-QA

Ronny Lorenz (ed.), *RNA Folding: Methods and Protocols*, Methods in Molecular Biology, vol. 2726, https://doi.org/10.1007/978-1-0716-3519-3_14,

1 Introduction

Many aspects of RNA structure and dynamics can be understood quantitatively at the secondary structure level. This includes equilibrium properties of the most stable structure, the existence and equilibrium distribution of alternative structures or structural elements, the (adiabatic) melting and refolding as a response to changes in temperature, the binding to other RNAs, and to a certain extent even the interaction with other types of molecules. These features can be efficiently modeled and computed because (1) the folding energy of an RNA structure is rather accurately approximated by the standard Turner energy model [23] and (2) efficient algorithms exist to compute the partition function of ensembles of RNA structures [15]. Despite this success, however, not all aspects of RNA folding that are important in practice are captured by equilibrium thermodynamics.

RNA folding is an inherently dynamic process, and the details of the folding trajectory are at least occasionally biologically relevant. Metastable states, for example, are sometimes the functional ones, as in the example of the hok/sok host killing system [8]. In living cells, a nascent RNA starts folding into stable structures while it is transcribed by an RNA polymerase. The structures formed in cotranscriptional folding have decisive functions, e.g., in the case of transcriptional riboswitches, where they decide whether the transcription is terminated or continued to produce a full-length RNA transcript. The structures formed cotranscriptionally are transient and will continue to refold as the transcript is elongated. The mechanisms underlying the biological function of such RNA elements thus can only be understood in terms of dynamics of the folding process.

Riboswitches and RNA thermometers that act on the level of transcription are of practical interest in synthetic biology. They form a class of fast-acting regulators for gene expression that can react to the presence of small molecules via suitable aptamer components or respond to environmental changes in temperature or ion strength that physically affect RNA structure formation. The rational design of such devices, however, requires not only a detailed understanding of the folding process but also computationally efficient methods to model and evaluate the folding dynamics. The size of RNA sequences and the timescales involved preclude 3D molecular dynamics simulations.

Since the standard Turner model assigns an energy to every RNA secondary structure, it is possible to simulate the dynamics of RNA at this level [6]. The calculation and analysis of a large number of trajectories are still computationally expensive and are infeasible at this level when (re)folding processes with timescales of seconds or even hours are to be studied. To be of practical use, e.g., in design applications, furthermore, computational models not only need to provide reasonably reliable predictions but also must be much faster and cheaper than experimental approaches. Such methods are indeed available. Instead of simulating the dynamics of RNA secondary structures directly, the idea is to use a coarse-grained representation of the underlying high-dimensional energy landscape and to approximate the (re)folding dynamics of the RNA by a dynamical system on the coarse-grained landscape [27].

A natural coarse-graining of an energy landscape is its barrier tree, which only considers the local minima (as representatives of low-energy basins) and the barriers between them [7]. By construction, the barrier tree does not retain any information on the proximity of different basins. Somewhat less coarse abstractions of landscapes, for instance, basin hopping graphs [10] or the gradient descent partition of the landscape discussed below, may serve as refine models. On a coarse-grained representation, we can define an (approximate) Markov process that describes the dynamics in terms of transitions between the coarse-grained macrostates. The main advantage of the coarse-grained representation is that the number of macrostates is small enough to explicitly compute the probability of occupancy for a given time instead of having to resort to simulating individual trajectories. As a consequence, the folding dynamics can be computed efficiently for arbitrary timescales – albeit of course only approximately and in terms of the coarse-grained macrostates.

Cotranscriptional folding cannot be modeled by a single energy landscape since the underlying RNA sequence changes with the addition of each nucleotide. The idea of BarMap [9] is to use a sequence of (coarse-grained) landscapes, one for each step of transcriptional elongation. The refolding dynamics between elongation steps is then modeled as a Markov process on the fixed landscape $\mathcal{L}$. Upon elongation, the landscape changes to $\mathcal{L}'$, the landscape of the RNA with one additional nucleotide at the 3'-end. Each structure x in $\mathcal{L}$, including the local minima of $\mathcal{L}$, corresponds to a structure x' in $\mathcal{L}'$ obtained by appending the newly added, unpaired nucleotide. On the other hand, x' belongs to some basin in $\mathcal{L}'$ that is represented again by a local minimum. In this way, each local minimum (basin) of $\mathcal{L}$ is assigned to a local minimum (basin) in $\mathcal{L}'$, thus defining the "bar map" of Hofacker et al. [9], which relates the barrier trees of consecutive landscapes. In particular, this allows to transfer the probabilities of the coarse-grained macrostates from one landscape to the next. This yields an approximate time course of the occupancy of coarse-grained macrostates for cotranscriptional folding without the need for expensive simulations of individual trajectories. The separation of computations on individual landscapes and the transition between them, furthermore, make it easy to explore the effects of variations in the speed of transcription on the formation of intermediate structures.

The BarMap software accompanying the publication of the method [9] is a collection of scripts to facilitate the kinetic analysis of several related RNA landscapes, for example, under changing environmental parameters such as temperature, or, more relevant here, for a stepwise elongated transcript. It makes use of Barriers [7] to coarse-grain the given RNA landscape and Treekin [27] to compute the folding dynamics at the level of macrostates represented by gradient basins of secondary structures. The script then stitches together the output of these tools to obtain a result for the full process. In contrast to other simulation tools, BarMap does not rely on the sampling of individual trajectories but offers an analytical solution based on the enumeration of representative states.

Despite its flexibility and extensibility, the original version of BarMap is a proof-of-concept rather than a ready-to-use tool. Its major drawbacks are:

1. The user has to call multiple scripts to get initial results, making the method unattractive for batch processing even for short input sequences.
2. The output is spread across multiple files, some of which are hardly human-readable, making the interpretation of the results a tedious and error-prone task.
3. There is no feedback that would allow the user to judge the quality of the produced results or help to improve on them.
4. Required software dependencies, in particular Barriers and Treekin, are developed for Linux. Although they can be compiled for other operating systems, this excludes many potential users.
5. There is no comprehensive documentation that guides the user step by step through a typical run, so getting started with the software is unnecessarily complicated.

It therefore takes considerable effort to use BarMap in the context of cotranscriptional folding. Its practical applications there have remained limited.

The intention of this contribution is to alleviate all of the mentioned deficiencies. We give detailed instructions to analyze a real-world example that the reader can easily reproduce on an average desktop computer. Further, we provide a pre-configured working environment including all required software to immediately jump into a cotranscriptional folding analysis. It is deployed inside a highly portable Docker image that can be loaded with a single command and requires no setup apart from a standard Docker installation, which is available on Windows, macOS, and Linux. The included scripts can produce preliminary output with just a single command and an input sequence, generating a full cotranscriptional folding simulation and post-processed plots in various formats including PDF and SVG. In addition, quality statistics are computed that allow the user, together with powerful helper scripts, to semiautomatically improve the simulation results if necessary. Finally, an integrating command line viewer for the output files is available that greatly simplifies the interpretation of the results and, together with the generated plots, provides a deep insight into the folding process.

2 Materials

Secondary Structures of RNA Molecules A given RNA molecule can, in theory, adopt any of an enormous number of possible structural conformations. It has been shown that, often, sufficiently precise predictions of an RNA's function can be made by limiting the analysis to its space of *secondary structures*, i.e., structures that are formed by intramolecular Watson–Crick base pairing according to the following rules: (i) each base can pair with at most one other base; (ii) base pairs cannot cross each other, i.e., there are no pseudo-knots; and (iii) paired bases have a minimum distance of three nucleotides. Note that tertiary structure elements, e.g., helix–helix interactions such as in kissing hairpins and other types of pseudo-knots as well as base triples, are not considered. The set of all possible secondary structures X for a given RNA sequence s is called the *structure ensemble* of s.

Probability of RNA Structures Each secondary structure $x \in X$ can be assigned an energy $\Delta G(x) \in (-\infty, \infty)$. Since x is an approximation for a *set* of 3D conformations, $\Delta G(x)$ is a Gibbs free energy and consists of both enthalpic and entropic contributions. As a consequence, $\Delta G(x)$ is explicitly temperature-dependent, a fact that is important, e.g., in the context of modeling RNA thermometers. In practice, $\Delta G(x)$ can be computed in very good approximation from the standard nearest neighbor energy model (Turner energy model), which comprises (temperature-dependent) energy parameters based on detailed thermodynamic measurements [13, 14, 23].

For an equilibrated RNA at fixed temperature T, the structure ensemble follows a Boltzmann distribution, i.e., the probability density of x is given by its *Boltzmann weight* $Z[x] := \exp(-\Delta G(x)/(RT))$, where R is the universal gas constant. The *partition function*

$$Z[Y] := \sum_{x \in Y} Z[x]$$

is the sum over all Boltzmann weights for a given subset of structures $Y \subseteq X$, and $Z := Z[X]$ refers to the partition function of the full ensemble. The probability of x is then given by $\Pr[x] = Z[x]/Z$

and for a set Υ of structures by $\Pr[\Upsilon] = Z[\Upsilon]/Z$. The lower the energy of a structure, the higher is its associated Boltzmann weight and thus its probability. We refer to the value $\Pr[\Upsilon]$ as the structural *coverage* of Υ with respect to the ensemble X because it measures how well the subset Υ covers the most probable (i.e., most important) structures of X.

RNA Energy Landscapes For a model of RNA folding kinetics, one needs, in addition to the structure space given by the ensemble X, a notion of *neighborhood* to define which structures a given structure $x \in X$ can refold into within an elementary simulation step. Here, the set of neighbor structures $N(x)$ of x is defined as the set of all structures obtained by removing or adding a single base pair from or to x. Together with the Gibbs free energy, these ingredients define the RNA energy landscape $\mathcal{L} := (X, N, \Delta G)$.

For any two structures $x, y \in X$, the *transition rate coefficient* $r_{x \to y}$ can be obtained by applying the Metropolis rule [16]:

$$r_{x \to y} := \begin{cases} \min\left\{ \exp\left(-\dfrac{\Delta G(y) - \Delta G(x)}{RT}\right), 1 \right\} & \text{if } y \in N(x) \\ 0 & \text{otherwise} \end{cases}$$

It specifies if and how fast structure x refolds into structure y. The pair $L := (X, r_{\cdot \to \cdot})$, which implicitly determines $N(x)$ (because $y \in N(x)$ if and only if $r_{x \to y} \neq 0$), represents the *Markov process* induced by the Metropolis rule on $\mathcal{L}$. L can also be interpreted as a graph where each structure is a node, labeled by its energy, and a weighted, directed edge connects each pair of adjacent structures (x, y) such that the edge weight corresponds to the rate coefficient $r_{x \to y}$. The weighted graph L is connected since each structure can refold into any other one, e.g., by first opening all of its base pairs and then closing all base pairs of the target structure, one after another.

Enumeration of the Energy Landscape: RNAsubopt The number of secondary structures $|X|$ increases exponentially with the sequence length $|s|$ of the given RNA [20]. In practice, enumerating the full ensemble is therefore infeasible even for small molecules. A first remedy is to only enumerate those structures with the lowest energies (and, thus, with the highest probability) up to a given threshold ΔG_{enum}. This can be achieved by applying Wuchty's algorithm [28] as implemented in the tool RNAsubopt of the ViennaRNA package [12]. The proper choice of ΔG_{enum} is nontrivial; it will be used as the main parameter in this work to obtain high-quality simulation results at the lowest possible resource cost. Besides losing possibly important transient structures, the major drawback of this approach is that it may disconnect the landscape, e.g., because all neighbors of a structure lying slightly below the

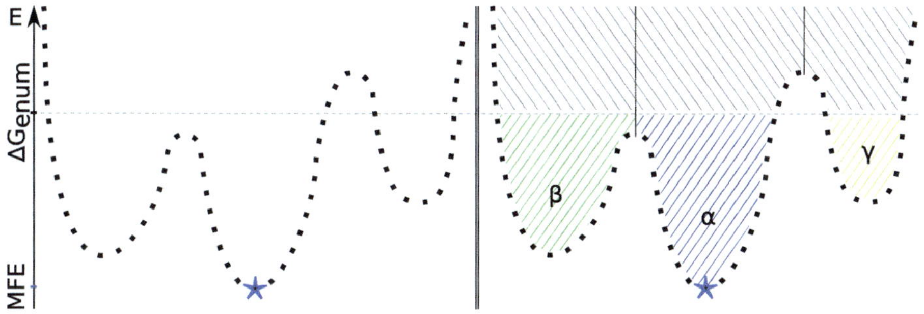

Fig. 1 Schematic and strongly simplified visualization of an RNA energy landscape to which a coarse-graining is applied. *left:* Individual structures are denoted as black squares, each being adjacent to its left and right neighbor. The *y*-axis displays their free energies. The MFE structure is marked with a blue asterisk. *right:* Coarse-grained version of the same energy landscape. The individual structures have been binned into gradient basins, denoted α, β, and γ here (hatched in green, blue, and yellow, respectively). Due to the enumeration threshold ΔG_{enum} (horizontal, dashed line), the upper part of the landscape (hatched in gray) is not part of the coarse-grained representation. As a result, γ is disconnected from the remaining basins

energy threshold are pruned, leaving an isolated state. Please refer to Fig. 1 for an example.

Coarse-Graining of the Energy Landscape: Barriers Despite the reduction of structures achieved by applying Wuchty's algorithm, the number of secondary structures is still far too high to use them directly as state set when performing kinetic folding simulations. Therefore, Barriers [7] performs a coarse-graining of the structure space X. It processes the *sorted* list of low-energy states $X' := \{x \in X \mid \Delta G(x) \leq \Delta G_{\mathrm{enum}}\}$ generated by RNAsubopt and extracts (i) the representative set $\widetilde{X}$ of local minima in X and (ii) the energy barriers between them. Barriers also (implicitly) assigns each $x \in X'$ to the local minimum $\gamma(x)$ that it reaches by performing a gradient descent in $\mathcal{L}$, thus binning the structures into well-defined macrostates. Macrostates are therefore simply sets of "microstates", i.e., individual RNA structures, that form a partition of the landscape. Figure 1 shows the result of applying the described procedure to an example.

For every pair of macrostates $\alpha, \beta \subseteq X$, the rate coefficient for the transition from α to β is computed as a weighted sum over the rate coefficients of each pair of microstates $x \in \alpha$ and $y \in \beta$:

$$\widetilde{r}_{\alpha \to \beta} := \sum_{x \in \alpha} \sum_{y \in \beta} \Pr[x|\alpha]\ r_{x \to y},$$

where $\Pr[x|\alpha]$ is the probability that the Markov chain L is in state x given we know that it is in the macrostate α to which x belongs. Assuming that basins are steep, the Markov process will approximately equilibrate within α before leaving the basin, justifying the approximation $\Pr[x|\alpha] = Z[x]/Z[\alpha]$. The approximation fails in particular for large, shallow basins, which are likely to appear in landscapes of sequences with extremely biased nucleotide distributions and unusually large fractions of unpaired bases in their ground state. The rate constants $\widetilde{r}_{\alpha \to \beta}$ define a Markov chain on the set of basins, which again can be seen as a graph $\widetilde{L} := (\widetilde{X}, \widetilde{r}_{\cdot \to \cdot})$ whose neighborhoods are given by $\widetilde{N}(\alpha) := \left\{ \beta \in \widetilde{X} \setminus \{\alpha\} \mid \exists x \in \alpha \; \exists y \in \beta : y \in N(x) \right\}$. Along with the basins and the corresponding rate matrix, Barriers computes several statistics on the number of structures and the partition functions $Z[\alpha]$ for the basins $\alpha \in \widetilde{X}$.

Analysis of the Folding Kinetics: Treekin Throughout, we consider the folding kinetics of a single RNA molecule as a Markov chain determined by the transition rates $r_{\cdot \to \cdot}$ at the level of individual secondary structures or $\widetilde{r}_{\cdot \to \cdot}$ at the level of energy basins as macrostates. Given a distribution $p(0)$ of initial states and the rate matrix $\mathbf{R} = (r_{ij})$ composed of the respective rate coefficients $r_{x_j \to x_i}$ or $r_{\alpha_j \to \alpha_i}$, the distribution after some time τ is simply given by $p(\tau) = \exp(\tau \mathbf{R})$. It can be computed using standard methods of linear algebra, requiring in essence the diagonalization of $\mathbf{R}$. The software Treekin [27] is the implementation used for this work. It can easily process rate matrices of a dimension up to a few thousands. The resulting output gives, for each time step, the population density of each state. The user can choose the length of the simulated time period, which corresponds to setting the transcription rate (cf. **Note 4.5**). Treekin expects a connected state space $(X, r_{\cdot \to \cdot})$, a precondition that is easily checked by means of a breadth-first search on $(X, r_{\cdot \to \cdot})$. As noted above, the full space of secondary structures is always connected. This is not necessarily true for the output of Barriers, since truncation of the landscape $(X, N, \Delta G)$ to low energies may consist of multiple connected components separated by energy barriers above the threshold energy ΔG_{enum}. The software collection of this contribution contains the script `barriers_keep_connected` that truncates any disconnected state from the rate matrix, thus generating Treekin-compatible input. As an alternative, heuristic methods such as `findpath` from the ViennaRNA package [12] or RNAEAPath [11] could be used to estimate energy barriers and, thus, approximate transition rates between the components of a disconnected landscape. However, this is beyond the scope of this contribution.

Cotranscriptional Folding: BarMap Several pre-computed, coarse-grained energy landscapes $\widetilde{\mathcal{L}}_1, \ldots, \widetilde{\mathcal{L}}_n$ and their associated rate matrices are combined by BarMap to obtain a single kinetics simulation. In our case, $\widetilde{\mathcal{L}}_i$ represents the growing RNA sequence of length i. To achieve this, BarMap first generates a mapping from

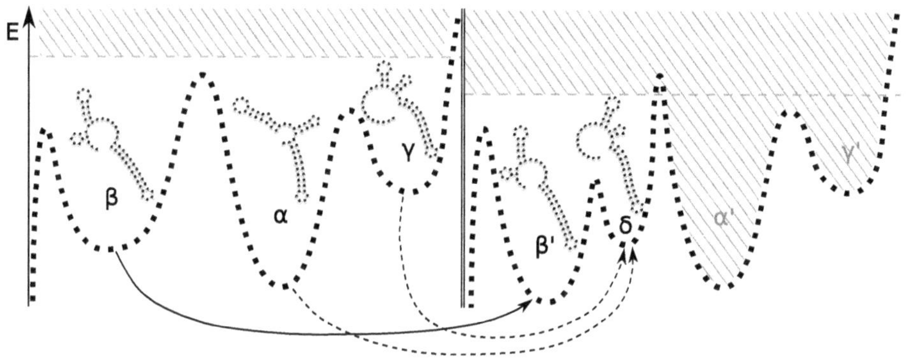

Fig. 2 Mapping of some exemplary basins for two coarse-grained, partially enumerated RNA energy landscapes. The left landscape is mapped to the right one. The dotted lines, again, indicate the RNA structures of the individual landscape, and each Greek letter denotes a basin. The structure of each basins' minimum is plotted above. Solid arrows mark exact mappings, while dashed arrows are approximate ones. Basin β from the left-hand landscape is mapped to the equivalent basin β' in the right-hand landscape, which has a lower local minimum due to the extension of the 3' hairpin by one base pair. Since the enumeration threshold (gray, horizontal, dashed line) of the right landscape is lower, the saddle between α' and β' is not enumerated. Thus, α' and γ', the equivalent basins of α and γ, are now disconnected and removed from the landscape. Instead, α and γ are mapped to a new basin δ due to their low base pair distance to the local minimum of γ

$\widetilde{X}_i$ to $\widetilde{X}_{i+1}$, $i \in \{1 \ldots n-1\}$, mapping each basin $\alpha_x \in \widetilde{X}_i$ by taking its representative minimum $x = \arg\min_{u \in \alpha_x} \Delta G(u)$, appending an unpaired base to obtain x', and then performing a gradient descent [3] on it. This yields $y := \gamma(x')$, a local minimum in X_{i+1} representing a basin $\beta_y \in \widetilde{X}_{i+1}$.

In practice, the structure y does not necessarily represent a basin in $\widetilde{X}_{i+1}$, e.g., because the low-energy part $X'_{i+1} \subseteq X_{i+1}$ was not enumerated up to sufficiently high energies to recover all previously connected minima (cf. Fig. 2). Barriers also applies heuristics to remove small shallow minima. In both cases, $y \notin X'_{i+1}$ may be the consequence. Furthermore, RNA energy landscapes typically contain adjacent structures with the same energy; thus, there is no guarantee that $y = \gamma(x')$ is the representative minimum of a basin computed by Barriers. To handle all this, BarMap uses an approximate approach and maps α_x to another $\beta_z \in \widetilde{X}_{i+1}$ such that the *distance* of minima y and z is minimal among all macrostates in $\widetilde{X}_{i+1}$. More precisely, the current implementation uses the *base pair distance* defined as $|(y \setminus z) \cap (z \setminus y)|$, i.e., the number of base pairs present in one structure, but not in the other. A basin mapped with a distance of 0 is said to be mapped *exactly*; otherwise, it is mapped *approximately*.

Next, Treekin is applied to the rate matrix of $\widetilde{\mathcal{L}}_1$ with a defined simulation duration. Then, the final populations of all basins in $\widetilde{X}_1$ are transferred to the respective basins in $\widetilde{X}_2$ using the mapping defined above. This process is continued until the final landscape $\widetilde{\mathcal{L}}_n$ is reached. As a result, a cotranscriptional folding simulation of the RNA molecule is obtained.

Measuring Simulation Quality A major problem of the original BarMap pipeline is that one cannot easily assess the quality of the produced output. To tackle this issue, our software provides rich statistics to evaluate:

1. The *ensemble coverage* of each generated landscape to ensure all highly probable (i.e., important) structures are contained in the simulation
2. The summed probability density of all exactly mapped (as defined in Paragraph "Cotranscriptional Folding: BarMap") macrostates – *exact coverage* for short – for each mapping step to ensure correctness of the mapping in the case of an equilibrated RNA
3. The fraction of *exactly mapped population* during the kinetics simulation to ensure highly populated states are mapped correctly during the simulation

Here, (1) measures the quality of the individual energy landscapes generated by Barriers, (2) measures the quality of the macrostate mapping generated by BarMap, and (3) measures the quality of the entire kinetics simulation. The coverage of a macrostate α is $\Pr[\alpha]$ as explained in Paragraph "Probability of RNA Structures". The total coverage of all enumerated basis is the ensemble coverage (1), which is directly controlled by the enumeration threshold of Wuchty's algorithm and serves as a measure of completeness of the coarse-grained landscape. The exact coverage of the mapping (2) is the sum of coverage of all macrostates that are mapped exactly to the next energy landscape by BarMap. Low values indicate that probable structures present in the previous landscape have not been enumerated in the next energy landscape. In such a case, the enumeration threshold needs to be increased for the next landscape, and the mapping has to be recomputed. Finally, the fraction of exact population (3) of a mapping step is given by the sum of the populations of exactly mapped macrostates at the end of this simulation step. Here, a low value also shows that an important state was not enumerated in the next energy landscape, which has to be re-enumerated with a higher threshold. The difference is that the inexactly mapped state could as well have a very low coverage but is highly populated for another reason (e.g., because it is an important transient state or a kinetic trap).

Software The following scripts are provided by our distribution:

- `barmap_gen_barmapfile` performs a full cotranscriptional folding run including sequence elongation, energy landscape generation, rate computation, landscape mapping, and output file filtering and uses a fixed energy threshold for enumerating all landscapes.

- `barmap_rebar` is used to re-enumerate individual energy landscapes, up to either an absolute energy or by a given energy difference. It also generates transition rate matrices and pre-computes eigenvalues and eigenvectors of these to speed up the remapping step and can be run in parallel.
- `barmap_remap` recomputes the macrostate mapping and reruns the full kinetics simulation.
- `barmap_filter_treekin` is used to post-process the final kinetics simulation output. It also prunes states below a given population threshold for clearer and more performant plotting.
- `barmap_merge_treekin` is used to post-process the final kinetics simulation similar to `barmap_filter_treekin` but additionally merges columns of states that are mapped onto each other in consecutive landscapes. This allows drawing a unichrome curve for a chain of macrostates mapped onto each other despite additional nucleotides being added.
- `barmap_show_run` i integrated command line viewer for the kinetics simulation output. It is used in conjunction with the generated plots to analyze the folding behavior of the sequence.
- `barriers_coverage` computes ensemble coverage and connectedness statistics for the given set of Barriers files.
- `barmap_exact_coverage` computes exact coverage values for each mapping step.
- `barmap_exact_population` computes exact population fractions for each mapping step.
- `barmap_plot_treekin` generates a kinetics plot from the (possibly merged or filtered) Treekin output file using the Grace package [22]. It also supports output of PDF documents, SVG vector graphics, and lossless PNG pixel graphics.

Many of the necessary computations can be accelerated by parallel execution (cf. **Note 4.4**).

The software distribution also contains utilities that are not directly important for the example demonstrated here. For a complete list, run `ls /BarMap_QA/bin` inside the Docker container or the corresponding directory in the source distribution.

System Requirements and Setup The software accompanying this work is deployed inside the Docker image `xilef1337/barmap-qa` hosted on Docker Hub. It is downloaded automatically when executing the commands provided in the Methods section. The only prerequisite is a working Docker (Community Edition) installation and at least 8 GB of main memory. It supports Microsoft Windows 10 (Pro, Enterprise, and Education editions), Apple macOS, as well as many Linux distributions (e.g., Ubuntu, Debian, Fedora, etc.).

Please refer to Docker Docs[1] on how to install Docker on your platform. **Note 4.1** contains instructions on how to avoid common issues encountered during the Docker setup.

Alternatively, a source distribution for Linux is provided on github.[2] It has been tested on the distributions Fedora 27 and 30 as well as Ubuntu 18.04 LTS. The user is responsible for providing all external dependencies as explained in the `README` file. Afterward, the shell script `install.sh` performs all necessary setup steps. We highly recommend to run the test suite by executing the command `make test` inside the distribution directory after the setup script completed successfully. To reproduce the data from this protocol, we suggest using the same versions of all external software, namely, ViennaRNA v2.4.13, Barriers v1.6.0, and Treekin v0.4.2, which is equivalent of using the docker image `xilef1337/barmap-qa:0.5.7`.

3 Methods

In this section, the application of the BarMap-QA pipeline to investigate the synthetic riboswitch RS10 [25] is described. In the original contribution, the authors aimed for designing riboswitches that regulate transcription termination. Therefore, six riboswitch constructs where experimentally investigated, and RS10 performed best, showing a threefold activation ratio [25]. A cotranscriptional folding analysis of RS10 should show an interesting switching behavior at least between two designed states: (i) the aptamer state, which is able to sense the ligand theophylline, and (ii) the terminator state, which disrupts the binding-competent aptamer conformation and triggers transcription termination. Following the subsequent step-by-step instructions, the reader will conduct a detailed kinetic analysis of the riboswitch RS10 using default parameter settings:

Step 1. We assume that the reader has set up Docker as described in Paragraph "System Requirements and Setup". To test your Docker installation, open a terminal (on Windows: a regular PowerShell window), and run:

```
docker run hello-world
```

You should get a message stating that Docker appears to be running correctly. In case of problems, please refer to **Note 4.1**.

[1] https://docs.docker.com/install/

[2] https://github.com/xileF1337/barmap-qa

Step 2. Update and run the `barmap-qa` Docker container. For Linux, we provide the convenience shell script `run_barmap-qa.sh`[3] that automatically updates the Docker image and starts a new container. Open a terminal running the Bash shell and execute the script. Provide a working directory that is mounted inside the container (here, the current directory "."):

```
./run_barmap-qa.sh .
```

for Windows and macOS, the Docker image can be updated by running

```
docker pull xilef1337/barmap-qa
```

inside a terminal. The container can be started using the following command:

```
docker run --rm -v C:\path\to\my\directory:/host \
-it xilef1337/barmap-qa
```

where `C:\path\to\my\directory` is the absolute path to the (existing) working directory of choice. The path to the current directory can be queried using the `pwd` command.

In case the container is started correctly, all generated output will be stored in the mounted working directory of the host machine and will, thus, still be available after the Docker container is shut down. The command prompt should look similar to:

```
barmap-qa@489979a0d4ed:host$
```

The number displayed in the prompt is a random container id and varies from run to run. The current directory, `host`, should contain all files present in the passed working directory of the host, as can be verified by running `ls`. Finally, create a new sub-directory and change into it:

```
mkdir 'kinetics_sim_rs10'
cd 'kinetics_sim_rs10'
```

[3] https://github.com/xileF1337/barmap-qa/

Step 3.
Now, an initial cotranscriptional folding simulation is performed. Store the RNA sequence in a variable, and run the all-in-one pipeline script to generate a complete set of output files:

```
rs10='AAGUGAUACCAGCAUCGUCUUGAUGCCCUUGGCAGCACUUCAGAAAUCUCUGAAGUGCUGUUUUUUUU'
barmap_gen_barmapfile -E 5 "$rs10"
```

The switch `-E` sets the enumerated energy range as described in Paragraph "Enumeration of the Energy Landscape: RNAsubopt". There is, however, a technical peculiarity to consider here: `-E` defines a *relative* threshold $\Delta\Delta G_{\mathrm{enum}}$ with respect to the MFE of the current sequence ΔG_{mfe} such that the *absolute* threshold is $\Delta G_{\mathrm{enum}} = \Delta G_{\mathrm{mfe}} + \Delta\Delta G_{\mathrm{enum}}$.

Here, a threshold of $\Delta\Delta G_{\mathrm{enum}} =$ 5 kcal mol^{-1} is used. The lower the value of `-E` is chosen, the faster the computation and the less resources (computation time and memory) will be consumed. The results of the kinetic simulation are automatically plotted and stored in both SVG and PDF format using the batch mode of the plotting tool Grace [22]. For a complete list of all generated files, we refer to **Note 4.2**.

Step 4. After generating initial results, the quality of the individual energy landscapes can be evaluated. Therefore, it is verified how well an enumeration applying the arbitrarily chosen threshold of 5 kcal mol^{-1} covers the space of all possible RNA secondary structures of RS10. This evaluation is performed automatically by `barmap_gen_barmapfile` but can also be called manually:

```
barriers_coverage -m $(ls *.bar | sort -n)
```

The switch `-m` adds asterisks (`*`) as markers to entries with low coverage. The notation `$(...)` triggers command substitution, and its content is evaluated and inserted. The contained sub-command `ls *.bar | sort -n` lists all Barriers files in the current directory and sorts them numerically such that the growing RNA sequences are processed in the order of their respective lengths. Results are printed on standard out:

```
file_name #conStr #struct cs/s [ ... ] fulRangeNrg EnsembleNrg conCovr fulCovr lowCv
    9.bar       4        4 100% [ ... ]   -0.003685    -0.004060  99.94%  99.94%
  [ ... ]                       [ ... ]
   21.bar      61       61 100% [ ... ]   -1.654210    -1.665671  98.16%  98.16% **
   22.bar      27       27 100% [ ... ]   -2.003425    -2.050692  92.62%  92.62% ****
   23.bar      27       27 100% [ ... ]   -2.003425    -2.061808  90.96%  90.96% ****
   24.bar      27       27 100% [ ... ]   -2.003425    -2.074347  89.13%  89.13% *****
   25.bar      27       27 100% [ ... ]   -2.003425    -2.458591  47.78%  47.78% *****
   26.bar      28       28 100% [ ... ]   -5.368760    -5.372944  99.32%  99.32% *
  [ ... ]
```

Note that columns or rows omitted due to space limitations are indicated with `[ ... ]` throughout the protocol. The rightmost column, labeled with header the `lowCv` , shows the markers activated by the `-m` option. The more markers are displayed, the lower the coverage is. The exact thresholds at which a certain number of asterisks are given as well as a key to the remaining listed information can be obtained from the help page of the script:

```
barriers_coverage --help
```

One sees at first glance that many energy landscapes suffer from a bad coverage. Specifically, Barriers files 22–25 (see example output above), 33–37, and 56–60 have less than 95% coverage and are thus marked with at least four asterisks. As an extreme example, file 25 has only 47.78% coverage, i.e., when this sequence folds into a random structure, in the majority of the cases, it will not be represented in the generated energy landscape. The potential to observe such a discrepancy increases the longer the sequence gets (cf. Fig. 3).

To improve on these results, the exploration threshold $\Delta\Delta G_{\mathrm{enum}}$ needs to be increased. For now, it suffices to increase the threshold for *all* sequences by rerunning the `barmap_gen_barmapfile` command and changing `-E` to 8 kcal mol^{-1} (cf. **Step 3**). Note that by rerunning the respective command, all the previously generated output will be overwritten.

Evaluation of the resulting energy landscapes utilizing again the `barriers_coverage` script now shows that an ensemble coverage of at least 97% has been achieved and all landscapes are thus sufficiently well approximated. For an equilibrated RNA molecule, a missing structure will, therefore, not appear with a probability higher than 3% in any intermediate landscape.

Step 5. We now proceed to the evaluation of the quality of the "bar map", i.e., the mapping from each coarse-grained energy landscape to the next one. As before, the quality measures for the mapping are generated automatically by `barmap_gen_barmapfile` or can be computed manually by executing:

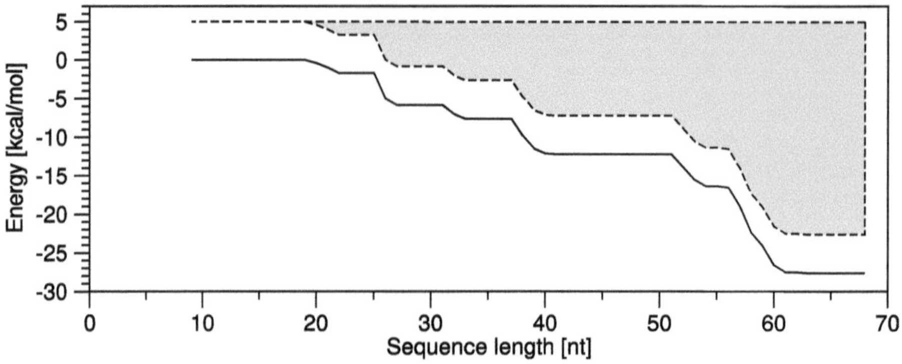

Fig. 3 Effect of an arbitrarily chosen $\Delta\Delta G_{\text{enum}}$ of 5 kcal mol^{-1} as exploration threshold. The solid black line depicts the energy of the most stable state at each sequence length. As expected, the RNA's MFE is dropping significantly with increasing length of the transcript. The dashed line above indicates the upper bound of a 5 kcal mol^{-1} energy band, and the gray shaded area highlights the increasing amount of potentially missed conformations

```
barmap_exact_coverage -m 'barmap.out'
```

Option `-m` again activates markers for bad mappings, and the file `barmap.out` contains the mappings of all Barriers files. For the previously generated results, enumerating all structures in the energy band of 8 kcal mol^{-1}, a dramatic error occurs when mapping Barriers file 25 to file 26. While both individual landscapes expose a high ensemble coverage, a randomly chosen RNA structure will be exactly mapped to the landscape of file 26 in only about 50% of the cases. In the remaining cases, the structure will be mapped approximately to the state of lowest base pair distance, introducing an error into the simulation. In contrast to the original BarMap implementation as described in Paragraph "Cotranscriptional Folding: BarMap", all mappings with a base pair distance less than three are counted as exact mappings in the BarMap-QA pipeline.

The approximately mapped minima are further investigated by rerunning the evaluation script and adding the `-a` option to the command. Note that single-letter options can be bundled using a single dash:

```
barmap_exact_coverage -ma 'barmap.out'
```

The extended output now summarizes the incorrectly mapped minima together with other useful information like the base pair distance (here seven base pairs) as well as the saddle height (here -1.70 kcal mol^{-1}) with respect to the global MFE basin:

```
from_file   to_file #exact #basin %exactBsn %%exactCov %totalCov %exactCov  lowCv
----------------------------------------------------------------------------------
   [ ... ]
25.bar ->   26.bar      21      61    34.43%      50.44%       99.97%    50.42% *****
  43.89%:         1 (((((((......)))).)))....       -1.70   0   8.00  saddle:  -1.70 kcal/mol
       ~>         4 ........................        0.00   1   2.70  dist:       7 bp
   3.95%:         4 ...((((......))))........       0.10   1   0.40  saddle:   0.50 kcal/mol
       ~>         4 ........................        0.00   1   2.70  dist:       4 bp
   0.47%:         8 ..(((.......)))..........       1.00   2   2.20  saddle:   3.50 kcal/mol
       ~>         4 ........................        0.00   1   2.70  dist:       3 bp
  [ ... ]
```

The global MFE basin of file 25, represented by a hairpin structure, is mapped to the open-chain conformation in file 26. This mapping completely destroys the characteristic structural feature of the MFE basin and is therefore a bad choice. Additionally, since often the MFE basin has a high Boltzmann weight and thus a high probability, it accounts for 43.89% of the incorrectly mapped probability. This mapping error is fixed by increasing the value of the parameter `-E` to 9 kcal mol^{-1} when rerunning the `barmap_gen_barmapfile` analysis. In the resulting Barriers file 26, the cause of the erroneous mapping becomes obvious (run `head -n6 26.bar`):

```
  AAGUGAUACCAGCAUCGUCUUGAUGC
1 ...........(((((.....)))))  -5.00   0   9.00   [ ... ]
2 ...((....))(((((.....)))))  -3.30   1   0.90   [ ... ]
3 ..((...))..(((((.....)))))  -2.30   1   0.90   [ ... ]
4 ((((((......)))).))).....   -1.70   1   5.20   [ ... ]
5 ........................    0.00   1   2.70   [ ... ]
```

Minimum 4 (-1.7 kcal mol^{-1}) is connected to its parent minimum (basin 1) via an energy barrier of 5.2 kcal mol^{-1}, such that the saddle structure with an energy of 3.2 kcal mol^{-1} ($= -1.7$ kcal mol^{-1} $+ 5.2$ kcal mol^{-1}) needs to be enumerated to connect this basin to the energy landscape. However, given the global MFE of the sequence from `26.bar`, -5 kcal mol^{-1}, a global exploration threshold of 8 kcal mol^{-1} was insufficient as it enumerates only structures up to 3 kcal mol^{-1}($= -5$ kcal mol^{-1} $+ -8$ kcal mol^{-1}), thus leaving the required structure disconnected in the previous run.

With the increased threshold, the fraction of the exactly mapped probability mass has been increased to more than 99% for most of the mapping steps, with only three steps having 97–98% of exact mappings. Consequently, when using this mapping for randomly chosen structures, there will be a mapping error in less than 3% of the cases.

Step 6. In the previous steps, the quality of the mapping has been verified, guaranteeing correctness if each sequence has sufficient time to equilibrate. However, since the whole point of cotranscriptional folding simulations is to observe effects resulting from the population of transient states, this guarantee is insufficient. Therefore, an approach to additionally assess the quality of a simulation run is required.

The script `barmap_gen_barmapfile` also generates population mapping statistics, which indicate how well the populations of the individual states of one energy landscape are mapped to the next one. Again, problematic transitions are marked automatically using asterisks. For the transition from file 26 to 27, one can observe a severe mis-mapping that transfers about 60% of the population incorrectly to the target landscape. For closer examination, the statistics can be recomputed with more detailed information by executing:

```
barmap_exact_population -ma barmap.out barmap.out.kin_t8_1e3
```

where, again, the `-m` switch enables the asterisk markers and `-a` emits every approximately mapped minimum. The issue is quickly identified:

```
from_file    to_file %exactPop lowPop
------------------------------------
  [ ... ]
  26.bar ->  27.bar     40.02 *****
       54.80%:    4 (((((((......)))).))).....   -1.70    1   5.20  saddle:    3.50 kcal/mol
            ~>    4 ..........................    0.00    1   2.70  dist:      7 bp
        4.87%:    6 ...((((......)))).........    0.10    4   0.40  saddle:    3.50 kcal/mol
            ~>    4 ..........................    0.00    1   2.70  dist:      4 bp
  [ ... ]
```

Minima 4 and 6, both exhibiting a leading hairpin structure, are mapped to the open chain. Similar to the problem described above, the exploration threshold $\Delta\Delta G_{\text{enum}} = 9$ kcal mol^{-1} is insufficient to connect and explore the saddle point of minima 4 and 6 at 3.5 kcal mol^{-1}: in file 27, the MFE is -5.8 kcal mol^{-1}, and therefore only structures up to 3.2 kcal mol^{-1} are connected to the landscape with $\Delta\Delta G_{\text{enum}}$ equal to 9 kcal mol^{-1}. This mapping error did not appear in the last run as the previous version of `26.bar` contained neither basin 4 nor basin 6.

Of course, further increasing the exploration threshold to 10 kcal mol^{-1} and running

```
barmap_gen_barmapfile -E 10 "$rs10"
```

solves this issue for the mapping between files 26 and 27. However, the problem is only postponed to a later transcription step, i.e., the one from file 31 to file 32, because the leader hairpin stays populated for a long time period and the MFE value decreases the longer the RNA sequence gets (cf. Fig. 3).

At a first glance, it seems reasonable to simply enumerate all Barriers files such that the saddle point at 3.5 kcal mol^{-1} is included. When looking at the low global MFE of -27.6 kcal mol^{-1} for the full-length sequence, however, it becomes apparent that this is not a feasible option. A (quite lengthy) run of RNAsubopt enumerating all structures in the energy range of 31.1 kcal mol^{-1} above the MFE reports a total of about 4.9 billion structures: by far too many to serve as input to Barriers.

! Input Size for Barriers
The run time and memory requirements of Barriers strongly depend on the number of input structures. As a rule of thumb, ten million of structures may easily be processed, while 50 million can be processed on a compute server with at least 30 gigabytes of main memory. To count the number of input structures, we refer to **Note 4.3**. Of course, these values also depend on the sequence length; our example RS10 consists of 68 nucleotides.

Returning to the enumeration of our transient states, we realize the population of these decreased from 54.80 to 19.13% already and will likely vanish within another few simulation steps, so it is sensible to consider a change in strategy. From now on, only some selected energy landscapes will be recomputed while reusing the results for the remaining landscapes. File 32 to file 50 are enumerated up to 3.5 kcal mol^{-1} when running:

```
barmap_rebar -a 3.5 {32..50}.bar
```

Parameter `-a` is used to specify the *absolute* enumeration threshold ΔG_{enum}. It is internally converted into the *relative* threshold $\Delta\Delta G_{\mathrm{enum}}$ depending on the sequence MFE, which is then used to call RNAsubopt and Barriers. Note that the number of connected minima in file 50 increased from 86 to 2668; thus, the execution of the command may take a few minutes. The kinetics simulation is recomputed by running:

```
barmap_remap 31.bar
```

Here, `barmap_remap` is given the filename of file 31 to start the mapping and simulation process at the corresponding sequence length and to connect the recalculated Barriers files. Since the quality for files 9–30 has already been verified, there is no need to waste resources by repeating these computations over and over again.

As can be verified by another run of `barmap_exact_population`, the quality of the recomputed part of the simulation is already quite good. However, the subsequent mapping step from file 50 to 51 only has an accuracy of about 92%. Most of the incorrectly mapped minima are connected via a saddle at 0.9 kcal mol^{-1}. A further accuracy improvement is achieved by enumerating files 51 to 55 up to a higher energy level:

```
barmap_rebar -a 0.9 {51..55}.bar
barmap_remap 50.bar
```

The next drop in exactness occurs when mapping the states of file 59 to file 60:

```
59.bar -> 60.bar     93.33 ***
     3.02%:   6 ((((((......)))).)))..((((...))))((((((((((....)))))))))).  -21.00 \
                1   5.90  saddle: -15.10 kcal/mol
        ~>   18 ............((((.....))))((...))(((((((((((....))))))))))) -20.50 \
                1   2.60  dist:    15 bp
     2.44%:   7 ((((((......)))).)))...(((...)))(((((((((((....)))))))))))  -21.00 \
                6   0.90  saddle: -15.10 kcal/mol
        ~>   18 ............((((.....))))((...))(((((((((((....))))))))))) -20.50 \
                1   2.60  dist:    13 bp
```

Minima 6 and 7 are connected to the MFE via a saddle structure with an energy of -15.5 kcal mol^{-1} in file 59. Therefore, file 60 and all subsequent files need to be recomputed. A remapping of the corresponding files to an absolute enumeration threshold of 15.1 kcal mol^{-1} is performed running:

```
barmap_rebar -a -15.1 {60..68}.bar
barmap_remap 59.bar
```

Note that the enumeration threshold of files 56–59 is still 10 kcal mol^{-1} and requires no further change, which significantly speeds up the analysis as, for example, the recalculation of only file 55 already requires 5:24 min on an Intel® Core™ i7-4770 (4×3.4 GHz) as well as 1.46 GB of main memory. The sequences in files 60 to 68 share a similar MFE, ranging from -26.6 to -27.6 kcal mol^{-1} (see Fig. 3). Here, the change in

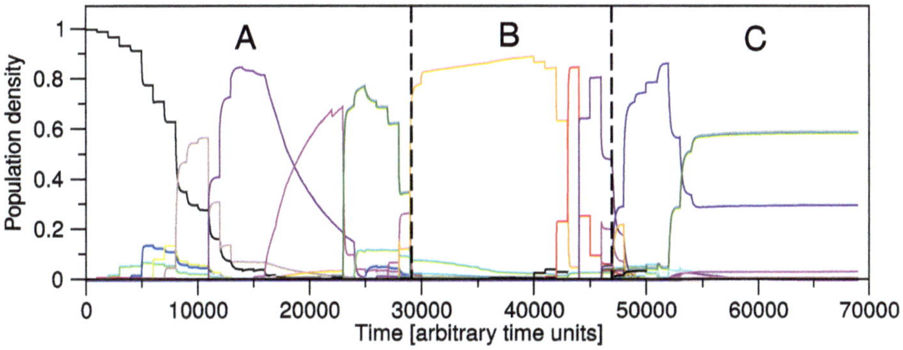

Fig. 4 Final results of the cotranscriptional folding simulation of RS10 with BarMap-QA. In conjunction with the text-based output of `barmap_show_run`, the line plot can be split into three functional parts: (**A**) the transcription and formation of the theophylline aptamer, (**B**) the ligand sensing window, and (**C**) the terminator formation. Although curves appear and disappear within the latter two, these steps are dominated by structures representing the binding-competent aptamer conformation at the 5'-end or the terminator hairpin at the 3'-end of the actual transcript in (**B**) and (**C**), respectively. Therefore, the actual switching between ligand sensing and terminator formation appears around 47,000 arbitrary time units

strategy paid off by making the calculation feasible. A final test of the mapping proofs its exactness. To compute the full kinetics, rerun `barmap_remap` without passing a file name.

Step 7. The final results can now be inspected. A line plot is automatically generated using the batch mode of Grace and stored in `barmap.out.kin_t8_1e3.merge.pdf`. An annotated version of this plot is shown in Fig. 4. For similar graphics of all intermediate refinement steps, we refer to **Note 4.7**. In such a plot, the *x*-axis shows the time (in arbitrary time units), and the *y*-axis indicates the population density ranging from 0 to 1. Each colored curve is a sequence of basins represented by locally minimal structures that are mapped onto each other by the generated mapping, i.e., the minima of one curve have a similar secondary structure.

Additionally, integrated simulation run information can be obtained by applying:

```
barmap_show_run barmap.out barmap.out.kin_t8_1e3 | less -S
```

The resulting text-based output summarizes, for each Barriers file, the population densities at the beginning and the end of the corresponding simulation step:

```
# final_population init_population min_index struct basin_mfe father_basin \
  barrier_height maps_to mapped_from
# '9.bar' (t0=1.0000e-01):
#                          AAGUGAUAC
   99.66% <= 100.00%   1 .........    0.00        0  5.80 ->    1  <= {}
# '10.bar' (t0=1.0001e+03):
#                          AAGUGAUACC
   98.77% <=  99.66%   1 ..........    0.00       0  7.30 ->    1  <= {1}
    1.23% <=   0.34%   2 ..((...)).   2.70        1  0.90 ->    3  <= {}
# '11.bar' (t0=2.0001e+03):
#                          AAGUGAUACCA
   96.58% <=  98.77%   1 ...........    0.00      0  7.40 ->    1  <= {1}
    2.24% <=   0.00%   2 ...((....))   2.30       1  0.30 ->    2  <= {}
    1.17% <=   1.23%   3 ..((...))..   2.70       1  0.90 ->    3  <= {2}
# '12.bar' (t0=3.0001e+03):
#                          AAGUGAUACCAG
   93.14% <=  96.58%   1 ............    0.00     0 10.00 ->    1  <= {1}
    5.73% <=   2.24%   2 ...((....)).   1.70      1  0.90 ->    2  <= {2}
    1.13% <=   1.17%   3 ..((...))...   2.70      1  0.90 ->    4  <= {3}
# '13.bar' (t0=4.0001e+03):
#                          AAGUGAUACCAGC
   90.97% <=  93.14%   1 .............    0.00    0  9.80 ->    1  <= {1}
    5.94% <=   5.73%   2 ...((....))..   1.70     1  0.90 ->    3  <= {2}
    1.99% <=   0.00%   3 ..((.......))   2.40     1  2.60 ->    2  <= {}
    1.10% <=   1.13%   4 ..((...))....   2.70     1  0.90 ->    4  <= {3}
# '14.bar' (t0=5.0001e+03):
#                          AAGUGAUACCAGCA
   77.31% <=  90.97%   1 ..............    0.00   0  9.80 ->    1  <= {1}
   13.81% <=   1.99%   2 ..((.......)).   1.10    1  2.70 ->    2  <= {3}
    7.22% <=   5.94%   3 ...((....))...   1.70    1  0.90 ->    4  <= {2}
    0.93% <=   1.10%   4 ..((...)).....   2.70    1  0.90 ->    5  <= {4}
[ ... ]
```

Note that each `t0` indicates the simulation start time in arbitrary time units (au) in the next landscape. Furthermore, for each populated basin, the representative structures and the corresponding MFE values are listed. In the example output above, the open chain represents the majority of all populated conformations. In Step 14, the corresponding density drops from 90.97 to 77.31%, and other, more structured states become populated. In conjunction with the line plot in Fig. 4, these information enable the user to easily analyze the generated results.

For the analyzed sequence RS10, the final kinetics simulation shows that as soon as the 42 nt long aptamer sequence is transcribed, the corresponding state is populated with a density of about 86% (cf. Fig. 4 at 33,000 au). Although the corresponding curve declines after 42,000 au, other states still containing the binding-competent aptamer conformation at their 5'-end are highly populated. This rapidly changes after 47,000 au when a sufficiently large part of the terminator hairpin is transcribed. From now on, the terminator structure gets extended step by step, and, thereby, the aptamer structure is suppressed. The exact timing and the speed of switching between the aptamer and the terminator state seem to be essential for functional transcription-regulating riboswitches [25, 26]. A close examination of the supposed switch of states after 51,000 au reveals that all the highly populated states contain a complete terminator hairpin and only the structural left-overs of the aptamer conformation differ. In other words, when asking for the fraction of states containing the terminator structure, the mass of all the highly populated states could be added up. However, this type of coarse-graining is beyond the scope of this protocol.

4 Notes

4.1 Docker Installation and Configuration

For Windows and macOS, download the Docker software from the corresponding official website. After the installation, you may need to reboot your machine. Ensure the Docker application is started before attempting to use the `docker run` command. On Windows, non-privileged users have to be added to the `docker-users` group to be able to start any Docker containers. On Linux, Docker can usually be installed via the system package manager (package `docker` on Fedora and Ubuntu). Please note that you may have to add a `docker` group (`sudo groupadd docker`) and add the current user to it (`sudo usermod -aG docker "$USER"`) if you get an error when using `docker run` . Additionally, it may be necessary to configure the system to automatically run the Docker daemon when the system is booting (`sudo systemctl enable docker`). On both Windows and Linux, log out and log in again after adding users to a group to apply the updated settings (or reboot the machine to be on the safe side). If you get "out of memory" or "segmentation fault" errors, add the option `-m 8G` to all `docker run` commands; it ensures that at least 8 GB of main memory are available to the container. Increase the memory limit if you apply the pipeline to longer input sequences.

4.2 Output Files

Table 1 lists all files generated by running the all-in-one script `barmap_gen_barmapfile`, which automatically performs a full BarMap-QA analysis without user interaction.

4.3 Enumeration of Structures

As described in Paragraph "Enumeration of the Energy Landscape: RNAsubopt", Wuchty's algorithm can be applied to enumerate all structures up to a given energy threshold ΔG_{enum}. However, the exponential dependence of the number of secondary structures on the sequence length imposes practical limitations upon the choice of this parameter, especially if the output is supposed to be

Table 1
Files that are generated when executing the command given in Step 3

File	Description
*.bar	Standard output of Barriers containing information about the respective coarse-grained landscape, e.g. representative structure and the barrier height
*.bar.log	Log file that summarizes the Barriers run
*.rates.bin	Binary rate matrix of the respective landscape generated by Barriers
*.evals.bin *.evecs.bin	Eigenvalues and -vectors of the respective rate matrix stored in a binary format. They are used in the final simulation step to speed up the calculation
*.rates.bin.log	Log file of the diagonalization process performed by Treekin
barmap.out	BarMap's state mapping indicating exact (->) and approximate (~>) mappings between consecutive landscapes
barmap.out.kin_t8_1e3	Full kinetics simulation output generated by multiple consecutive Treekin simulations over all landscapes
barmap.out.kin_t8_1e3.filt	Kinetics simulation output filtered only for highly populated states
barmap.out.kin_t8_1e3.filt.[pdf | svg]	Plot of the filtered Treekin simulation in SVG and PDF format
barmap.out.kin_t8_1e3.merge	Filtered and merged (cf. above) kinetics simulation output
barmap.out.kin_t8_1e3.merge.[pdf | svg]	Plot of the merged Treekin simulation in SVG and PDF format

processed by computationally demanding methods such as Barriers. Here, we show how to determine the number of structures for a given value of ΔG_{enum} such that the user can make a qualified decision about the feasibility of continuing the analysis.

As an example, we consider again the riboswitch RS10, for which a saddle state at $\Delta G_{\mathrm{enum}} = 3.5$ kcal mol^{-1} had been identified. To connect the respective minimum to the global MFE basin located at 27.6 kcal mol^{-1}, $\Delta\Delta G_{\mathrm{enum}}$ needs to be set to 31.1 kcal mol^{-1}. Running

```
RNAsubopt -e 31.1 <<< "$rs10" | perl -wlne 'print $. unless $. % 1e6'
```

reports about 4.9 billion structures. In detail, the one-line Perl script counts the number of enumerated conformations, indicating its progress every one million structures. % is the modulo operator and $. stores the current line number. The switches passed to the

Perl interpreter enable warnings (`-w`), automatically add newline characters to printed lines (`-l`), enable line-wise processing of the standard input stream (`-n`), and specify that the script to be executed follows directly on the command line (`-e`).

4.4 Parallelization

The script `barmap_rebar` supports multi-threading. The number of threads to launch can be specified using the switch `-t`. This switch is also supported by `barmap_gen_barmapfile`, which passes it to all other scripts it is calling:

```
barmap_gen_barmapfile -t 4 -E 5 "$rs10"
```

Note that running multiple instances of Barriers with a high number of sequences requires a lot of main memory. We recommend keeping an eye on the spawned threads using a system monitor and setting a per-process memory limit using the shell built-in `ulimit`. For example,

```
ulimit -m $((8 * 10**6))
```

sets the maximum resident set size (i.e., the amount of occupied main memory) for each sub-process of the current shell to 8 GB. The notation `$(( ... ))` triggers arithmetic expansion of its content and `**` is the exponentiation operator. The value to `-m` is assumed to be given in kilobytes.

4.5 The Influence of Transcription Rates

Obviously, how close the distribution of structures of a given RNA sequence is to the equilibrium distribution depends on the time available for equilibration per step before the sequence is elongated again, i.e., the rate of transcription. With high transcription rates (i.e., many added nucleotides (nt) per time unit), a transient state may still be highly populated when the RNA chain is extended, and thus a different part of the energy landscape – which could hardly be reached from the ground state – may become accessible. As a consequence, the quality analysis of the kinetics simulation needs to be repeated if the latter is performed with a different transcription rate.

The BarMap pipeline has no option to directly set the transcription rate, but, instead, the end time (and thus the duration) of the simulation for each elongation step can be defined in *arbitrary time units (au)*. The question how exactly they correspond to real time is left to the user. Recently, 1 au nt^{-1} has been mapped to 5×10^{-6} s nt^{-1} [19], an estimate also applicable to BarMap [1]. Thereby, a transcription rate of 200 nt s^{-1} translates into 1000 au s^{-1}, which is equivalent to 0.005 s nt^{-1}. These are only rough estimates, and the true transcription rate varies for different types of RNA molecules [2], environmental conditions such as temperature [18], or growth phase [24]. We therefore advice the reader to tune this parameter to known experimental data, ideally from the biological system of interest.

To change the time per elongation step, the `-e` option of `barmap_remap` can be set:

```
barmap_remap -e 4e3
```

This changes the default duration of 1×10^3 au to 4×10^3 au, which corresponds to a transcription rate of 50 nt s^{-1} following the approximations above. Additionally, BarMap-QA adjusts the simulation time of the last landscape to be ten times as long as the simulation time of all other landscapes. This allows the user to get an impression of the equilibrated structure distribution after the transcription process has finished. Note that when running `barmap_remap`, only the final BarMap simulation step is repeated. Therefore, previous results are kept, and new files with the prefix `barmap.out.kin_t8_4e3` are created.

4.6 Post-processing of the Kinetics File

BarMap-QA provides two types of plots, marked with either the `.filt` or the `.merged` suffix, which result from two different post-processing procedures. The first one is a simple filter that removes any state that does not acquire a minimal population from its output and also allows to limit the total number of states in each simulation step. Manually calling the script `barmap_filter_treekin` allows the user to adjust the minimum population threshold using the `-p` switch. The filtered output can then be plotted using `barmap_plot_treekin`, which supports multiple output formats (currently PDF, SVG, and PNG). Strikingly, the color of the population curves change every time a new Barriers file is processed (i.e., if the next nucleotide is added to the RNA sequence), which can be helpful during the analysis of the plot. The second option is provided by the script `barmap_merge_treekin`. It offers the same filtering options but additionally merges the states of successive Barriers simulations in the output according to the generated BarMap file. As a result, the population curves have a single color, which often provides a more pleasant visual experience. The output can, again, be processed with `barmap_plot_treekin`.

4.7 Refinement Visualization

A visualization of each intermediate refinement steps can be obtained by inspecting the corresponding `barmap.out.kin_t8_1e3.merge.pdf`. Figure 5 depicts a summary of all refinement steps described in the methods section.

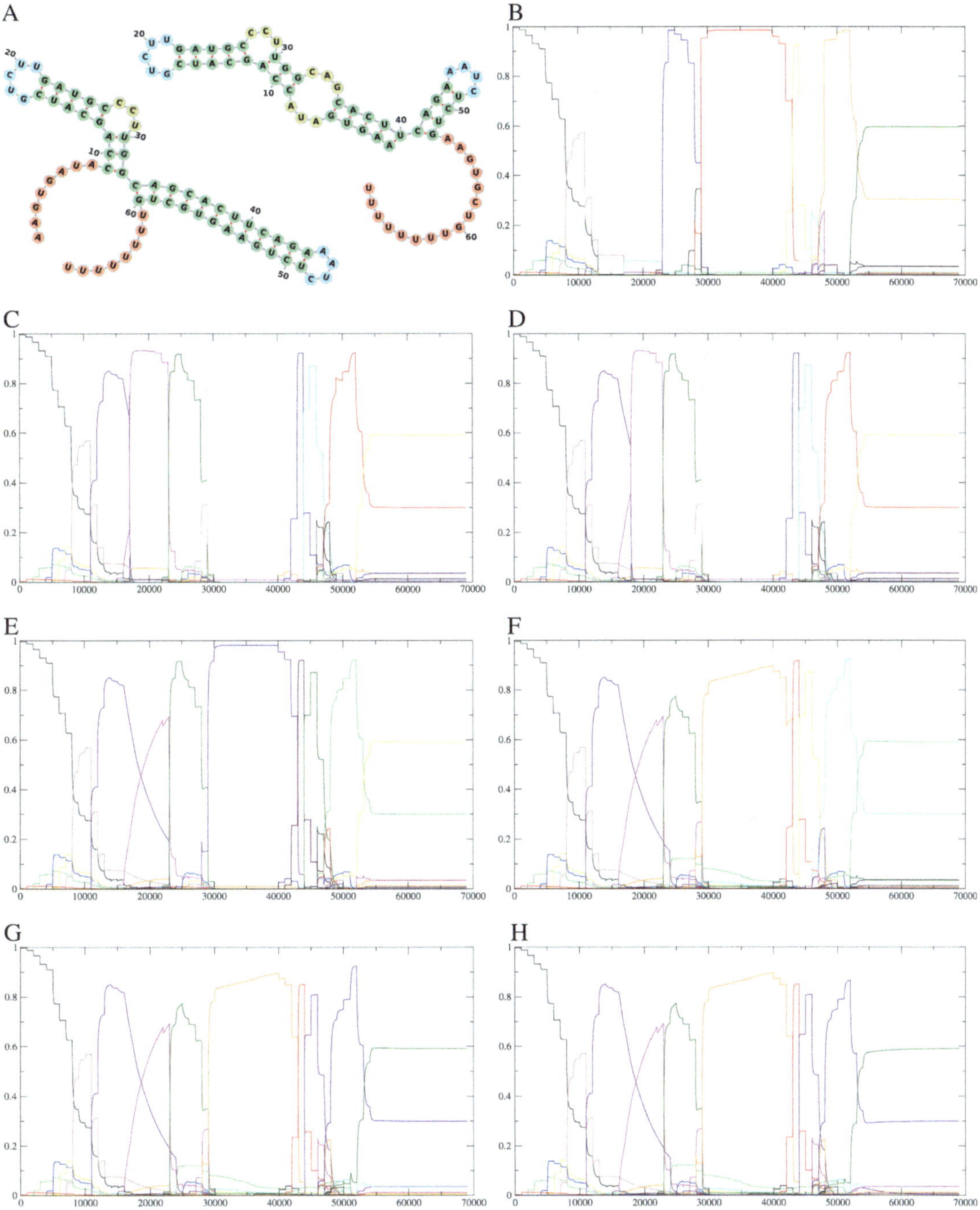

Fig. 5 Refinement steps of the folding kinetics analysis of the transcriptional theophylline riboswitch RS10. (**A**) Secondary structure plots of RS10 in its *on* state (*top right*) and its *off* state (*bottom left*). While the *on* state displays the aptamer's binding pocket consisting of two characteristic interior loops, the *off* state features a stable hairpin loop followed by a poly-U tail, which in conjunction constitute an intrinsic terminator. (**B**) Initial kinetics plot after enumerating all landscapes by 5 kcal mol^{-1}. (**C**–**E**) Kinetics plots after increasing the relative enumeration threshold to 8, 9, and 10 kcal mol^{-1}, respectively. (**F**) Kinetics plot after enumerating Barriers files 32–50 up to 3.5 kcal mol^{-1}. (**G**) Kinetics plot after enumerating Barriers files 51–55 up to 0.9 kcal mol^{-1}. (**H**) Kinetics plot after enumerating Barriers files 60–68 up to -15.1 kcal mol^{-1}

4.8 Extending BarMap-QA

BarMap is designed to be applicable to a broad range of problems and thus very flexible. Due to the semiautomatic architecture of our pipeline, this flexibility is transparently propagated despite the added functionality. For instance, the user can provide custom rate matrices accounting for ligand binding or changing environmental temperatures and may still compute the quality statistics as described above.

The current BarMap implementation does not support variations in the rate of transcription along the input RNA – an effect that has been reported in literature [4, 17]. Such a functionality would be desirable, for example, to understand the effect of pause sites [8, 21] on cotranscriptional RNA structure formation. We expect that this functionality will become available in a future release of BarMap.

pourRNA, a memory-efficient and parallelizable alternative to Barriers, has been released recently [5]. Integrating this new tool into BarMap-QA might speed up the computation but remains as a future task.

4.9 Feasibility of Cotranscriptional Folding Simulations

Apart from the length of the input sequence, the feasibility of a BarMap analysis also depends on the sequence's folding characteristics. An RNA molecule with a smooth energy landscape with a dominant global minimum will be much easier to simulate than bistable sequences exhibiting highly probable alternative conformations or "folding traps". The reason is that persistent intermediate folding states require an enumeration up to a constant energy level during the entire elongation process, leading to a combinatorial explosion of the number of secondary structures. In contrast, if the most likely folding paths quickly lead to the global minimum, only a small number of structures need to be enumerated to obtain high-quality simulation results even for long RNA sequences.

As an example for a sequence that is more difficult to analyze, we encourage the reader to experiment with the nonfunctional riboswitch candidate RS8 [25], whose sequence[4] is very similar to RS10.

Trademark Notices Windows® is a registered trademark of Microsoft Corporation in the USA and other countries. This work is an independent publication and is neither affiliated with nor authorized, sponsored, or approved by Microsoft Corporation or other companies. macOS® is a trademark of Apple Inc., registered in the USA and other countries. Linux® is the registered trademark of Linus Torvalds in the USA and other countries. Docker is a registered trademark of Docker, Inc. in the USA

[4] AAGUGAUACCAGCAUCGUCUUGAUGCCCUUGGCAGCACUUCACUCCUAGUGGAGUGAAGUGCUGUUUUUUUU

List of Abbreviations This work makes use of the following abbreviations:

au	arbitrary time units
E. coli	*Escherichia coli*
MFE	minimum free energy
nt	nucleotide
RNA	ribonucleic acid

Acknowledgements

This work was supported in part by the German Research Foundation (DFG; grants STA 850/15-2 and MO 634/9-2).
We thank Ivo L. Hofacker for fruitful discussions on variations of the BarMap approach, and Anne Hoffmann and Jens Steuck for their IT support to test our software.

References

1. Badelt S, Hammer S, Flamm C, Hofacker IL (2015) Thermodynamic and kinetic folding of riboswitches. Methods Enzymol 553:193–213. https://doi.org/10.1016/bs.mie.2014.10.060
2. Bremer H, Dennis PP (2008) Modulation of chemical composition and other parameters of the cell at different exponential growth rates. EcoSal Plus 3(1) (2008). https://doi.org/10.1128/ecosal.5.2.3
3. Day L, Souki OAE, Albrecht AA, Steinhöfel K (2016) Random versus deterministic descent in RNA energy landscape analysis. Adv Bioinformat 2016:9654921. https://doi.org/10.1155/2016/9654921
4. Dennis PP, Ehrenberg M, Fange D, Bremer H (2009) Varying rate of RNA chain elongation during RRN transcription in Escherichia coli. J Bacteriol 191(11):3740–3746. https://doi.org/10.1128/jb.00128-09
5. Entzian G, Raden M (2019) pourRNA—a time- and memory-efficient approach for the guided exploration of RNA energy landscapes. Bioinformatics. https://doi.org/10.1093/bioinformatics/btz583
6. Flamm C, Fontana W, Hofacker IL, Schuster P (2000) RNA folding at elementary step resolution. RNA 6:325–338. https://doi.org/10.1017/s1355838200992161
7. Flamm C, Hofacker IL, Stadler PF, Wolfinger MT (2002) Barrier trees of degenerate landscapes. Zeitschrift für Physikalische Chemie 216(2/2002). https://doi.org/10.1524/zpch.2002.216.2.155
8. Gerdes K, Wagner EG (2007) RNA antitoxins. Curr Opin Microbiol 10:117–124. https://doi.org/10.1016/j.mib.2007.03.003
9. Hofacker IL, Flamm C, Heine C, Wolfinger MT, Scheuermann G, Stadler PF (2010) BarMap: RNA folding on dynamic energy landscapes. RNA 16(7):1308–1316. https://doi.org/10.1261/rna.2093310.
10. Kucharík M, Hofacker IL, Stadler PF, Qin J (2014) Basin hopping graph: a computational framework to characterize RNA folding landscapes. Bioinformatics 30(14):2009–2017. https://doi.org/10.1093/bioinformatics/btu156. https://academic.oup.com/bioinformatics/article-lookup/doi/10.1093/bioinformatics/btu156

11. Li Y, Zhang S (2012) Predicting folding pathways between RNA conformational structures guided by RNA stacks. BMC Bioinformat 13(3):S5. https://doi.org/10.1186/1471-2105-13-S3-S5
12. Lorenz R, Bernhart SH, Siederdissen CHZ, Tafer H, Flamm C, Stadler PF, Hofacker IL (2011) ViennaRNA Package 2.0. Algorithms Mol Biol 6:26. https://doi.org/10.1186/1748-7188-6-26
13. Mathews DH, Sabina J, Zuker M, Turner DH (1999) Expanded sequence dependence of thermodynamic parameters improves prediction of RNA secondary structure. J Mol Biol 288(5):911–940. https://doi.org/10.1006/jmbi.1999.2700
14. Mathews DH, Disney MD, Childs JL, Schroeder SJ, Zuker M, Turner DH (2004) Incorporating chemical modification constraints into a dynamic programming algorithm for prediction of RNA secondary structure. Proc Natl Acad Sci U S A 101(19):7287–7292. https://doi.org/10.1073/pnas.0401799101
15. McCaskill JS (1990) The equilibrium partition function and base pair binding probabilities for RNA secondary structure. Biopolymers 29: 1105–1119
16. Metropolis N, Rosenbluth AW, Rosenbluth MN, Teller AH, Teller E (1953) Equation of state calculations by fast computing machines. J Chem Phys 21(6):1087–1092. https://doi.org/10.1063/1.1699114
17. Mills DR, Dobkin C, Kramer FR (1978) Template-determined, variable rate of RNA chain elongation. Cell 15(2):541–550. https://doi.org/10.1016/0092-8674(78)90022-3
18. Ryals J, Little R, Bremer H (1982) Temperature dependence of RNA synthesis parameters in Escherichia coli. J Bacteriol 151:879–887
19. Sauerwine B, Widom M (2013) Folding kinetics of riboswitch transcriptional terminators and sequesterers. Entropy 15(8):3088–3099. https://doi.org/10.3390/e15083088
20. Stein PR, Waterman MS (1979) On some new sequences generalizing the Catalan and Motzkin numbers. Discrete Math 26(3):261–272. https://doi.org/10.1016/0012-365X(79)90033-5.
21. Szlachta K, Thys RG, Atkin ND, Pierce LCT, Bekiranov S, Wang YH (2018) Alternative DNA secondary structure formation affects RNA polymerase II promoter-proximal pausing in human. Genome Biol 19:89. https://doi.org/10.1186/s13059-018-1463-8
22. The Grace contributors: Grace – a WYSIWYG 2D plotting tool for the X Window System and M*tif (2015). http://plasma-gate.weizmann.ac.il/Grace/. Online; Accessed 10 Sep 2019
23. Turner DH, Mathews DH (2010) NNDB: the nearest neighbor parameter database for predicting stability of nucleic acid secondary structure. Nucleic Acids Res 38(Database issue): D280–D282. https://doi.org/10.1093/nar/gkp892
24. Vogel U, Jensen KF (1994) The RNA chain elongation rate in Escherichia coli depends on the growth rate. J Bacteriol 176:2807–2813. https://doi.org/10.1128/jb.176.10.2807-2813.1994
25. Wachsmuth M, Findeiß S, Weissheimer N, Stadler PF, Mörl M (2013) De novo design of a synthetic riboswitch that regulates transcription termination. Nucleic Acids Res. 41: 2541–2551
26. Wachsmuth M, Domin G, Lorenz R, Serfling R, Findeiß S, Stadler PF, Mörl M (2015) Design criteria for synthetic riboswitches acting on transcription. RNA Biol 12(2):221–231. https://doi.org/10.1080/15476286.2015.1017235.
27. Wolfinger MT, Svrcek-Seiler WA, Flamm C, Hofacker IL, Stadler PF (2004) Efficient computation of RNA folding dynamics. J Phys A Math Gener 37(17):4731–4741. https://doi.org/10.1088/0305-4470/37/17/005
28. Wuchty S, Fontana W, Hofacker IL, Schuster P (1999) Complete suboptimal folding of RNA and the stability of secondary structures. Biopolymers 49(2):145–165. https://onlinelibrary.wiley.com/doi/10.1002/(SICI)1097-0282(199902)49:2%3C145::AID-BIP4%3E3.0.CO;2-G

Chapter 15

The Multiscale Ernwin/SPQR RNA Structure Prediction Pipeline

Bernhard C. Thiel, Simón Poblete, and Ivo L. Hofacker

Abstract

Aside from the well-known role in protein synthesis, RNA can perform catalytic, regulatory, and other essential biological functions which are determined by its three-dimensional structure. In this regard, a great effort has been made during the past decade to develop computational tools for the prediction of the structure of RNAs from the knowledge of their sequence, incorporating experimental data to refine or guide the modeling process. Nevertheless, this task can become exceptionally challenging when dealing with long noncoding RNAs, constituted by more than 200 nucleotides, due to their large size and the specific interactions involved. In this chapter, we describe a multiscale approach to predict such structures, incorporating SAXS experimental data into a hierarchical procedure which couples two coarse-grained representations: Ernwin, a helix-based approach, which deals with the global arrangement of secondary structure elements, and SPQR, a nucleotide-centered coarse-grained model, which corrects and refines the structures predicted at the coarser level.

We describe the methodology through its application on the *Braveheart* long noncoding RNA, starting from the SAXS and secondary structure data to propose a refined, all-atom structure.

Key words RNA structure, Structure prediction, SAXS, Multiscale modeling, Coarse-grain

1 Introduction

The knowledge of the three-dimensional structure of RNA molecules is fundamental for the thorough understanding of their biological function. In this regard, several computational methods have been proposed during the recent years to predict the structure from the sole knowledge of the nucleotide sequence, incorporating if possible experimental information obtained with techniques such as X-ray spectroscopy, NMR, or cryo-EM, among others. In particular, small-angle X-ray scattering (SAXS) provides structural information of a macromolecule in solution, which constitutes a more realistic scenario under a less invasive scheme. It is therefore desirable to provide a computational approach which models large RNA structures making use of this information, and several programs

Ronny Lorenz (ed.), *RNA Folding: Methods and Protocols*, Methods in Molecular Biology, vol. 2726,
https://doi.org/10.1007/978-1-0716-3519-3_15,

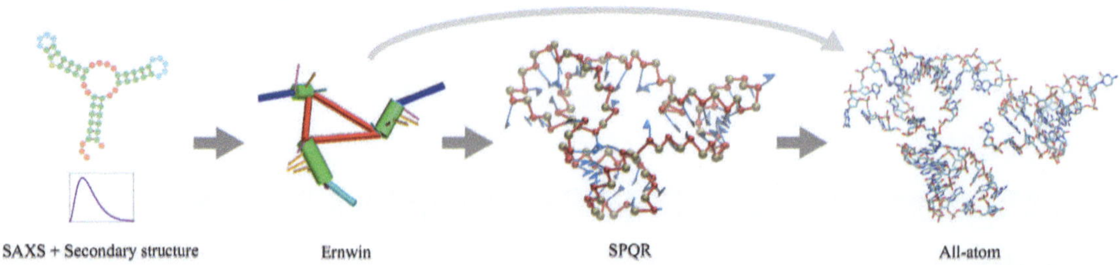

Fig. 1 Ernwin-SPQR pipeline schematized. Starting from the secondary structure and SAXS data, a low-resolution 3D model is built by Ernwin. From here, an atomistic structure can be proposed, which is prone to have clashes and topological links between its elements. A better result can be achieved by refining with the SPQR model at nucleotide level before reconstructing the final all-atom representation

attempt this [1–7]. Here, we describe a multiscale scheme, able to integrate the global arrangement of the RNA domains with local features expressed in a realistic geometry at a nucleotide level, through the prediction of the three-dimensional structure of a lncRNA already studied by some of us [8]. The method, illustrated in Fig. 1, bridges the results of Ernwin, a low-resolution model, with SPQR, a nucleotide-level description.

Ernwin [9] describes an RNA molecule as a set of secondary structure elements. Stems are represented by cylinders which are linked through junctions and internal loops. Ernwin also supports an additional potential based on the pair distance distribution function derived from the SAXS data [8]. An all-atom reconstruction can also be obtained from the proposed structures, although it is likely that some refinement will have to take place in order to ensure that bonds, angles, dihedrals, and positions assume realistic values.

SPQR (SPlit and conQueR) [10], on the other hand, is a coarse-grained (CG) representation for the simulation and modeling of RNA at a nucleotide resolution. The model represents each nucleotide as two elements: a point particle situated at the phosphorous atom which corresponds to the phosphate group and a structured, anisotropic particle which stands for the nucleoside. Altogether, they constitute a set of five dependent particle coordinates. The interactions among these particles include stacking, canonical and noncanonical base-pairing, and base-phosphate interactions, while the backbone conformations are able to specify the sugar pucker and glycosidic bond angle states.

Another important feature is that SPQR supports the steering of a set of fragments toward a template configuration by minimizing the $\mathcal{E}$RMSD metric between them. This procedure allows to apply secondary structure restraints [11] to enforce a particular arrangement of nucleotides as in viral capsids [12] or to construct an atomistic description in explicit solvent starting from a CG conformation [13].

In the following, we will describe briefly the installation steps, give an overview over the basic usage, and illustrate the prediction and refinement pipeline through the modeling of the *Braveheart* long noncoding RNA, a system previously studied in [8].

2 Materials

2.1 Source Code

- Ernwin: https://github.com/ViennaRNA/ernwin
- SPQR: https://github.com/srnas/spqr
- Tutorial Files: https://github.com/ViennaRNA/ERNWIN-SPQR-Tutorial

2.2 Installation

2.2.1 Ernwin

Ernwin requires Python and has been tested on Linux (Ubuntu 16.04, 20.04, and 22.04 with Python 3.10). We recommend using a virtual environment for installing it and its dependencies.

Ernwin can be installed either from the cloned git repository or from the Python package index via pip. Its dependency forgi contains Cython code which requires a compiler and header files for Python and NumPy before installation:

```
sudo apt install gcc g++ python3-dev python3-numpy
python -m pip install ernwin
```

If you like to create your own coarse-grained fragments ("stats") as described in Subheading 3.3.1, you should install DSSR [14] or MC-Annotate [15] to improve the detection of the secondary structure in PDB files (*see* **Note** 3). If the DSSR or MC-Annotate executable is in your `PATH`, it will be used automatically by the scripts described in Subheading 3.3.1.

To perform all-atom reconstructions (as described in Subheading 3.3.3), some resources are needed. Extract the coarse-grained RNA files from the Ernwin repository into one directory (e.g., called `ernwin_cgs`) by running the following command in the target directory:

```
tar -xzf /path/to/ernwin/RESOURCES/CGS.tar.gz
```

Additionally, download the corresponding MMCIF files from the Protein Data Bank. To do so, first, download the representative set of RNA structures [16] version 3.36 at 4 Å cutoff from

```
http://rna.bgsu.edu/rna3dhub/nrlist/download/3.36/4.0A/csv
```

Then, in a new directory, e.g., called `ernwin_pdbs`, give this file to the `download_pdb_files.sh` script provided in the `RESOURCES` directory of the Ernwin repository:

```
path/to/ernwin/RESOURCES/download_pdb_files.sh
↪  path/to/nrlist_3.36_4.0A.csv
```

The experimental SAXS data for Braveheart used for our example can be downloaded from the SASBDB at https://www.sasbdb.org/data/SASDHE3/.

2.2.2 SPQR

The core SPQR programs are written in C. In addition, the simulation package contains interaction energy files and a set of Python and Bash tools which help to prepare and analyze simulations or systematically run a sequence of simulations which are needed for the refinement of large structures. The code has been tested in Mac OS X 10.14.6 and Ubuntu Linux 16.04 and later versions. In addition, the scripts are available for Python 2 and 3.

SPQR can be compiled on any standard C compiler. The way it is built is quite easy and allows to generate a number of binaries which can be used once the path of the interaction energy files is specified. Download the SPQR source from the master branch of the git repository:

```
git clone https://github.com/srnas/spqr
```

Go to the `spqr/src/` directory and run

```
autoconf
./configure
make
make install
```

This will create the binaries in `spqr/bin/` and energy files in the file `spqr/interactions/intrac.btb`. With this, the installation is complete. The tools used for the refinement of structures refined in this tutorial require any version of Python 3 and are contained in `spqr/tools/` together with other C-written analysis programs. Python 2 scripts are included in the directory `spqr/tools/python2/`.

2.3 Units and File Formats

All distances in Ernwin and SPQR are measured in ångstrom. Ernwin energies are in arbitrary units, and SPQR energy units correspond to reduced units, where a simulated annealing starts typically from a temperature of 25 energy units.

Ernwin uses the forgi [17] file format to store coarse-grained RNA structures. Files in this format often use ".cg" or ".coord" as extension. Through forgi, Ernwin can read most common formats for RNA secondary structure, where the Vienna format (fasta files with a dotbracket string below the sequence) is the preferred input format. It can read MMCIF and PDB files but can write only PDB files.

The following is an example of a file in forgi format representing a structure with two stems and an internal loop:

```
name minimal_example
length 9
seq GCAGC&GCGC
seq_ids A:1 A:2 A:3 A:4 A:5 B:1 B:2 B:3 B:4
define s0 1 2 8 9
define i0 3 3
define s1 4 5 6 7
connect s0 i0
connect s1 i0
info Energy 51.2729353666 ROG,AME(i),
coord s1 2.7367535917493999 6.6531844066631471
↪   -2.8776818573773837 3.0382061066493598 9.1842524365562443
↪   -3.1815074805785324
coord s0 0.0000000000000000 0.0000000000000000
↪   1.0000000000000000 0.0000000000000000 2.5670000000000002
↪   1.0000000000000000
twist s1 0.2267985120684020 -0.1426614042130008
↪   -0.9634366396767013 -0.3820438841325927 -0.0651550606912912
↪   -0.9218445035162905
twist s0 1.0000000000000000 0.0000000000000000
↪   0.0000000000000000 0.7759115102200425 0.0000000000000000
↪   -0.6308417617026102
sampled i0 5XYI_2:i_50 2 49 49
sampled s1 4V50_BA:s_89 4 513 514 524 525
sampled s0 5T83_A:s_5 4 16 17 33 34

vres i0 6.45142706 5.26841793 -6.59969566
vbase i0 6.45142706 5.26841793 -6.59969566
vsugar i0 6.45142706 5.26841793 -6.59969566
vbackbone i0 6.45142706 5.26841793 -6.59969566
```

SPQR works with conformations stored in its own file format with extension `mc`, in order to keep the full precision and reproducibility of the results. In addition, the PDB format is available for human readability, and the scripts for their visualization in VMD [18] and conversion among them are available in the `tools/` directory. In the PDB format, each nucleotide is characterized by five particles, which stand for the oriented nucleobase (`BASE`, `XVEC`, and `YVEC`, the sugar bead `SUGR`, and the phosphate bead `PHOS`) as shown in the example below:

```
ATOM     10 BASE   C 0   2     53.7613 110.402 50.6610    A3NA
ATOM     11 XVEC   C 0   2     53.7337 110.835 49.7602
ATOM     12 YVEC   C 0   2     52.9166 110.875 50.9139
ATOM     13 SUGR   C 0   2     56.9731 109.741 49.0783
ATOM     14 PHOS   C 0   2     57.2721 107.274 52.8009
```

Additionally, the `BASE` particle contains, between columns 54 and 57, four characters which specify the state of the glycosidic bond angle (A for anti, H for high anti, and S for syn), the sugar pucker (3 for C3' endo and 2 for C2'endo), and two characters which state whether the position and the backbone state can be changed (A for flexible and N for fixed). In the example, the cytosine nucleotide can move freely, but its pucker and glycosidic states are fixed.

A simulation run creates the directory `configs/` where the checkpoints and the simulation trajectories will be stored in mc format and, optionally, in PDB format, which is the default for the final structure.

3 Methods

3.1 Overview over the Pipeline

Our pipeline allows to predict RNA 3D structures with additional experimental constraints from SAXS experiments. It starts with an RNA sequence and secondary structure as input (*see* **Note 1** about the prediction of secondary structure from the sequence). Ernwin then builds an initial model out of coarse-grained fragments taken from RNA PDB structures, where each fragment corresponds to a secondary structure element. Starting from this initial model, a Monte Carlo (MC) simulation samples a trajectory of different structures with an energy function that can optionally direct the sample toward the global shape shown by experimental SAXS data. At regular intervals in the simulation, all-atom structures are reconstructed from the coarse-grained model. These structures contain locally correct geometries within the secondary structure elements, but it is very likely that they exhibit gaps and clashes between the nucleotide atoms as well. Even more, the superposition of fragments can lead to entangled conformations that are hard to fix in an all-atom representation. Therefore, after filtering the all-atom reconstructions for the best candidates (e.g., by using CRYSOL [19] to compare them again to the SAXS data), we submit those candidates to further refinement with SPQR.

The SPQR refinement consists of a series of restrained Monte Carlo and simulated annealing simulations, summarized in four steps:

1. A short energy minimization, to push away clashes and reconstruct broken bonds along the backbone
2. A check of topological artifacts and removal, if they are present
3. A short steered simulation which pulls the structure back to the original model
4. An energy minimization

After a more realistic model is obtained, one can annotate the unrefined Ernwin structure and check for a possible nucleotide-level realization of the proposed contacts between the secondary structure elements. The procedure to follow for each contact is to crop a set of nucleotides containing the interacting loops and simulate them under SPQR in order to explore its conformation space. This step may require many independent simulations, but given the small size of the selected fragments, it results in easy-to-parallelize and quick run. Once a predicted contact is found, the spatial arrangement of the interacting nucleotides can be imposed on the full-refined SPQR structure by means of $\mathcal{E}$RMSD restraints. These restraints, as mentioned before, enforce the relative distances and orientations between the nucleotides of a template fragment into the corresponding set of the refined structure. The final product is a full-refined structure as faithful as possible to the Ernwin prediction, which can be converted into an atomistic model for further purposes.

The basic usage of each method will be briefly described in the following subsection. Then we will illustrate how to combine these approaches using the *Braveheart* long noncoding RNA [8].

3.2 Basic Usage

3.2.1 Ernwin

Ernwin is based on sampling RNA molecules by assembling coarse-grained fragments that are based on secondary structure elements (stems and loops). *See* Fig. 2 and Thiel et al. [17] for a detailed description of this model underlying our approach.

Ernwin is completely run and controlled via options and input files given through the command line. Ernwin expects a single argument, the input RNA file, typically as fasta file and can have a lot of options (preceded by `--`). Files can be given as absolute or relative path.

For a simple simulation with only default options, run `ernwin.py` followed by the name of a fasta file. This will create an output folder in the current directory, with the name of the RNA from the fasta file:

```
ernwin.py example.fa
```

Ernwin is written in a way that it consists of multiple components, where each component has its own options. Thus, the

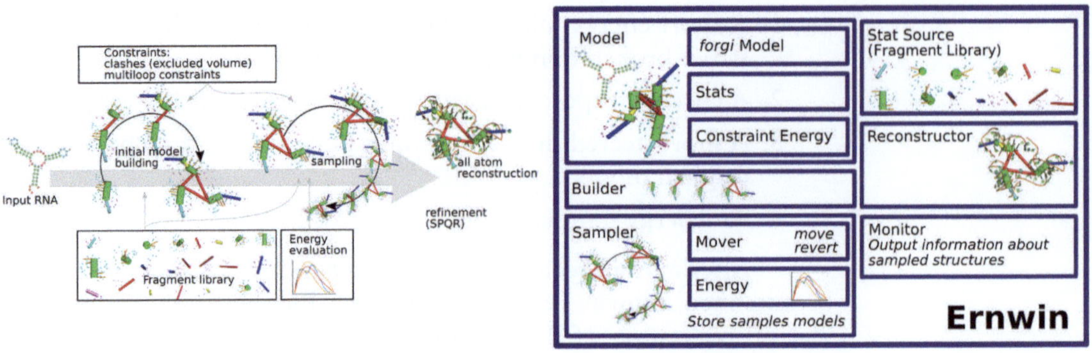

Fig. 2 Overview over the Ernwin model. **Left:** The flow of the RNA through the program: from the input secondary structure, an initial 3D model is assembled from coarse-grained, rigid fragments. This model must fulfill certain constraints. Then from this model, a Monte Carlo sampling is started. In regular intervals, but typically not at every step of the trajectory, an all-atom structure is reconstructed from the coarse-grained model, which can be refined by SPQR. **Right:** An architectural view of Ernwin, illustrating its different components and modules

number of available options can be quite overwhelming. You can run `ernwin.py --help` to display a help message explaining all options. In this section, we will give an overview over the components of Ernwin (as illustrated in Fig. 2), highlighting the most important options and explaining their relevance. A practical example how Ernwin is used will follow in the next chapter.

First of all, Ernwin **loads the RNA input file**, using the forgi library [17], and it inherits some options from forgi for file loading, which are not relevant when loading fasta files. Then, from the secondary structure, **an initial 3D model is built**, taking into account the constraint energies (clashes due to excluded volume and multiloop constraints). An example where a different constraint energy is chosen using the `--constraint-energy-per-ml` option will be given in Subheading 3.3.2. The algorithm for creating an initial model samples one element after the other in an order given by the minimum spanning tree, goes back one or more elements in case of clashes, and is therefore not 100% fair, i.e., does not sample all possible structures equally likely. As the initial building is typically followed by fair Monte Carlo sampling, this is usually fine. With `--num-builds`, you can create more than one independent initial model, and Ernwin will sample a trajectory for each of them, either in a serial way (default) or in parallel using one subprocess each if `--parallel` is given. After the initial 3D models have been built, **sampling** can start. The `--iter` option is used to set the number of iterations for sampling. A good choice for the number of iterations usually depends on size of the RNA molecule, in particular on the number of interior and multiloops. For small RNAs (less than 100nts), usually between 2500 and 5000 iterations are enough; for longer molecules, we recommend 10000 or more iterations.

Building the initial structure and subsequent sampling using a Monte Carlo simulation uses the "stat source" component of Ernwin to **sample geometries ("stats") of the structure elements**. The angles and distances with which the elements are parametrized were described in detail by Kerpedjiev et al. [9]. The default list of stats to sample from can be replaced with the `--stats-file` option or amended with the `--fallback-stats-files` option. *See* Subheading 3.3.1 for an example.

The main sampling algorithm consists of two components: the **Mover**, which takes one or more elements and replaces their geometry with a different one sampled from the stat source, and the **Energy Function**, which assigns a score to the new structure. Then the new structure is accepted or rejected with a probability based on the difference in energy. Details on the different move types implemented in Ernwin will be presented elsewhere [20]. The default energy based on the radius of gyration, loop-loop interactions, and A-Minor interactions is optimized for predicting structures when no experimental SAXS data is available and was described in detail by Kerpedjiev et al. [9]. In Subheading 3.3.4, we will give an example how a different energy can be used to sample the *Braveheart* lncRNA with the help of SAXS data.

Finally, a **reconstructed all-atom structure** can be saved in regular intervals, e.g., every 100 sampling steps, using `--reconstruct-every-n 100`. The reconstructor takes all-atom fragments corresponding to the sampled coarse-grained fragments from the full PDB/MMCIF files and inserts them into the target structure. To extract the fragment from the original PDB/MMCIF file, Ernwin needs two directories, one containing the PDB/MMCIF files (and given using `--source-pdb-dir`) and another one containing the corresponding forgi files (`--source-cg-dir`), which are used for storing the definition of fragments in the PDB files. *See* Subheading 3.3.3 for a detailed example.

3.2.2 SPQR

Here, we briefly describe the main features of a single SPQR simulation run. In the refinement process, however, many simulations and file conversions will be executed automatically in a procedure described in the Methods section. An atomistic PDB structure can be converted to SPQR format by running

```
python pdb2spqr.py -i input.pdb -o spqr-output.pdb
```

For a single simulation, it is required that the initial condition is contained in the `spqr_inits` directory. The `params.spqr` parameter file must also be present in the simulation directory. This file contains the simulation parameters, and an example can be found in the `spqr/examples` directory. The main parameters are the temperature TEMPERATURE, the total number of MC sweeps MC_STEPS, and the number of sweeps between saved configurations MC_TRAJ_STEPS. The path of the energy file must also be specified here through the variable ENERGS_PATH. In addition, the simulated annealing simulations must specify the initial temperature SA_TINI, the annealing factor SA_TFAC, and the maximum number of runs SA_NT.

Then, a simulation is run by

```
SPQR_<bin> [ -i <job_index> ]
```

where <job_index> serves as a simulation identifier and operates on the random seed, while <bin> stands for the type of binary and method to be used. For example, MC performs constant temperature Monte Carlo simulations, eMC allows the incorporation of an energy function proportional to the difference of $\mathcal{E}$RMSD with respect to a set of target fragments, SA performs a simulated annealing simulation, and eSA runs an annealing under $\mathcal{E}$RMSD restraints.

During the simulation run, the trajectories and checkpoints are saved in the directory configs/ in binary format and, optionally, as PDB files if such is specified in the parameter file. Trajectories are stored in the file confs.p<XX>.mc where XX is the job index. Checkpoints are named chk.<YY>.<XX>.mc, where YY stands for the MC step and XX for the job index. These files can easily be converted to a PDB file by using the MC2PDB script. In addition, the last conformation is saved in the local directory under the name final.p<XX>.pdb.

Additionally, the $\mathcal{E}$RMSD restraints require the ermsd_frags.lst file, which defines the groups of nucleotides that will be enforced, followed by the coordinates of the template structure for each group in the PDB format of the SPQR representation. The functional form of the $\mathcal{E}$RMSD energy is given by $U_{\mathcal{E}} = \frac{1}{2} K_{\mathcal{E}} (\mathcal{E}\mathrm{RMSD})^2$, where $K_{\mathcal{E}}$ is a harmonic constant that can be specified for each group. As an example, if there are three groups composed of the nucleotides 2 to 8, 20 to 25, and 11, 12, 40, and 41, with $K_{\mathcal{E}} = 50$ for all of them and a dimensionless cutoff of 10 [21], the starting lines of the $\mathcal{E}$RMSD file must be

```
REMARK ERMSD PARAMS 3 10
REMARK ERMSD GROUP 50 2 3 4 5 6 7 8
REMARK ERMSD GROUP 50 20 21 22 23 24 25
REMARK ERMSD GROUP 50 11 12 40 41
```

which are followed by the respective coordinates of the nucleotides of the groups in SPQR format, in the same order as they were introduced. It is worth to remember that only the relative positions of the particles matter; each group can be extracted from an independent template. The ℰRMSD restraints produce an output file, ermsd_obs.dat, which contain the ℰRMSD as a function of the MC step and the corresponding energy.

3.3 Modeling of Full-Length Bvht (SASDHE3)

In 2020, we have used Ernwin to predict possible all-atom RNA 3D structures for the long noncoding RNA *Braveheart* [8]. Here, we will describe an improved version of the pipeline we used for those predictions step by step, which can be followed by the files provided in the Tutorial Files.

3.3.1 Creating Additional Fragments

As a fragment-based method, Ernwin is limited by the available RNA 3D structures. Consequently, it can happen that the input secondary structure contains elements for which few or no fragments exist. The *Braveheart* secondary structure, in particular (see Tutorial Files, building/bvht.fa), contains several interior loops with unusually long unpaired regions, making simulations with the default list of stats (fragments) problematic. These loops, such as the biologically relevant AGIL motif, still exist after we have introduced additional base pairs with RNAfold [22] into the SHAPE-based published secondary structure of *Braveheart* [23]. The lack of fragments in the knowledge base for these motifs can be seen by the warning and error messages displayed after starting the simulation:

```
ernwin.py --iter 1 bvht.fa --step-save 1
WARNING:fess.builder.builder.load_sampled_elements[46]: No (useable) 3D
↪  structure found in the input file. Starting to build an initial model.
WARNING:fess.builder.stat_container._possible_stats_inner[245]: Only 38
↪  angle-stats found for interior loop with 0 and 4 unpaired nucleotides
[...]
WARNING:fess.builder.stat_container._possible_stats_inner[245]: Only 0
↪  angle-stats found for interior loop with 14 and 3 unpaired nucleotides
ERROR:fess.builder.stat_container._possible_stats_inner[280]: Trying key
↪  interior loop with 13 and 3 unpaired nucleotides instead of interior loop
↪  with 14 and 3 unpaired nucleotides for angle-stat
WARNING:fess.builder.stat_container._possible_stats_inner[245]: Only 1
↪  angle-stats found for interior loop with 13 and 3 unpaired nucleotides
```

The first warning is harmless in this case, as we started with a fasta file which does not contain any 3D structure. Next comes a bunch of warnings about secondary structure elements, for which Ernwin does not have enough fragments in its knowledge base (stats file). This warning appears whenever less than 100 fragments are found for a secondary structure element. For many applications, having only around 30 stats for an element is still acceptable. However, the fact that no fragments at all were found for the interior loop with 14 unpaired nucleotides on one side and 3 on the other side (the AGIL motif) is a problem. If no fragments are found, Ernwin tries to fall back to fragments corresponding to slightly smaller loops. As we are interested in the pair distance distribution (for the SAXS energy) and all-atom reconstructions, this is not what we want. So we have to supplement the built-in fragments with additional fragments.

This can be done using any atom-level RNA 3D structure prediction program, which allows for secondary structure constraints (**Note** 2). Create a bunch of artificial structures only for the loop we need plus some adjacent base pairs. Some structures we created with simRNA [24] for the AGIL loop can be found in the Tutorial Files (`building/3_3_1/`). Assuming you have all the resulting predicted PDB files in a directory called "predicted_AGIL_loop," convert all of them to coarse-grained forgi files using the forgi library:

```
rnaConvert -T forgi --to-file predicted_AGIL_loop/*.pdb
```

This creates for each connected component of RNA chains in each of the PDB files, a *.cg file, with the original file name followed by the identifier of the PDB chain. *See* **Note** 3 for details about the conversion from PDBs to other file formats with forgi.

Then, run the script `create_stats.py` to create a file containing all the stats (coarse-grained fragments) from your sampled RNA structures:

```
create_stats.py *.cg > CUSTOM.stats
```

With this stat file as fallback, you can try sampling again:

```
ernwin.py --iter 1 bvht.fa --step-save 1 --fallback-stats path/to/CUSTOM.stats
```

3.3.2 Multiloop Constraints and Move Sets

Due to the complexity of *Braveheart* and its multiloops, finding an initial structure from which we can start sampling can take a lot of time. To reduce the build time of the initial model and increase the probability of finding acceptable moves from one conformation to the next during sampling, we can relax the multiloop constraint energy using `--constraint-energy-per-ml MAX16[1SFJ1]` as explained in Subheading 3.2. (*See* also **Note** 4.)

3.3.3 All-Atom Reconstruction

Filtering with CRYSOL (*see* Subheading 3.3.6) and refinement with SPQR require all-atom reconstructions of the structures predicted by Ernwin as input. To enable all-atom reconstruction, all PDB/MMCIF files from which a fragment was extracted (for custom stat files as well as for Ernwin's default stats file) have to be put into one directory (or sym-linked into one directory) for the all-atom reconstruction module to find them. The path to this directory has to be given on the command line. The MMCIF files corresponding to the default stats file of Ernwin can be downloaded as described in the section about the installation of Ernwin (2.2.1).

As different external annotation tools used by our scripts can sometimes assign different secondary structures to the same PDB file, the secondary structures corresponding to each PDB/MMCIF file have to be stored as well. These are the cg files which we used in Subheading 3.3.1 as input to the `create_stats.py` script and the files extracted from the `RESOURCES` directory of Ernwin (*see* Subheading 2.2.1). Again, all cg files have to be in one directory, which can be the same as the PDB/MMCIF's directory or a different one.

To speed up repeated reconstructions of structures that share the same fragments, use the `--reconstruction-cache-dir` option. After all files have been prepared, you can run Ernwin with the following options to create an all-atom reconstruction every 100 steps (this example does not use any SAXS data and samples just a short trajectory):

```
ernwin.py --iter 1000 bvht.fa --step-save 25 --fallback-stats
↪  path/to/CUSTOM_STATFILE.stats --reconstruct-every-n 100
↪  --source-pdb-dir /path/to/ernwin_pdbs --source-cg-dir
↪  /path/to/forgi/files/ernwin_cgs  --reconstruction-cache-dir
↪  ~/.cache/ernwin
```

3.3.4 Using SAXS Data to Guide the Sampling

There are two options how to include SAXS data in the Ernwin sampling via pseudoenergies. The first approach (PDD energy) simply is an exponential potential on the area between expected and observed pair distance distribution functions, while the second one is an ensemble-based approach (EPD energy) using the reference ration method on each bin of the pair distance distribution function. While the first one is arguably easier to interpret and more descriptive, the latter interacts better with other (not SAXS-based) energy contributions.

A standard tool for obtaining an experimental pair distance distribution function from experimental SAXS intensities is GNOM [25], and Ernwin can read GNOM output files directly if they use ångstrom as unit. If distances are in nm (as is often the case), Ernwin will misinterpret the file. For compatibility with other programs and in case the GNOM output has the wrong units, Ernwin can also read csv files with the pair distance distribution function. Such files can be easily generated manually from the output of any programs that calculates a pair distance distribution function. Again, all units have to be in ångstrom for Ernwin to work correctly. The following shows the beginning of an example csv file with a pair distance distribution function in the csv format understood by Ernwin:

```
distance,count,error
0.0000E+00, 0.0000E+00, 0.0000E+00
0.6743E+00, 0.2116E-01, 0.2566E-03
0.1349E+01, 0.4455E-01, 0.4572E-03
0.2023E+01, 0.7010E-01, 0.6033E-03
0.2697E+01, 0.9775E-01, 0.6974E-03
0.3371E+01, 0.1273E+00, 0.7429E-03
```

The path to this file has to be given with the command line option, but it is only used if a corresponding energy is set via the command line:

```
--energy PDD[R] --pdd-file PDDFILE.csv
```

The option in square brackets given to the PDD energy specifies the level of detail at which we calculate the pair distance distribution function of the sample structures. We recommend to always use "R," which stands for 1 point per virtual residue. This level of detail is sufficient for large molecules like *Braveheart* and allows for quick evaluation on the coarse-grained level.

As the `energy` option overwrites all default energy contributions, we have to add the loop-loop interaction energy and the A-Minor energy manually to the energy definition: `--energy EPD[R],LLI,AME` in the case of large, extended molecules with isolated hairpins and `--energy EPD[R],SLD,AME` for compact molecules with hairpins closer to each other. The SLD energy (shortest loop distance) and AME (A-Minor energy) are described in [9], while the LLI (loop-loop interaction) was introduced in [8] to avoid circularization of the bvht RNA by isolated hairpins trying to come close to other hairpins.

The EPD (or PDD) energy can be scaled toward your needs using a number before (prefactor) and after (adjustment) the short name "EPD." The prefactor is simply a weight for the energy compared to other contributions to the energy function. The default is 30, but it is often desirable to reduce the weight of the SAXS data compared to other energies, e.g., by using a prefactor of 5. In the case of the EPD energy, the adjustment is used to scale the experimental error which will be used as standard deviation for the target distribution around each bin of the pair distance distribution function. It has a hard-coded minimum of 5. A value of 25 would increase the tolerance around the experimental data, which is often recommended.

It is also possible to do simulated annealing, by defining the adjustment as `25_-1_500`, which means start with an adjustment of 25 and decrease it by 1 every 500 steps.

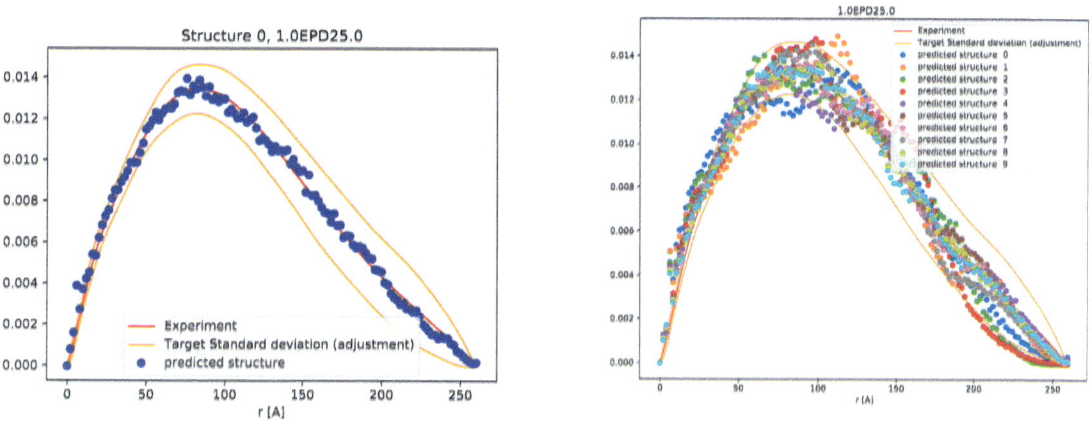

Fig. 3 Visualization of the EPD energy of Ernwin for *Braveheart* SAXS data. The lines above and below the experimental SAXS data indicate the standard deviation used by the EPD energy, which is based on the experimental error and the adjustment provided via the command line and is applied to each bin individually. The dots represent the pair distance distribution function of sampled structures and indicate the location of the bins. The left image shows a single structure, while the right images show multiple structures distributed over the course of a trajectory

On the other hand, the pdd energy is calculated as $p * e^{a*x}$ where p is the prefactor, a is the adjustment, and x is the area between the experimental and the simulated pair distance distribution function (binned to the stepsize).

If the experimental pair distance distribution function is defined with unequally spaced distances, it might be necessary to give a target stepsize via the command line with the `--pdd-stepsize 2` option, where 2 would stand for 2Å.

To ensure that Ernwin correctly loads the pair distance distribution function and that the chosen energy parameters work as intended, we provide the script `plot_ernwin_pdd.py`. It understands the same energy and pdd-related options as Ernwin, takes one or more RNA structures as input, and creates a plot for every combination of RNA structure and PDD or EPD energy:

```
plot_ernwin_pdd.py --energy 1EPD25[R] --pdd-file
↪  path/to/gnom.out path/to/simulation_01/step*000.coord
```

Figure 3 shows an example of the plots created by this script.

3.3.5 The Final Ernwin Command

After having put together all the command line options and input files for Ernwin, the final command line looks like this (to sample four trajectories in parallel and store the results in a folder with suffix "simulation_results"):

```
ernwin.py bvht.fa --iter 10000 --num-builds 4 --parallel
↪  --fallback-stat  path/to/CUSTOM.stats --step-save 50
↪  -reconstruct-every-n 100 ---source-pdb-dir
↪  /path/to/ernwin_pdbs --source-cg-dir
↪  /path/to/forgi/files/ernwin_cgs  --reconstruction-cache-dir
↪  ~/.cache/ernwin --constraint-energy-per-ml MAX16[1SFJ1]
↪  --energy 5EPD25_-1_500[R],LLI,AME --pdd-file
↪  /path/to/SAXS_PDD.csv -o _simulation_results
```

3.3.6 Filtering

The reconstructed PDB files can be filtered for agreement with the experimental SAXS data using CRYSOL 3.0 [19]. CRYSOL is a standard tool for predicting SAXS data from PDB files and for evaluating the agreement between experimental SAXS data and a PDB file. In contrast to the coarse-grained view on the pair distance distribution function implemented in Ernwin, CRYSOL works at a higher (atomic) level of detail, accounts for the effect of the hydration shell, and takes intensities and errors of the original SAXS profile into account in a more statistical rigorous way. We have observed that the Ernwin energy helps to guide sampling toward structures with a good fit to the experimental SAXS profile, but not every structure with a low Ernwin energy has a good χ^2 value when evaluated with CRYSOL, such that we recommend this additional filtering step.

Since version 3.0, CRYSOL can batch-process multiple PDB files:

```
crysol_30 bvht_simulation_results/simulation*/*.pdb
↪  path/to/SASDHE3.dat
```

We can then grep for the χ^2 value of CRYSOL in the CRYSOL output files and sort them in descending order. By including a space at the end of the grep pattern, we exclude all structures with a χ^2 of 10 and above and allow for sorting without having to further process the string:

```
grep "Chi^2: " bvht_simulation_results/simulation_0*/*.fit |
↪  awk '{print $NF,$0}' | sort -nr
```

Be careful when reporting and interpreting the final χ^2 value in a statistical way, because our pipeline involves multiple testing, as we screen a huge number of predicted structures.

The final result is a set of all-atom PDB structures. Three samples can be found in the Tutorial Files (`refinement/samples`) for further refinement.

3.3.7 Refinement

Starting from the PDB structures predicted by Ernwin, we illustrate the procedure described in Subheading 3.1 using the scripts provided with the SPQR package. In this manner, the subsequent MC simulations and annealing runs are already performed by a set of Bash scripts. The parameters chosen for the length and temperature of the simulations are also contained in a human-readable fashion. If the refinement procedure fails, an error message will specify in which step the convergence criteria was not reached and which parameters should be tuned for fixing this.

We illustrate the procedure on the first example in the Tutorial Files, `refinement/samples/bvht1.pdb`.

The refinement script `SPQR_REFINE` is found in the `spqr/tools` directory, as is the rest of the scripts used in the following. Provided that the initial structure is given in `bvht1.pdb` and the fasta sequence with secondary structure in Vienna format is in the file `bvht.fa`, the command is

```
./SPQR_REFINE -i bvht1.pdb -o bvht1 -t bvht.fa
```

The `-o` option specifies the label of the output files. The results will be stored in the directory `refSPQR_bvht1`. The refined structure will be saved as `at_refSPQR_bvht1.pdb` in the all-atom representation, while `refSPQR_bvht1.pdb` and `refSPQR_bvht1.mc` are its counterparts in SPQR representation and full-precision format for further SPQR simulations, respectively. Independent runs can be executed with a different random seed `SEED` by adding the `-r SEED` flag.

Additionally, the directories `01_CLASHRM`, `02_ERMSDMI`, and `03_ENERGMI` contain the simulations of the **steps 1**, **3**, and **4** described in Subheading 3.1. **Step 2** is executed only if topological artifacts are detected between the secondary structure elements. The directory `01_CLASHRM_LR` contains the simulations which take care of their removal. An illustration of the refinement procedure is shown in Fig. 4.

3.3.8 Search for Tertiary Contacts

One can go a step further and search with more care for a realization of possible tertiary contacts. The annotation with Ernwin specifies the possible contacts between elements of the secondary structure, which can be obtained by running

```
rnaConvert.py bvht1.pdb > bvht1.cg
python ernwin_annot.py bvht1.cg
```

In particular, we would like to focus on

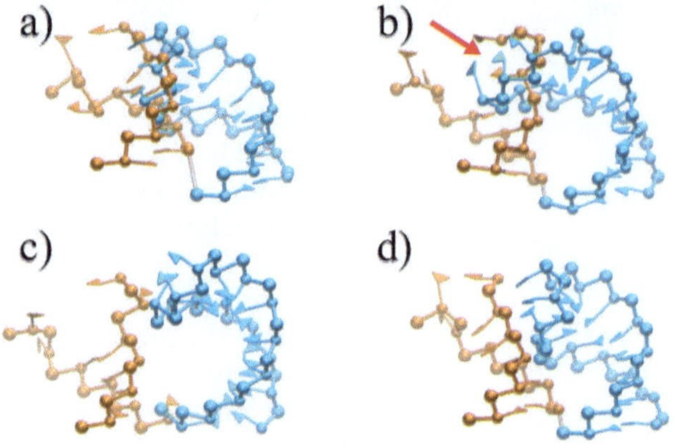

Fig. 4 Example of SPQR refinement on fragment of *Braveheart* lncRNA, on stem s0 (orange) and hairpin h11 (cyan), on file `bvht3.pdb`. (**a**) The reconstructed structure by Ernwin is converted to SPQR format. In this case, several clashes and an entanglement are present. (**b**) The clashes are removed in **step 1**, and the bonds have realistic values. However, the hairpin pierces the duplex toward the region pointed by the arrow. (**c**) The link between these elements is identified and removed in **step 2**. Still, the structure is distorted from the original geometry. (**d**) Finally, the structure is gently pulled back toward its initial conformation, and the energy is minimized. Sample file `frag_bvht3.pdb` contains this fragment for a much faster relaxation and simulation

```
array(['h4', 's18'], dtype='|S3')
```

which is a long-range contact between the hairpin 4 and stem 18, defined in bvht1.cg as

```
define h4 229 233
define s18 201 207 216 222
```

Thus, a good choice is to work with the residues between 201 and 238, which ensures that the secondary structure elements in between are complete. From the refined `refinement/csearch/spqr-bvht1.mc` file, the fragment is cropped using

```
./MCEDIT -i spqr-bvht1.mc -o crop_bvht1.mc -L 201-238 -s
↪  frag.fa -p 1
```

The selected fragment is saved in `crop_bvht1.mc`. The option `-s` makes it possible to restrain the pucker and glycosidic bond angle of the nucleotides which belong to the stems specified in the secondary structure while giving complete freedom inside the rest of the loops. The flag `-p` indicates the index of the first residue for consistency with the Ernwin numbering.

With these files, a simulated annealing run can be performed over the conformation, provided the right secondary file is used as a restraint. For this, create the $\mathcal{E}$RMSD restraint file using the secondary structure of the fragment `refinement/csearch/frag.fa` by running

```
./MC2PDB -i crop_bvht1.mc -o crop_bvht1.pdb
python spqr2ermsd.py -i crop_bvht1.pdb -f frag.fa -o
↪  ermsd_frags.lst -r
```

and create and annotate an Ernwin format structure of the fragment:

```
python SPQR_BBACKMAP.py -i crop_bvht1.pdb -o at_crop_bvht1.pdb
rnaConvert.py at_crop_bvht1.pdb > crop_bvht1.cg
python ernwin_annot.py crop_bvht1.cg
```

Then run the simulated annealing with the parameters contained in the file `refinement/csearch/params.spqr` file in the Tutorial Files and using the cropped structure as initial condition by copying it to `spqr_inits/init.mc`, by executing

```
./SPQR_eSA
```

which can be parallelized easily by adding different values to the argument `-i`. The path to the energy files `ENERGS_PATH` must be updated in the simulation parameter file, just as in the example `params.spqr`. The binary tool `SPQR_ENERG` can be used to evaluate the energy and annotate the snapshots both in PDB and in mc format, among other functionalities. Once the data is produced, the generated trajectory stored in the `configs/confs.p00.mc` file can be annotated according to the secondary structure elements used by Ernwin, by running

```
./SPQR_ENERG -i crop_bhvt1.mc -e crop_bhvt1.cg -T
↪  configs/confs.p00.mc
```

The output reports the contacts found in each frame. In this case, the line

```
41 [ s0 h1 (G1,G30 SH) ]
```

states that in frame 41 there is a sugar-edge/Hoogsteen pair between two guanines belonging to the s0 and h1 loops, respectively, which correspond to s18 and h4 in the full fragment.

The annotation also generates a PDB file for each contact, which can be used as a template for enforcing its arrangement on the full structure. For example, if the output tctc_s0_h1.p00.lst is selected, it must renamed using

```
cp tctc_s0_h1.p00.lst ermsd_frags.lst
```

and its first lines must read

```
REMARK ERMSD PARAMS 1 100
REMARK ERMSD GROUP 10 200 201 202 203 204 205 206 207 208 209
↪ 210 211 212 213 214 215 216 217 218 219 220 221 222 223 224
↪ 225 226 227 228 229 230 231 232 233 234 235 236 237
```

before the ATOM lines, keeping in mind that the nucleotide numbering starts from zero. Since this step depends strongly on the random seed and initial condition, we have included both files in the Tutorial Files in refinement/enforce. The ermsd_-frags.lst file, together with the parameter file contained in the Tutorial Files (refinement/enforce/params.spqr), is enough to run a single simulation on the refined spqr-bvht1.mc initial condition through the command

```
./SPQR_eSA -i 1
```

Finally, the backmapping to full atom resolution can be performed on the final structure of the previous simulation using

```
SPQR_BBACKMAP.py -i final.p01.pdb
```

This is a crude way of generating an atomistic structure, by placing an atomistic nucleotide template on each of the CG nucleotides, which is specific for the states of sugar pucker and glycosidic orientation. Further refinement requires MD simulations and the employment of the same templates, as it has been already shown in other systems [13]. In this example, a contact between G201 and G230 is formed and detected, which is finally backmapped as shown in Fig. 5.

4 Notes

1. Many tools exist to predict RNA secondary structures from sequences, optionally with additional constraints from experimental data, such as SHAPE probing. For our structure prediction pipeline, we most commonly use the Vienna RNA package [22] with the program RNAfold to get a secondary structure from the sequence. This program allows for the use of

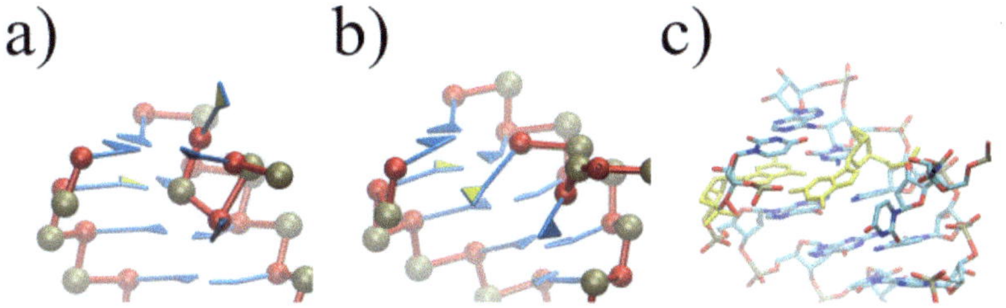

Fig. 5 Contact search and atomistic backmapping. Starting from the refined structure in (**a**), the two nucleotides G201 and G230, marked in yellow, form a tSH contact found during by the search procedure, as shown in (**b**). The back-mapped structure is shown in (**c**)

hard constraints (e.g., for base pairs we are certain of due to NMR experiments) and soft constraints (e.g., based on SHAPE probing activities) [26].

2. Creating fragments that fulfill a certain secondary structure exactly can be quite tricky with most atom-level RNA 3D structure prediction programs, because many programs only allow the user to enforce base pairs with constraints, but not to disallow base pairs. Predicting fragments with the exact correct number of unpaired nucleotides often requires some tricks, such as playing with the types of constraints provided by the program or modifying the sequence to make base pairs impossible. (The latter is fine for Ernwin, as Ernwin can use fragments independent of the sequence and can replace the bases during all-atom reconstruction.)
3. To convert a PDB/MMCIF file to other file formats with forgi, we recommend that you install either MC-Annotate [15] or DSSR [14] in a path findable by forgi to detect the secondary structure in the PDB/MMCIF files. If both are installed, use the script `forgi_config.py` to choose your preferred tool. If none is installed, forgi will use a built-in heuristic which is less powerful than the aforementioned tools. As an alternative, the secondary structure can be enforced via the command line:

```
rnaConvert -T forgi --to-file input.pdb --chain A
↪  --pdb-secondary-structure '(((...)))...(((...)))...'
```

Make sure you properly quote/escape the brackets in the dotbracket string on your shell.

4. To relax the junction constraint energy, the option `--constraint-energy-per-ml MAX16[1SFJ1]` can be used. The junction-based energy SFJ (**S**earching for the best **f**ragment to close the **j**unction) is proportional to the gap (in terms of distance and angle) that is left in the multiloop if the best fitting fragment is used to bridge the gap that is left in the multiloop after sampling all but the last segment (which can't

be sampled because it is fully determined by other segments). The numbers "1" in front and after the energy are needed to ensure that the energy is scaling linearly with the gap. The prepended term "MAX16" means that a maximal gap of 16Å is allowed (making the energy infinite if the gap is larger and 0 otherwise). This value can of course be varied freely, but we recommend not to use a value below 8Å to give the junctions enough flexibility for sampling.

If building an initial structure with the junction constraints is too difficult (takes too long), it is possible to use the SFJ energy as an exponential potential for sampling instead of using it as constraint:

```
ernwin.py bvht.fa --iter 500 --energy SFJ
↪   --constraint-energy-per-ml MAX100[1SFJ1] --track-energies
↪   1SFJ1
```

The forgi files created with such a simulation can be used as input for follow-up simulations with the SFJ energy as constraint energy. If the initial 3D structure fulfills the constraints, then it will be taken as initial model without resampling.

A similar approach can be applied to the clash constraints.

To relax the junction constraint energy by using a distance-based instead of a fragment-based one, use `--constraint-energy-per-ml JDIST`. More details on the effect of using different junction constraint energies will be given elsewhere [20].

Acknowledgements

S. P. acknowledges Prof. Judit Lisoni for infrastructure support.

References

1. Cheng CY, Chou F-C, Das R (2015) Modeling complex rna tertiary folds with Rosetta. Methods Enzymol 35–64. https://doi.org/10.1016/bs.mie.2014.10.051
2. Ponce-Salvatierra A, Astha, Merdas K, Nithin C, Ghosh P, Mukherjee S, Bujnicki JM (2019) Computational modelling of RNA 3d structure based on experimental data. Biosci Rep 39:BSR20180430. https://doi.org/10.1042/BSR20180430
3. Li B, Cao Y, Westhof E, Miao Z (2020) Advances in RNA 3D structure modeling using experimental data. Front Genet 11: 574485. ISSN 1664-8021. https://doi.org/10.3389/fgene.2020.574485
4. Miao Z, Adamiak RW, Antczak M, Boniecki MJ, Bujnicki J, Chen S-J, Cheng CY, Cheng Y, Chou F-C, Das R et al (2020) RNA-puzzles round iv: 3d structure predictions of four ribozymes and two aptamers. RNA 26(8):982–995. https://doi.org/10.1261/rna.075341.120
5. Mazzanti L, Alferkh L, Frezza E, Pasquali S (2021) Biasing rna coarse-grained folding simulations with small-angle x-ray scattering data. J Chem Theory Comput 17:6509–6521. https://doi.org/10.1021/acs.jctc.1c00441
6. Chojnowski G, Zaborowski R, Magnus M, Bujnicki JM (2021) RNA fragment assembly

with experimental restraints. bioRxiv 2021–02. https://doi.org/10.1101/2021.02.08.430198
7. Bernetti M, Hall KB, Bussi G (2021) Reweighting of molecular simulations with explicit-solvent SAXS restraints elucidates ion-dependent RNA ensembles. Nucleic Acids Res 49(14):e84–e84
8. Kim DN, Thiel BC, Mrozowich T, Hennelly SP, Hofacker IL, Patel TR, Sanbonmatsu KY (2020) Zinc-finger protein CNBP alters the 3-d structure of lncRNA braveheart in solution. Nat Commun 11(1). https://doi.org/10.1038/s41467-019-13942-4
9. Kerpedjiev P, Zu Siederdissen CH, Hofacker IL (2015) Predicting RNA 3D structure using a coarse-grain helix-centered model. RNA 21(6):1110–1121. https://doi.org/10.1261/rna.047522.114
10. Poblete S, Bottaro S, Bussi G (2018) A nucleobase-centered coarse-grained representation for structure prediction of RNA motifs. Nucleic Acids Res 46(4):1674–1683. https://doi.org/10.1093/nar/gkx1269.
11. Poblete S, Bozic A, Kanduc M, Podgornik R, Guzman HV (2021) RNA secondary structures regulate adsorption of fragments onto flat substrates. ACS Omega 6(48):32823–32831
12. Poblete S, Guzman HV (2021) Structural 3D domain reconstruction of the RNA genome from viruses with secondary structure models. Viruses 13(8):1555. https://doi.org/10.3390/v13081555
13. Poblete S, Bottaro S, Bussi G (2018) Effects and limitations of a nucleobase-driven backmapping procedure for nucleic acids using steered molecular dynamics. Biochem Biophys Res Commun 498(2):352–358. https://doi.org/10.1016/j.bbrc.2017.12.057
14. Lu X-J, Bussemaker HJ, Olson WK (2015) DSSR: an integrated software tool for dissecting the spatial structure of RNA. Nucleic Acids Res. https://doi.org/10.1093/nar/gkv716
15. Gendron P, Lemieux S, Major F (2001) Quantitative analysis of nucleic acid three-dimensional structures. J Mol Biol 308(5):919–936. https://doi.org/10.1006/jmbi.2001.4626
16. Leontis NB, Zirbel CL (2012) Nonredundant 3d structure datasets for RNA knowledge extraction and benchmarking. In: Nucleic Acids and Molecular Biology. Springer, Berlin, pp 281–298
17. Thiel BC, Beckmann IK, Kerpedjiev P, Hofacker IL (2019) 3d based on 2d: Calculating helix angles and stacking patterns using forgi 2.0, an RNA Python library centered on secondary structure elements. F1000Research 8:287. https://doi.org/10.12688/f1000research.18458.2
18. Humphrey W, Dalke A, Schulten K (1996) VMD—visual molecular dynamics. J Mol Graph 14:33–38.
19. Franke D, Petoukhov MV, Konarev PV, Panjkovich A, Tuukkanen A, Mertens HDT, Kikhney AG, Hajizadeh NR, Franklin JM, Jeffries CM, Svergun DI (2017) ATSAS 2.8: a comprehensive data analysis suite for small-angle scattering from macromolecular solutions. J Appl Crystallogr 50(4):1212–1225. https://doi.org/10.1107/s1600576717007786
20. Thiel BC, Bussi G, Poblete S, Hofacker IL (2021) Sampling globally and locally correct RNA 3D structures using Ernwin, SPQR and experimental SAXS data. In preparation
21. Bottaro S, Di Palma F, Bussi G (2014) The role of nucleobase interactions in RNA structure and dynamics. Nucleic Acids Res 42(21):13306–13314. ISSN 0305-1048. https://doi.org/10.1093/nar/gku972
22. Lorenz R, Bernhart SH, Höner zu Siederdissen C, Tafer H, Flamm C, Stadler PF, Hofacker IL (2011) ViennaRNA package 2.0. Algorithms Mol Biol 6(1):26. https://doi.org/10.1186/1748-7188-6-26
23. Xue Z, Hennelly S, Doyle B, Gulati AA, Novikova IV, Sanbonmatsu KY, Boyer LA (2016) A G-rich motif in the lncRNA braveheart interacts with a zinc-finger transcription factor to specify the cardiovascular lineage. Mol Cell 64(1):37–50. https://doi.org/10.1016/j.molcel.2016.08.010
24. Boniecki MJ, Lach G, Dawson WK, Tomala K, Lukasz P, Soltysinski T, Rother KM, Bujnicki JM (2015) SimRNA: a coarse-grained method for RNA folding simulations and 3d structure prediction. Nucleic Acids Res 44(7):e63–e63. https://doi.org/10.1093/nar/gkv1479
25. Svergun DI (1992) Determination of the regularization parameter in indirect-transform methods using perceptual criteria. J Appl Crystallogr 25(4):495–503. https://doi.org/10.1107/s0021889892001663
26. Ronny Lorenz, Ivo L. Hofacker, and Peter F. Stadler. Lorenz R, Hofacker IL, Stadler PF (2016) RNA folding with hard and soft constraints. Algorithms Mol Biol 11(1):1–13. ISSN 1748-7188. https://doi.org/10.1186/s13015-016-0070-z

INDEX

Ronny Lorenz (ed.), *RNA Folding: Methods and Protocols*, Methods in Molecular Biology, vol. 2726,
https://doi.org/10.1007/978-1-0716-3519-3,

S

T

MIX
Papier aus verantwortungsvollen Quellen
Paper from responsible sources
FSC® C105338

If you have any concerns about our products,
you can contact us on
ProductSafety@springernature.com

In case Publisher is established outside the EU,
the EU authorized representative is:
Springer Nature Customer Service Center GmbH
Europaplatz 3, 69115 Heidelberg, Germany

Printed by Libri Plureos GmbH
in Hamburg, Germany